INTRODUCTION TO
HUMAN GEOGRAPHY
USING ArcGIS ONLINE

BY J. CHRIS CARTER

Esri Press
REDLANDS | CALIFORNIA

Cover credits: feiyuezhangjie/Shutterstock.com, Avatar_023/Shutterstock.com

Esri Press, 380 New York Street, Redlands, California 92373-8100
Copyright 2019 Esri
All rights reserved.
23 22 21 20 19   3 4 5 6 7 8 9 10

Printed in the United States of America

Library of Congress Cataloging-in-Publication Data
Names: Carter, J. Chris, author.
Title: Introduction to human geography : using ArcGIS online / J. Chris
   Carter.
Description: Redlands, California : Esri Press, [2019] | Includes
   bibliographical references and index.
Identifiers: LCCN 2018047382 (print) | LCCN 2018055844 (ebook) | ISBN
   9781589485198 (electronic) | ISBN 9781589485181 (pbk. : alk. paper)
Subjects: LCSH: Human geography--Data processing. | Geographic information
   systems. | ArcGIS.
Classification: LCC GF50 (ebook) | LCC GF50 .C36 2019 (print) | DDC
   304.20285--dc23
LC record available at https://urldefense.proofpoint.com/v2/url?u=https-
3A__lccn.loc.gov_2018047382&d=DwIFAg&c=n6-cguzQvX_tUIrZOS_4Og&r=qNU49__SCQN30XC-f38qj8bYYMTIH4VCOt-
Jb8fvjUA&m=FXsY31O4qU-aG-
6qKD4iIeTNWOI4uB9hKu7x6pOyzRs&s=upFIsjv5n_E9SSQMTSkCprR90WPG90tKOOXBMlvSEjA&e=

Ask for Esri Press titles at your local bookstore or order by calling 1-800-447-9778. You can also shop online at www.esri.com/esripress. Outside the United States, contact your local Esri distributor or shop online at eurospanbookstore.com/esri.

Esri Press titles are distributed to the trade by the following:

In North America:
Ingram Publisher Services
Toll-free telephone: 800-648-3104
Toll-free fax: 800-838-1149
E-mail: customerservice@ingrampublisherservices.com

In the United Kingdom, Europe, the Middle East and Africa, Asia, and Australia:
Eurospan Group
3 Henrietta Street
London WC2E 8LU
United Kingdom
Telephone 44(0) 1767 604972
Fax: 44(0) 1767 6016-40
E-mail:eurospan@turpin-distribution.com

Contents

Preface

Purpose and organization of the book

This book introduces undergraduate university students and AP human geography high school students to the essential concepts and theories of human geography. While many people think of place-name memorizing geography bees when they sign up for a geography class, in reality it is a subject that helps us understand the organization of human society and its impact on our planet. Each chapter focuses on a key area of human geography, covering essential concepts, and is illustrated with real-world data and examples.

Although many human geography textbooks also cover essential concepts, this one leverages the power of Esri's ArcGIS® Online, a browser-based geographic information system with interactive maps and data. Unlike most geography textbooks, with static map figures printed in the book, most map figures in this book include a link that enables students and teachers to further explore data, allowing them to examine places of interest in more detail. A student or instructor can view a map in the textbook on, say, level of development, and then open the ArcGIS Online link and click on individual countries to see how life expectancy, education, and income each contribute to its development ranking.

Each chapter also includes ArcGIS Online exercises, posted in the Esri Press group of the Learn ArcGIS organization, where students can apply geographic concepts to real-world map data. Students will interact with and run spatial analysis on datasets from global organizations such as the United Nations and World Bank, national agencies such as the US Census Bureau and Environmental Protection Agency, and private companies such as Esri®. In many cases, exercises are designed so that students can focus on data within their own city or neighborhood, making analysis more directly related to the place in which they live.

Instructors can use these exercises in a variety of ways. In classrooms with computers, exercises can be completed during class hours, with discussions of questions done in small groups or with the entire class. In other cases, instructors can assign the exercises as homework and discuss results later during class time. The wide range of maps included in the exercises also allows instructors to customize their own exercises.

ArcGIS Online accounts

If you have an existing ArcGIS Online account, you can use it to complete the exercises for this book. If you need a new ArcGIS Online account, it's recommended that you sign up for a free Learn GIS organizational account here: http://learngis.maps.arcgis.com/home/index.html. Some instructors may choose to add students to their campus organizational accounts, where they can set credit limits, add and remove students, and set other parameters. Instructors should contact their campus ArcGIS Online administrator for assistance.

Go to the book resource page at esri.com/Human-Geography for links to access the exercises and data.

Acknowledgments

This book would not be possible without the support of my family, especially my wife Alejandra, who graciously sacrificed weekend and summer family time as I devoted many hours to writing chapters and developing exercises. I would also like to thank Esri staff members Jennifer Bell, Sirisha Karamchedu, Joseph Kerski, Veronica Rojas, and Lauren Scott, among others, who provided essential feedback on this project.

Chapter 1
Introduction

What is geography?

In the news, immigrants risk their lives to reach safety and opportunity in new lands, while some citizens in destination countries worry about losses of jobs and their cultures. In parts of the world, parents struggle to feed multiple children, while in other places employers struggle to fill positions as populations shrink. The decline of manufacturing employment in the developed economies of Europe and North America has devastated many towns and left myriad workers unemployed or with wages well below their previous salaries. At the same time, a burgeoning working class has developed in much of Asia as farmers leave the fields and take up jobs in urban factories. Struggles for political power depend on how voting districts are drawn, while citizens hotly debate the influence of religion in public life and the benefits and challenges of linguistic and ethnic diversity. These topics, and many more, are the subject of human geography. But with that said, what makes geography distinct from other disciplines that also study these issues?

Students often associate geography with identifying countries, cities, rivers, mountains, and other features on a map. Although the ability to find features such as these on a map is of use to geographers, it is not the focus of geography. Geography exists as a distinct academic discipline because of its focus on space, and for this reason, it is considered a spatial science. When geographers use the words *space* and *spatial*, it is in the context of geometric space, not outer space. It is concerned with the three-dimensional location of features on the surface of the earth. To put it simply, the key questions that geographers ask are, *Where* are things located and *why* are they there?

These questions give geographers a unique understanding of how the world is organized and how human and physical features interact to create unique places and regions. They look at the spatial patterns, or distributions, of everything from plant species to unemployment. Geographers further study the spatial relationship between different phenomena, such as how political attitudes and religious beliefs overlap in particular places. The concepts of origin, diffusion, and spatial interaction are also important elements of geography. The world religions of Christianity, Judaism, and Islam originated in the Holy Land of the Middle East and then diffused across the globe, transforming societies as they spread to new locations. Finally, geography looks at human-environment interaction, or how humans influence and change the environment, as well as how the environment shapes humans in terms of where we live, what we eat, and much more. By understanding spatial distributions and the processes that drive them, geographers help us understand the world in which we live. This knowledge allows us to make predictions and decisions on how to address a wide range of pressing social and environmental issues. Each of these concepts is discussed in more detail later in this chapter.

Geographic inquiry is thus wide-ranging and focuses on big issues, with the goal of understanding the causes and potential solutions to economic development and employment, food production, urban congestion, population explosions and busts, religious and ethnic conflict, climate change, plant and animal extinctions, and other contemporary challenges (figure 1.1, figure 1.2).

ArcGIS Online mapping service

Given that the guiding principle of geography is understanding where things are located and why they are there, the map is an essential tool. While people have used maps for millennia, in recent decades maps have evolved from being static and drawn on paper to being dynamic and digital. This book examines a wide range of geographic issues, drawing heavily on the power of Esri's ArcGIS Online digital mapping service (figure 1.3).

ArcGIS Online is a powerful cloud-based system that allows users to explore and analyze thousands of geographic datasets. Traditional data, in the form of text and spreadsheets, becomes immensely more useful by adding a spatial component via maps.

For instance, by mapping a conventional list of customers' addresses, it becomes possible not only to visualize where customers live but also to identify neighborhoods where few or no customers reside. Analytical tools can further enhance an understanding of customers by mapping statistically significant hot spots, where clusters of customers live, and cold spots, where few customers live. By detecting these patterns, it is then possible to look at underlying social, economic, and environmental characteristics of the hot spots and cold spots. Additional data can be added to the map, which may indicate the cold spot is due to concentration of a distinct immigrant group. On the basis of this geographic information, a site-specific marketing campaign can be developed to appeal to this group.

In this book, most maps are produced with data from ArcGIS Online. This allows students and instructors to not just view maps in a static, printed, format, but to explore them in more detail in ArcGIS Online.

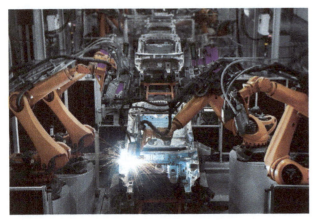

Figure 1.1. The spatial pattern of economic development, such as where industry locates, is one issue explored by geographers. The automobile industry has gone through dramatic shifts in recent decades as manufacturing has moved away from Detroit to factories in the southern United States, Asia, and Latin America. These shifts, as well as technological change such as the increased use of robotics, impact the quantity and types of employment. Photo by Xieyuliang. Stock photo ID: 587205803, Shutterstock.

Figure 1.2. Geographers also study demographic patterns. Fertility rates vary greatly from place to place. Large families can still be found in much of Africa and the Middle East, while smaller families now dominate North America, Europe, and much of Asia. Photo by Avatar_023. Stock photo ID: 23509669, Shutterstock.

In addition, each chapter includes ArcGIS Online exercises, where you will explore geographic datasets with sophisticated analytical tools.

Given that this book is built around ArcGIS Online, before moving on to more detail on the discipline of geography, it is important to first understand how maps function and how new digital technologies are reshaping the way geographers study the world.

Geographic tools and data

Geospatial technology

The traditional tools that geographers have used throughout history have gone through a dramatic transformation with the development of geospatial technologies. These are digital technologies developed in recent decades that allow geographers to collect data about the earth and run sophisticated analyses. With global positioning systems (GPS), remote sensing, and geographic information systems (GIS), vast quantities of data about human and natural features can be collected with great precision and analyzed with sophisticated techniques. Most people are not even aware that these geospatial technologies have become an integral part of our lives. Your cell phone can track your location with GPS, while Google Maps provides vast quantities of satellite imagery and geographic data on roads,

businesses, parks, public buildings, and more. Based on this information, it is possible to determine where you are, then calculate the fastest route from your location to a coffee shop, or to find not just any local coffee shop but a coffee shop with a high customer rating.

GPS is a powerful technology that identifies the location of a receiver unit (such as your cell phone) on the surface of the earth. Created by the US Department of Defense to aid in precision targeting and navigation, the system relies on three components: a receiver unit, a constellation of satellites, and ground-based tracking stations (figure 1.4). A system of twenty-four satellites circles the globe, and the precise location of each satellite is tracked by ground stations. GPS receivers communicate with satellites by sending and receiving radio waves. The time it takes for radio waves to travel between the receiver and a satellite is used to calculate the distance between them. With a minimum of three satellites, a two-dimensional location (latitude and longitude) on the earth's surface is determined. With at least four satellites, a three-dimensional location (latitude, longitude, and altitude) is determined. Based on this system, a GPS receiver works only when it has a clear line of sight to satellites and thus is of limited use indoors. However, many cell phones use technology that compensates for this limitation by using Wi-Fi and cell tower connections with known latitude and longitude coordinates to determine location.

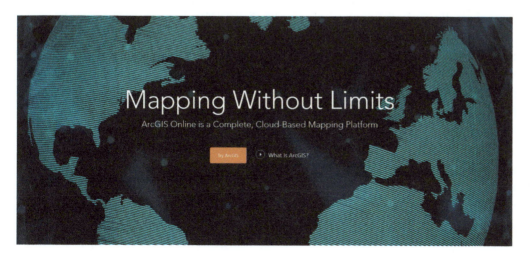

Figure 1.3. This book is designed around ArcGIS Online. Access ArcGIS Online at https://www.arcgis.com. Image by Esri.

Figure 1.4. GPS systems consist of a receiver unit, ground control stations, and a constellation of satellites. Ground control stations track the precise location of satellites. Location is determined by measuring the time it takes radio signals to travel between a receiver unit and satellites with known locations. Image by Art Alex. Stock vector ID: 532342483, Shutterstock.

The most common use of GPS is for navigation. You use GPS technology every time you use Google Maps on your phone to identify where you are and where you need to go. GPS also assists navigation for aircraft, ships, and ground vehicles. But GPS receivers are also powerful tools used for field data collection. Many GPS units allow for the collection of data as points, lines, and areas. An urban arborist can collect point data on trees, taking note not only of each tree's location but also of information on the species, height, health, and more. A surveyor can collect line data on property boundaries and roadways, with associated information on owners, condition, material, and so on. A biogeographer can collect data on areas of illegal logging, noting where the logging has occurred as well as the time period and type of species that is being stolen.

Another important geospatial tool is remote sensing. Remote sensing consists of images of the earth's surface, typically taken from satellites or aircraft

(figure 1.5). Passive remote sensing instruments mounted on these platforms read reflections of the sun's radiation or heat emitted from the earth's surface. Different types of features, such as asphalt, cement, water, soils, rocks, and vegetation types, all reflect radiation differently, thus giving features a unique spectral signature. Active remote sensing instruments emit energy, such as with a laser or microwaves, which bounces off features, showing their location and shape.

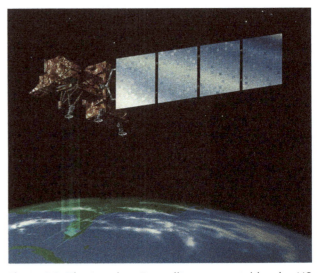

Figure 1.5. The Landsat-7 satellite, operated by the US Geological Survey. Satellites and aircraft are common sources of remote sensing imagery. Image by NASA.

One of the most common uses of remotely sensed imagery is for basemaps, as used in digital maps such as ArcGIS Online. However, imagery goes well beyond simple basemaps. By analyzing the spectral signature of features, areas can be classified, such as in a thematic map of land use/land cover that shows urban areas, forests, different crop types, and more (figure 1.6). Remote sensing is also used for economic research by looking, for example, at the number of cars in retail parking lots and viewing tanker railcars at oil refineries. In environmental monitoring, it is used to track oil spills and determine the health of forests. Local governments use remote sensing to study urban growth and transportation needs. International aid and human rights organizations use it to help evaluate the condition of refugee settlements or to identify areas with mass graves from war crimes. In public health, remote sensing helps evaluate areas of mosquito infestation. As these examples show, remote sensing data is used in myriad professional and technical fields.

GIS is a powerful tool for creating, storing, and analyzing geographic data. GIS combines spatial data (i.e., the location of things) with attribute data (i.e., characteristics of things), essentially bringing the power of maps and spreadsheets together. GIS data is stored and viewed as layers, where each layer is a specific theme (figure 1.7). For instance, a municipal GIS database can have a layer of city trees with their location as well as attribute information on tree species, health, and height. Another layer can have sewer systems with attribute information on diameter and age. Another layer can have parcels with attributes on ownership, land-use zoning, and type of structure.

Figure 1.7. A geographic information system consists of layers of data, which can include land use, roads, parcels, buildings, vegetation, topography, and much more. Image by Naschy. Stock vector ID: 526267657, Shutterstock.

Satellite Remote Sensing Imagery

Remote sensing data can be used for **economic analysis** by counting cars in commercial areas, as seen in a mall in Riverside, CA.

False color infrared images, which show vegetation in red, are used to identify land uses and monitor the **health of vegetation**, as seen around Des Moines, Iowa.

Figure 1.6. Satellite remote sensing imagery. Log in to your ArcGIS Online account to explore these maps. High resolution imagery of mall: https://arcg.is/1L5PWX. False color infrared imagery: http://arcg.is/2m4ByRF. Data sources: World Imagery basemap, Esri, DigitalGlobe, GeoEye, Earthstar Geographics, CNES/Airbus DS, USDA, USGS, AEX, Getmapping, Aerogrid, IGN, IGP, swisstopo, and the GIS User Community. Infrared vegetation. USA NAIP Imagery: False Color. Esri; data sources: Esri, USDA Farm Service Agency.

GIS is a powerful tool for studying spatial distributions and spatial relationships. By looking at a layer of mosquito habitat and comparing it to a layer of recent urban growth, public health officials can analyze and predict how many malaria infections are likely to occur. With a layer of household income, a layer of ethnicity, and a layer of population density, a company can find the best location to sell a product targeting an ethnic niche. For environmental analysis, a layer of roads and a layer of tree species can be used to predict where logging is likely to occur.

Because of the myriad uses of geospatial technologies, there are many employment opportunities for people with these skills. Private companies, such as insurers, market researchers, and environmental consultants, need people who can collect data and map it with geospatial technologies. Government agencies, such as in urban and community development, environmental protection, public health, public works, and economic development, need people with these skills as well. Nonprofit organizations that provide social services, protect the environment, and improve health and economies locally and internationally also hire many people with backgrounds in geospatial technology.

Data sources

Geographic data can be produced in a wide variety of ways. Private companies produce much data, as do governments and researchers at universities and think tanks.

Private companies often collect data on customers, such as their home addresses and purchasing history. With this data, they can produce maps showing the types of products and services people buy in different parts of cities. A more detailed picture of population can be mapped by adding census data collected by governments, which is based on household surveys and can include the number of people, race and ethnicity, income, education, and many other variables. Phone interviews and mail surveys can also be used to collect data and map people's attitudes and opinions on public issues.

Geospatial technologies, such as GPS and airborne remote sensing, are also important sources of data. As mentioned previously, GPS units are used in the field to collect data on any number of things, such as the location of potholes in streets, graffiti locations, buildings in rural villages, well sites, vegetation clusters, and bird nests. Remote sensing technology uses satellites and aircraft to collect data on larger areas. With this technology, data on crop types and health, urban growth, deforestation, illegal construction, and more can be collected.

Field analysis of the cultural landscape is also commonly used by geographers. By going into the field and making observations of the cultural landscape, from how people move and interact in particular parts of the city to types of buildings and land uses in different locations to peoples' perceptions of neighborhoods, geographers collect and map a wide range of data.

Data quality and metadata

With myriad sources of geographic data, users must be very careful when evaluating data quality. Many times, a GIS user will find interesting data that appears useful for a work project or class paper. However, without investigating the quality and source of the data, the user may end up with inaccurate or misleading analysis results.

The most common types of data quality issues include spatial, temporal, and attribute accuracy; completeness; and data source reliability.

Spatial accuracy

Are features in the correct location, and with what degree of precision? For instance, is a hospital mapped at the correct street address, or did it get placed at a similar address in the wrong city? Is a property boundary mapped at a survey level of precision down to centimeters, or is it mapped at a coarser scale, such as meters? If you are building a perimeter wall around a property, a dataset mapped with an accuracy of meters will not suffice.

Temporal accuracy

When was the data created? A map showing voting patterns by county can be very helpful in understanding attitudes toward social issues. However, the map user needs to know if the data is current or if it was created too long ago to be of use.

Attribute accuracy

Are the values in attribute fields correct? For instance, does a map of average income by ZIP code have the correct values? Poorly built databases may have errors, or the numbers presented may have wide margins of error that must be accounted for when interpreting patterns.

Completeness

Are all features included, or are some missing? For example, when mapping home burglaries, is data available for all parts of the city? If not, there may be a false impression that no burglaries occur in one area, while in reality, the absence of burglaries may be due to missing data.

Data source

The origin of the data can indicate level of quality. For instance, a dataset made by the US Census Bureau should be based on high data quality standards. A dataset made by an unknown blogger or for a class project may not be as reliable.

Data quality and other important information is part of a spatial dataset's metadata. Metadata is information about a dataset. It can include data quality, as discussed, as well as information on data collection methods, who produced the data, projection and coordinate systems, and more. When evaluating spatial data, it is important to review the metadata.

Go to ArcGIS Online to complete exercise 1.1, "Introduction to ArcGIS Online."

Map basics

To work well with geospatial technologies, it is important to understand maps and the various ways in which data is presented with them. Different map types are available for conveying different varieties of data, while map scale can influence levels of detail and the types of spatial processes observed. Map projections can influence the user's perceptions of size, shape, and direction when reading maps, and various coordinate systems are used to describe where features are located. Count and rate data are often misunderstood by novice map users, while classification schemes can have a significant impact on how people interpret data. Each of these issues is discussed in more detail below.

Map types

Maps can be classified into two broad categories: *reference maps* and *thematic maps*. Reference maps have a wide range of general information on them. For instance, US Geological Survey topographic maps have information on natural and cultural features such as elevation, roads, public buildings, water features, and political boundaries. Many online maps, such as Google Maps, also have general reference information on roads, businesses, public institutions, entertainment, and more. When you create a new map in ArcGIS Online, you are presented with a topographic reference map as a basemap (figure 1.8).

Reference Map

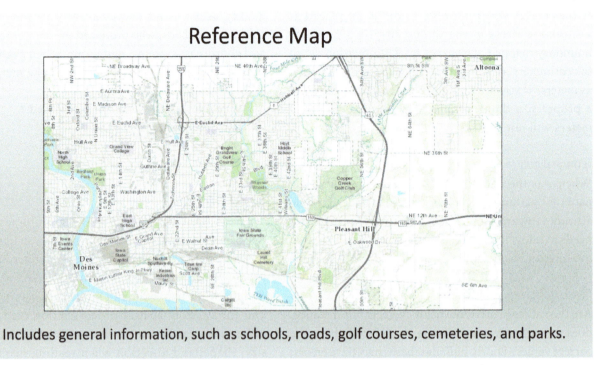

Includes general information, such as schools, roads, golf courses, cemeteries, and parks.

Figure 1.8. Reference map. The topographic basemap in ArcGIS Online includes basic reference information. Data sources: World Topo Map. HERE, DeLorme, Intermap, increment P Corp., GEBCO, USGS, FAO, NPS, NRCAN, GeoBase, IGN, Kadaster NL, Ordnance Survey, Esri Japan, METI, Esri China (Hong Kong), swisstopo, MapmyIndia, © OpenStreetMap contributors, and the GIS User Community.

Thematic maps, in contrast, focus on a single topic, or theme. This type of map may show population density, average income, dominant language, soil type, annual precipitation, or any number of other physical or cultural features. When you add layers to ArcGIS Online (excluding basemaps), such as Living Atlas of the World layers, you are adding thematic maps. Thematic maps can be represented in several different ways, including choropleth maps, graduated circle maps, isoline maps, dot density maps, flowline maps, and cartograms.

A common type of thematic map is the *choropleth map*. Choropleth maps use shades or colors to represent values of a variable within an area, such as census tracts, cities, counties, or states (figure 1.9).

Like choropleth maps, *graduated circle maps* also represent values of a variable within an area. However, instead of using shades or colors to distinguish values, circles of different sizes are used. A large circle represents a high value, while smaller circles represent lower values (figure 1.9).

Thematic Maps: Choropleth and Graduated Circle

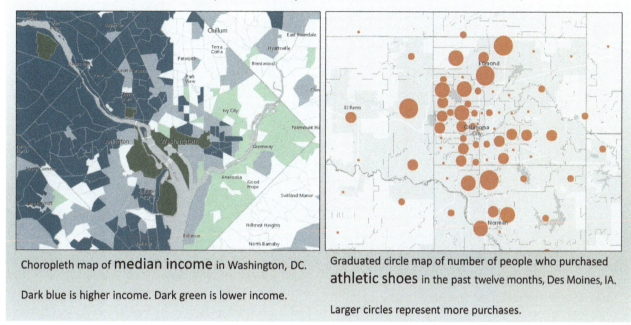

Choropleth map of **median income** in Washington, DC.

Dark blue is higher income. Dark green is lower income.

Graduated circle map of number of people who purchased **athletic shoes** in the past twelve months, Des Moines, IA.

Larger circles represent more purchases.

Figure 1.9. Thematic maps: Choropleth and graduated circle. Choropleth maps use colors or shades within areal features to represent data. Graduated circle maps use circles of different sizes to represent data. Log in to your ArcGIS Online account to explore these maps. Choropleth map of median income: https://arcg.is/1WiuC4. Graduated circle map of market potential for regular exercise routines: https://arcg.is/1jjKHz. Maps by author. Data sources: 2016 USA Median Household Income, Esri, US Census Bureau. 2016 USA Adults That Exercise Regularly, Esri and GfK US, LLC, the GfK MRI division.

Isoline maps consist of lines that connect points of the same value. Typically, these are used to map continuous surfaces, where data values change often over the earth's surface, such as with temperature or elevation (figure 1.10).

Dot density maps use dots to represent a specified value within a geographic feature (figure 1.10). If the population of a county is 10,000 people, then a dot density map where one dot equals 1,000 people would have ten dots randomly placed within the county borders.

Thematic Maps: Isoline and Dot Density

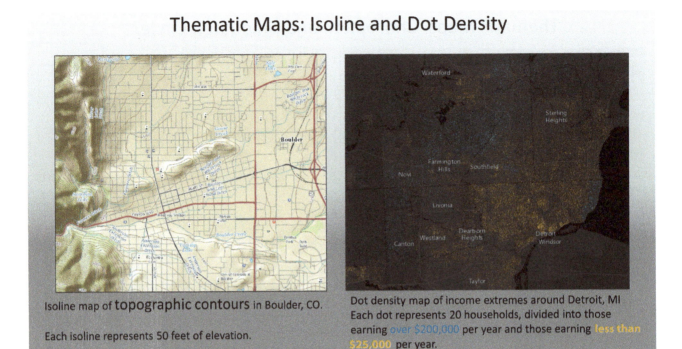

Isoline map of **topographic contours** in Boulder, CO.

Each isoline represents 50 feet of elevation.

Dot density map of income extremes around Detroit, MI Each dot represents 20 households, divided into those earning over $200,000 per year and those earning less than $25,000 per year.

Figure 1.10. Thematic maps: Isoline and dot density. Elevation contours on a topographic map are a type of isoline. Dot density maps use dots to represent values, such as number of households. Log in to your ArcGIS Online account to view these maps. USGS National Map with topographic isolines: https://arcg.is/91zf1. Dot density map of income extremes: https://arcg.is/m8DHL. Data sources: USGS National Map by Esri–USGS The National Map: National Boundaries Dataset, National Elevation Dataset, Geographic Names Information System, National Hydrography Dataset, National Land Cover Database, National Structures Dataset, and National Transportation Dataset; US Census Bureau–TIGER/Line; HERE Road Data. Income Extremes by Lisa Berry–Esri.

Flowline maps use lines of varying thickness to show the direction and quantity of spatial interaction between places. Thicker lines represent larger quantities, while thinner lines represent smaller quantities. These maps are often used to represent trade and migration flows between countries (figure 1.11).

Cartogram maps distort the area of features based on the value of a variable. A cartogram of population will show places with more people as larger and places with fewer people as smaller. In figure 1.11, state populations are shown for three time periods. The size of each state varies according to its population size. Note how western states, such as California, change in size in each time period.

Thematic Maps: Flow Line and Cartogram

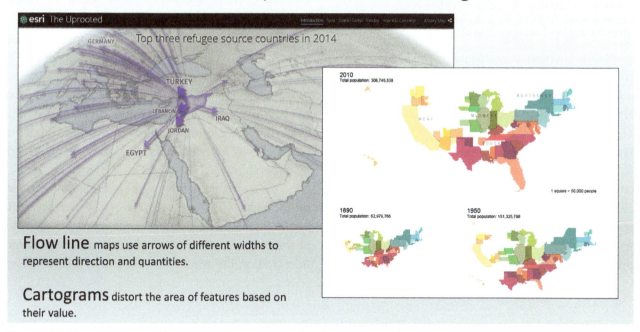

Flow line maps use arrows of different widths to represent direction and quantities.

Cartograms distort the area of features based on their value.

Figure 1.11. Thematic maps: Flowline and cartogram. The flowline map shows Syrian refugee flows in 2014. View the Syrian refugee flow map at https://storymaps.esri.com/stories/2016/the-uprooted/index.html. Cartogram from US Census. Image sources: The Uprooted by Esri Story Maps Team; data sources: UNHCR, Airbus Defense and Space. Cartograms of State Populations in 1890, 1950, and 2010 by US Census Bureau; data sources: Census 2010 tables.

Map scale

Scale is another issue to be aware of when creating and interpreting maps. Real estate companies often produce maps with no scale or with distorted scales to make desirable places seem closer. For instance, a real estate map may include the location of a new housing development, with lines showing freeways, beaches, and parks, giving the impression that they are all nearby. However, with no given scale, these places are often drawn to appear much closer than they really are.

Properly produced maps include a clearly defined map scale that indicates the ratio of map distance to real-world distance. The scale allows map readers to measure the size of features and the distance between them. Map scale is represented verbally, graphically, or as a ratio or fraction.

Verbal scale: 1 inch equals 1 mile
Graphic scale: 0 50 100mi
Ratio scale: 1:24,000
Fraction scale: 1/24,000

In the case of ratio and fraction scales, the units remain the same on both sides of the scale. Using the examples noted, 1 inch on the map represents 24,000 inches in the real world.

Maps are often described as being *large scale* or *small scale* (figure 1.12). A large-scale map refers to a larger fraction or ratio, while a small-scale map refers to a smaller fraction or ratio. For instance, 1:24,000 is a larger ratio than 1:100,000, so it is a larger scale map.

Large-scale maps are more "zoomed in." They cover a smaller area and include more detail. A city map is a larger-scale map than a country map. Small-scale maps are "zoomed out" and cover a larger area with less detail. A country map is a smaller-scale map than a city or neighborhood map.

To remember the difference between large- and small-scale maps, either think in terms of ratios or fractions, or use this trick: your neighborhood looks larger on a large-scale map (because it is more zoomed in), while your neighborhood looks smaller on a small-scale map (because it is more zoomed out).

While map scale is important for measuring size and distance and determining the level of detail shown, it is also important to understand scale in terms of how it affects the spatial patterns observed by geographers.

This is often referred to as the *modifiable areal unit problem* (MAUP). In essence, the unit of measurement used for analysis, be it countries, states, counties, cities, or some other area, can strongly influence the patterns observed on the map. For instance, at a state scale,

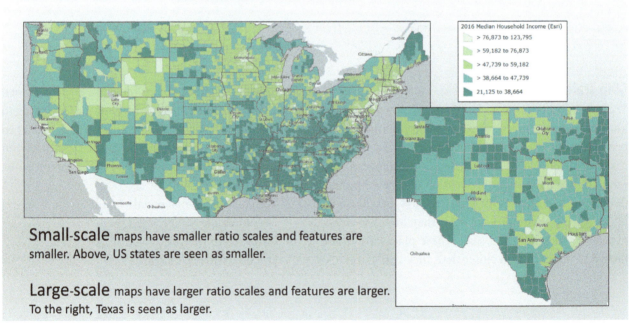

Small-Scale and Large-Scale Maps

2016 Median Household Income (Esri)
- > 76,873 to 123,795
- > 59,182 to 76,873
- > 47,739 to 59,182
- > 38,664 to 47,739
- 21,125 to 38,664

Small-scale maps have smaller ratio scales and features are smaller. Above, US states are seen as smaller.

Large-scale maps have larger ratio scales and features are larger. To the right, Texas is seen as larger.

Figure 1.12. Small-scale and large-scale maps. Large-scale maps are more zoomed in than small-scale maps. Explore this map at https://arcg.is/rjL8K. Maps by author. Data sources: 2016 USA Median Household Income by Esri. Esri, US Census Bureau.

the "red state/blue state" divide in US presidential elections clearly shows states such as Texas as solidly red (Republican). But by changing the scale of analysis, new spatial patterns emerge. At a county scale, large urban areas within Texas appear as blue (Democratic) patches (figure 1.13). So, while a state level of analysis is useful in understanding the Electoral College for presidential elections, a county-scale analysis is more useful for understanding House of Representative and local election results.

There is no single "proper" scale of analysis for all geographic questions. Rather, the proper scale depends on the question being asked. If the US government has funds available to help states tackle high unemployment, then analyzing unemployment rates at a state level makes sense. On the other hand, if a city government wants to identify neighborhoods with high unemployment rates, then the proper scale of analysis would be urban neighborhoods.

Geographers are interested in spatial patterns at a wide range of scales, always keeping in mind how patterns and processes interact between global and local levels. These interactions have become even more essential to understand due to globalization, the process whereby places become increasingly interconnected through communication networks, transportation technology, and political policies.

For instance, global patterns of manufacturing output and employment show dramatic shifts from developed countries to developing countries, especially

Scale of Analysis and the Modifiable Areal Unit Problem

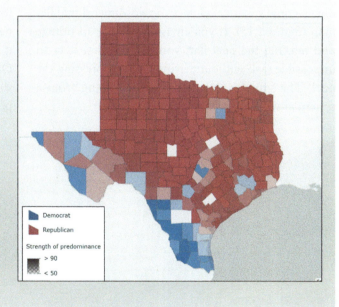

The areal unit used in a map heavily influences observed spatial patterns.

Data aggregated at a state scale illuminates different patterns than data aggregated at a county scale.

Texas is heavily red (Republican), when mapped at the state level. It has voted Republican in all presidential elections since 1980.

But by changing the areal unit to counties, one can see areas of blue (Democrat).

Figure 1.13. Scale of analysis and the modifiable areal unit problem. Explore these maps at https://arcg.is/yDHKy.

Maps by author. Data sources: State level–Federal Election Commission. Texas counties–Texas Office of the Secretary of State.

China and other Asian states. This shift has had a profound impact on development patterns at a global scale, most obviously with the economic, political, and military rise of China. However, these global processes also play out at a more local scale. The shutdown of automobile factories in Detroit has had a devastating impact on that city (figure 1.14).

Myriad impacts, such as massive population decline, abandonment of entire neighborhoods, increases in crime, and municipal fiscal crises have played out locally, all because of global shifts in manufacturing production. At the same time, local-scale impacts in China have transformed many cities, with greater wealth and opportunity combined with pollution of air, water, and soils.

Thus, when deciding the proper scale for creating a map, it is essential to first have a clear idea as to what processes—from global to local—you want to address.

Map projections

Map projections are necessary to transform a three-dimensional spherical globe to a two-dimensional flat map (figure 1.15). If you envision peeling an orange and making the peel flat, you can see that it is an impossible task without tearing and compressing the peel. The same problem arises when going from a spherical world to a flat map.

Map projections cannot preserve all spatial elements of a map: area, shape, distance, and direction. Just like when flattening an orange peel, something must give. Maps projections that preserve area are known as *equal-area projections*. These projections show the correct area, such as the square miles of countries and states, but shape, distance, and direction will be incorrect. Projections that preserve shape are known as *conformal projections*. With these projections, the shape of features, such as country or state boundaries, are correct, but area, direction, and distance measurements will be off.

The Mollweide projection is a good example of an equal-area projection (figure 1.16). Area is preserved, so for example, the square miles within each country are accurate. However, shape, distance, and direction are distorted.

A popular map projection that illustrates the tradeoff between area and shape is the Mercator projection (figure 1.16). This projection is conformal, so shape is preserved, but area is dramatically distorted toward the poles. For example, Greenland appears to be the same size as the entire continent of Africa, while it is actually about fourteen times smaller. ArcGIS Online uses the Web Mercator projection, which is a slightly modified version of the traditional Mercator projection.

Figure 1.14. Abandoned Packard automobile factory in Detroit, Michigan. Geographic processes are linked from the global to the local scales. Global shifts in manufacturing have had devastating impacts on some local **areas.** Photo by Atomazul. Stock photo ID: 154954085, Shutterstock.

Map Projection

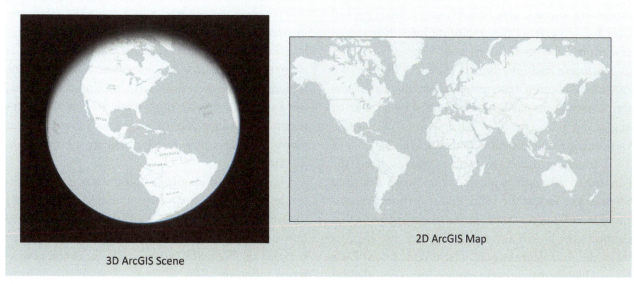

3D ArcGIS Scene

2D ArcGIS Map

Figure 1.15. Map projection. When transforming a spherical representation of the world to a flat representation, distortions are unavoidable. Distortions can be in area, shape, distance, and direction. Images by Esri.

Equal Area and Conformal Map Projections

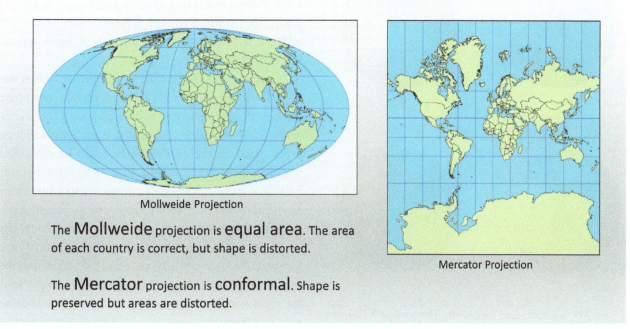

Mollweide Projection

The **Mollweide** projection is **equal area**. The area of each country is correct, but shape is distorted.

The **Mercator** projection is **conformal**. Shape is preserved but areas are distorted.

Mercator Projection

Figure 1.16. Equal-area and conformal map projections. These examples represent an equal-area projection and a conformal projection. Explore the Mollweide projection at http://arcg.is/2m4Q8so. Explore the Mercator projection at http://arcg.is/2l4zdBT. Maps by EsriedtmCF.

Coordinate systems

Given that the major focus of geography is on where things are located, geographers use various types of coordinate systems that facilitate identification of places on the surface of the earth.

Latitude and longitude is the most well-known geographic coordinate system. It allows all locations on the surface of the earth to be identified by measuring angles north and south of the equator and east and west of the prime meridian (figure 1.17).

Latitude is measured from 0 degrees along the equator to 90 degrees north at the North Pole and 90 degrees south at the South Pole. Longitude is measured from 0 degrees at the prime meridian, a line that connects the North and South Poles, to 180 degrees west and 180 degrees east. The International Date Line, which demarcates the change from one calendar day to the next, is located approximately along the 180-degree meridian.

Whereas the equator, which splits the earth into northern and southern hemispheres, is a natural location for starting latitude measurements, there is no natural place to begin longitude measurements. Different prime meridians have been used over time, but by the late 1800s, due largely to Great Britain's maritime dominance in the nineteenth century, most maps began using the prime meridian at Greenwich, England.

Latitude and longitude coordinates can be written in decimal or degree/minutes/seconds formats (figure 1.18). For example, the White House, located between the 38th and 39th northern parallels and between the 77th and 78th western meridians, is written as follows:

Decimal degrees: 38.8977° N, 77.0366° W
Degrees/minutes/seconds: N 38° 53' 49.5456",
    W 77° 2' 11.562"

Another commonly used method for describing the location of a place is with street addresses, whereby each address refers to a specific building in a specific place. The location for the White House, as a street address, is 1600 Pennsylvania Ave NW, Washington, DC 20500.

One unusual and innovative coordinate system has been developed by What-3-Words. With this coordinate system, the entire world is divided into 3 × 3 meter grids, each of which is assigned three words. Thus, every place on the earth's surface can be identified with just three words within three meters of accuracy. This has some advantages compared to traditional coordinate systems. First, many places do not have an official street address, which severely restricts the usefulness of a street address system in identifying locations. Second, while latitude and longitude describe specific locations, they are too long and complicated for most people to remember. In contrast, it is quite easy to remember three words. With this system, the location of the White House is described as "sulk.held. raves." With the What-3-Words app, businesses and governments can deliver goods and services to precise

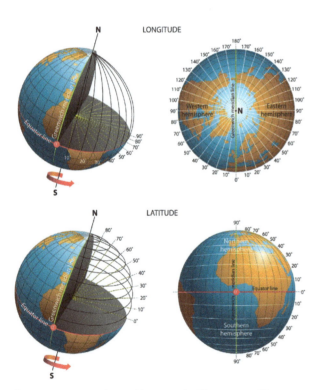

Figure 1.17. Latitude and longitude. This image illustrates longitude lines running from zero degrees at the Greenwich prime meridian to 180 degrees and latitude lines running from the equator to the North and South Poles. Image by NoPainNoGain. Stock vector ID: 326090990. Shutterstock.

locations, from the proper building entrance on a large corporate campus to a remote home in rural Kenya. In 2016, the postal service of Mongolia, where few streets have official names, began using this system nationwide.

Many other types of coordinate systems are used throughout the world. When you take additional classes on geography and geographic information systems, you will be able to delve more deeply into them.

Counts vs. rates

Another issue to keep in mind when creating and reading maps is the difference between counts and rates. As the name implies, *counts* are a count of the number of features in an area. A population count map will show the number of people in an area, such as a city, while a terrorist activity count map will show the number of terrorist incidents, such as within a country.

Rates compare one variable to another. In geography, it is common to calculate rates on the basis of population or area. A wheat production map can show the amount of wheat within a county divided by the area in square miles of the county, resulting in wheat production per square mile. Likewise, the number of people with influenza within a state can be divided by the total population of the state, giving the influenza rate per 100,000 people.

Understanding the difference between counts and rates is essential. If a political party targets the Hispanic community and is looking for a good location for a get-out-the-vote campaign, a map showing counts and a map showing rates can lead to very different location decisions (figure 1.19). For instance, there may be census tracts with a very high proportion of Hispanic people (i.e., a high rate). This high rate may appear to indicate a good location for the campaign. However, while 90 percent of the population may be

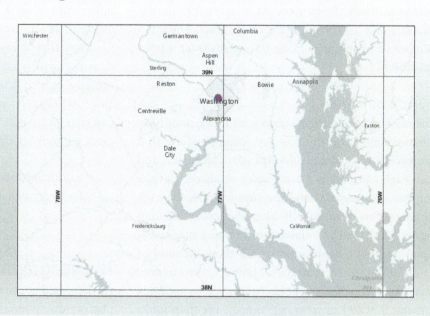

Latitude and Longitude: Location of the White House

Located between the **38**th and **39**th northern parallels and the **77**th and **78**th western meridians.

Decimal degree notation:
38.8977° N
77.0366° W

Degrees/minutes/seconds notation:
N 38° 53' 49.5456"
W 77° 2' 11.562"

Figure 1.18. Latitude and longitude: Location of the White House. This map shows the location of the White House in relation to 1-degree latitude and longitude grid lines. Explore this map at http://arcg.is/2m4WPKR. Map by author. Data sources: Esri, HERE, Garmin, NGA, USGS, NPS.

Counts vs. Rates

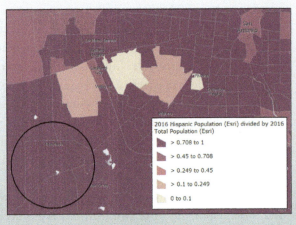 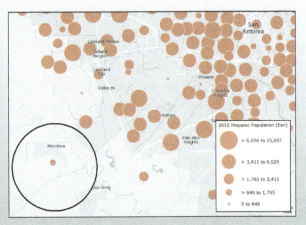

Rate map: Hispanic Population **divided by** Total Population

Count map: Hispanic Population

Is Macdona, TX a significant Latino neighborhood? The answer varies depending if Latino **rates** are mapped or if Latino **counts** are mapped.

Figure 1.19. Counts vs. rates. When creating and interpreting maps, very different impressions result from classifying data by rates and by counts. Explore these maps at https://arcg.is/0H1uvO. Maps by author. Data sources: 2016 USA Diversity Index. Esri, US Census Bureau.

Hispanic, when mapping counts, it may turn out that there are only 100 people in the census tract. The small number of people may make the census tract a poor location in reality.

Map classification

The classification scheme used with a map can have a major impact on the way it is interpreted. With a choropleth map, data is divided into categories, and then each category is given a color or shade. But the number of categories and the cutoff points for each category can dramatically alter the look of a map (figure 1.20). In the following example, a map using equal interval classification would show incomes of $160,000 in the top category. However, the quantile classification scheme would include all households earning $79,894 or more. Obviously, the map looks very different depending solely on the chosen classification scheme (figure 1.21). One scheme gives the impression that wide swaths of the Seattle region are

upper income, while the other scheme makes the prevalence of upper income areas look much more limited.

Note that changing the map classification scheme does not involve changing any of the data. The data remains exactly the same. All that changes are the cutoff points for each color category. Cartographers can thus easily manipulate the perception that a map gives without falsifying data in any way.

Go to ArcGIS Online to complete exercise 1.2: "Map basics with ArcGIS Online."

The geographic perspective

As discussed at the beginning of this chapter, geography is a discipline that, at its core, asks where things are located and why they are there. Broadly speaking, geography can be seen from a spatial perspective and an ecological perspective. The spatial perspective

Classification Schemes

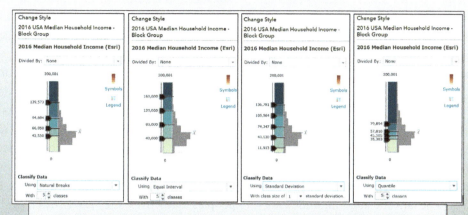

- Natural Breaks. Data is divided into categories based on natural groups within the data.
- Equal Interval. Data is divided so that each category has the same range of values.
- Standard Deviation. Data is divided by into categories by standard deviations above and below the mean.
- Quantile. Data is divided so that groups contain an equal number of values.

Figure 1.20. Classification schemes. Different classification schemes using the USA Median Household Income layer. Note how the category cutoff points can change dramatically depending on the classification scheme used. Image by author. Data Source: Esri, US Census Bureau.

Is Seattle Rich or Poor?

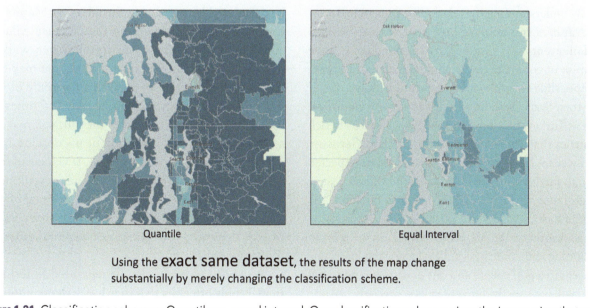

Quantile   Equal Interval

Using the **exact same dataset**, the results of the map change
substantially by merely changing the classification scheme.

Figure 1.21. Classification schemes: Quantile vs. equal interval. One classification scheme gives the impression that most of Seattle is affluent, while the other shows affluent areas as much more limited in scope. Explore this map at http://arcg.is/2m5n4B3. Maps by author. Data sources: 2016 USA Median Household Income by Esri; Esri, US Census Bureau.

examines spatial distributions and processes, while the ecological perspective offers a holistic view that incorporates both human actions and environmental opportunities and constraints. This section dives deeper into the fundamental concepts that constitute the geographic perspective.

Space

Location and distance are key components of geographic inquiry and can be viewed in both absolute and relative terms.

Absolute location describes a fixed point on the surface of the earth. The latitude and longitude coordinate systems, as well as street address systems, refer to absolute location.

Relative location is another way of describing where things are and is arguably more significant for much geographic research. Relative location describes where a feature is located in relation to another feature. For example, the location of a house can be described as *1 mile from* the freeway, *close to* shopping, *far from* the beach, or *adjacent* to a park. Each of these terms describes where the house is located relative to other important landscape features.

By understanding the relative location of features, geographers can analyze how spatial relationships explain events. For instance, by knowing the relative location of countries in the Middle East and Europe, it is possible to understand migration flows out of war-torn Syria. Syrians will flee to nearby countries, such as Turkey, Lebanon, and Jordan, as well as to rich countries that are not too far away, such as Germany and Sweden. Many fewer migrants would be expected to go to farther away to Canada or the United States, which have a relative distance that is far from the Middle East.

As another example, relative location is useful in explaining real estate prices. Two identical houses, one adjacent to a golf course and one close to an industrial park, will have vastly different values, precisely because of their location relative to different land uses.

Closely related to location is the concept of distance. As with location, distance can be measured in absolute and relative terms. Absolute distance can be measured in traditional units, such as miles and feet or kilometers and meters. Relative distance looks at distance in terms of a surrogate value such as cost or difficulty.

Absolute distance is commonly measured by geographers in two ways (figure 1.22). Euclidean distance measures the distance between two points in a straight line. When people use the common vernacular "as the crow flies," they are referring to Euclidean distance. Drawing a straight line from your house to school would give you the Euclidean distance. However, in peoples' daily lives, they rarely travel in straight lines. For this reason, Manhattan distance, also called *network distance*, is also used in geographic analysis. Manhattan distance (named after the rectangular layout of Manhattan streets) is the distance between two places along a grid. When you travel from home to school, you probably don't fly in a straight line. Most likely, you follow a street grid, which results in a longer total distance travelled.

Distance can also be measured in relative terms as cost distance. This can include cost in time or cost in difficulty of travel. For instance, cost distance can be calculated by measuring Euclidean or Manhattan distance and then weighting the distance value to account for the difficulty of travel. When walking from your house to the grocery store, you may have two options. Option one may be a flat route of 0.75 miles, while option two may be only 0.5 miles but include a steep hill. Because of the hill, you may add a cost value (either consciously or unconsciously) to give that distance a greater weight. If you decide that walking over the hill is twice as difficult as walking on the flat route, you can multiply the hill route by two (0.5 miles × 2 = 1.0 mile). Based on this calculation of cost distance, you would decide to take the flat 0.75-mile route.

Cost distance can also be measured in terms of time. People often say that they live "twenty minutes" from school rather than saying they live eight miles from school. Geographers use cost distance when calculating drive times. Different types of roads have

Measuring Absolute Distance

Absolute distance can be measured by Euclidean distance, "as the crow flies,"
or by
Manhattan (Network) distance,
along a street grid.

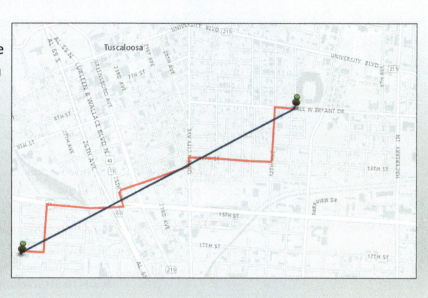

Figure 1.22. Measuring absolute distance. Euclidean distance in green (1.48 miles) follows a straight path between two points. Manhattan or network distance in red (1.93 miles) follows the street grid. The red line can also be measured as cost distance in terms of time. The cost in time will vary on the basis of traffic conditions, so that at midnight it may be 8.5 minutes, while at 5:30 p.m. it may be 12 minutes. Map by author. Data sources: City of Tuscaloosa, Esri, HERE, Garmin, INCREMENT P, NGA, USGS.

different speed limits or are made of different materials. A vehicle travelling for twenty minutes will go much farther on a state highway than on a narrow dirt road. For this reason, different road types can be weighted differently for calculating travel time. Also, traffic conditions can vary by time of day, resulting in a cost distance that varies not only over space but also over time.

Go to ArcGIS Online to complete exercise 1.3: "Location and distance."

Spatial patterns

Features on the earth's surface arrange themselves in *spatial patterns*. Analyzing these patterns allows geographers to elucidate not only how human and physical features are arranged but also the processes behind their formation.

A commonly used description of spatial patterns is density. Density is the number of features per unit area, as in the number of people per square mile or number of trees per square kilometer. Density is useful for illustrating spatial patterns that would not be seen using raw numbers alone. For example, the population of California is about 39 million people, while the population of Singapore is only 5.5 million. With no additional information, one may get the impression that California is more crowded than Singapore. But when information on area is added, that impression quickly changes. California consists of 163,696 square miles, while Singapore is made up of just 278 square miles. So, in reality, Singapore has a much higher population density than California (figure 1.23).

Figure 1.23. Population density: Singapore. Singapore has one of the highest population densities in the world, with 5.5 million people living in just 278 square miles. Photo by joyfull. Stock photo ID: 138766448. Shutterstock.

Spatial patterns can also be viewed in terms of clustering, randomness, and dispersion (figure 1.24). As the name implies, clustered features are found grouped near each other. Clusters are often identified with hot spot analysis or with a heat map (figure 1.25). Randomly distributed features have no particular spatial pattern. Dispersed features are those that repel each other. They are certainly not clustered and are even farther from each other than if the distribution were random.

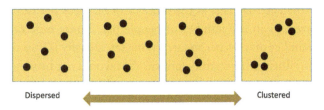

Figure 1.24. Spatial patterns can be seen as dispersed, random, or clustered. Image by author.

Analysis of these types of spatial patterns has many applications. For example, if home burglaries are found to be clustered in a specific neighborhood, police can increase patrols in that area, while detectives and community groups can focus on what the

underlying causes of the crime cluster are. It may turn out that a prolific burglar lives nearby, or youth from a local high school may be committing crimes after school. If home burglaries are not clustered, but have a more random pattern, then other causes may be at play, such as burglaries being crimes of opportunity, where criminals take advantage of homes with open windows.

Diseases often cluster as well. If cancer rates are found to cluster in a neighborhood, then health researchers may search for environmental causes of the disease, such as a nearby toxic waste site. If cancer cases are randomly distributed around a city, then environmental factors are less likely to be the cause.

Dispersed features can include shopping malls or chain restaurants in an urban region. Mall owners may intentionally maintain a distance from competing malls to avoid competition, while owners of a restaurant chain may space their stores so that they do not cannibalize sales from each other.

Spatial patterns can also be analyzed by measuring the center of features. With a map of consumer purchasing behavior, a business may want to find a new store location that lies at the center of its specific market segment. Likewise, geographers can study shifts in population by mapping the center of US population over time.

Spatial relationships

Mapping the *spatial relationships* of two or more features can offer insight into why particular patterns exist. Whereas spatial distributions describe how features are clustered or dispersed, spatial relationships depict where features are located in relationship to other types of features. For instance, geographers study the distance between different types of features or whether different feature types overlap (figure 1.26). If there is a disease cluster, geographers can examine the distance between the cluster and factories that emit toxic effluent. If the cluster is nearby, then the effluent may be the cause of the disease. They can also study whether the disease cluster overlaps with the residences of workers in a specific type of occupation. It may turn

Mapping Clusters as Hot spots

Residential burglary clusters in Long Beach, CA.

ArcGIS hot spot analysis of residential burglaries shows statistically significant hot spots and cold spots.

Red represents hot spots with more burglaries, while blue represents cold spots with fewer burglaries.

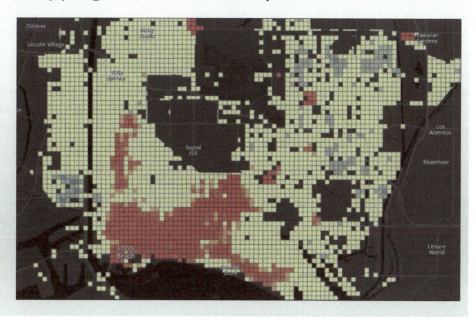

Figure 1.25. Mapping clusters as hot spots. Hot spot analysis can uncover clusters of crime, different demographic groups, disease, natural hazard events, and much more. Map by author. Data source: Long Beach Police Department.

out that the cluster is not due to nearby toxic effluent but rather that many residents in the disease cluster work in a mine that uses toxic chemicals.

Figure 1.26. Spatial relationships. Geographers study the spatial relationship between features, such as how far apart they are (left) or whether they overlap (right). Image by author.

Statistical tools are often used to study spatial relationships. With spatial correlation, it is possible to analyze the strength and direction of spatial relationships (figure 1.27), be they positive, negative, or unrelated. A positive relationship is when both variables change in the same direction, as when places with high unemployment also have high rates of alcohol consumption.

A negative relationship is when an increase in one variable leads to a decrease in another, as when areas with high unemployment have lower traffic fatalities due to people driving less. When there is no pattern of increase or decrease between two variables, they are unrelated, as when places with high unemployment have no correlation with the number of earthquakes.

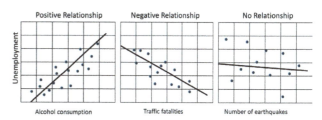

Figure 1.27. Spatial correlation. Variables in the same place can be plotted to see if they have positive, negative, or no relationship. Image by author.

With quantitative analysis, the phrase "correlation does not imply causation" is commonly used to describe the case where variables can be correlated, but one variable does not cause the other to change. To build on an earlier example, a cancer cluster may be located near a toxic effluent site, leading some people to infer that cancer risk increases because of proximity to the site. But, in reality, there may be a third variable that is not being considered. Even though the cancer cluster correlates with distance to toxic effluent, it may turn out that the cancer is due to where residents of the cluster work. It may be that many residents of the cluster work in a mine that uses toxic chemicals and that exposure to those chemicals is causing the disease. Cancer may have a strong correlation with proximity to toxic effluent, but the proximity is not the cause.

It is therefore important to consider multiple explanations when looking at correlations and to use previous research and theory when determining which variables to include in an analysis. When mapping heart disease by county and determining which factors contribute to it, current scientific research says that variables such as smoking, diet, and physical inactivity are contributing factors. Rates of smoking, rates of high cholesterol from poor diet, and average hours of exercise per capita can be mapped on top of a heart disease map. With spatial statistical analysis, the strength of each variable can be analyzed in relation to rates of heart disease. It may become clear that some counties have high rates of heart disease primarily due to high rates of smoking, while others may have high rates due to a lack of physical activity.

Go to ArcGIS Online to complete exercise 1.4: "Spatial patterns and spatial relationships. An analysis of homicide patterns in Chicago."

Places and regions

Many people are drawn to geography because they love to explore and learn about the great diversity of the world. From quaint towns along the coast of Italy to the busy markets of Casablanca, and from the scenic valley of Yosemite to the steppes of Kenya, the world offers a vast range of interesting landscapes to experience. To better study and understand the unique characteristics of the world's landscapes—the locations and spatial patterns of human and physical features—geographers use the concepts of place and region.

Places

Places are locations with a set of physical and/or human features that make them unique from other locations. Because of their uniqueness, they typically have names that can be found on general reference maps. For example, Yellowstone National Park is a place that is distinguished by natural geysers; wolves and other wildlife; and a complex tourism infrastructure of lodging, restaurants, roads, and hiking trails. Manhattan is a place that is distinguished by dense, high-rise buildings; an economy focused on areas such as finance, law, and advertising; and landmarks such as Central Park and Times Square. Venice Beach in Los Angeles is a place with eccentric boardwalk venders and entertainers.

The unique combination of features within a place, when experienced by people, create what is known as a *sense of place*. Sense of place comes from an emotional reaction that forms as humans interact with places. Some places have a very strong sense of place, evoking either a positive or negative reaction in people. For many, a place such as Paris has a very strong, positive sense of place. History, architecture, street layout, cafés, pedestrian activity, and parks combine to create a unique sense of place that people are strongly attached to (figure 1.28). Because of this, demand for housing is strong and tourism is a flourishing industry there. In contrast, a neighborhood in a big city that is trash strewn, largely abandoned, and littered with remnants of drug use may have a strong negative sense of place that repels people.

Placelessness is the antithesis of sense of place. Some places lack uniqueness, offering homogenous landscapes that differ very little from other places. Many urban areas in North American cities consist of wide arterial streets lined by fast-food restaurants and gas stations. The architecture and design of these places remain basically the same, whether it is in Los Angeles, Miami, Atlanta, or Phoenix. Some argue that

Figure 1.28. Café in Paris. Paris is a city with a strong sense of place that comes from its unique combination of human and physical features. Photo by Stefano Ember. Stock photo ID: 237982663. Shutterstock.

cookie-cutter suburban residential development is also placeless. By this argument, these developments consist of large swaths of homogenous suburban homes that lack design tied to local history or culture (figure 1.29).

When geographers talk about places, they typically focus on the wide range of characteristics that compose the place—the people, the built environment, the natural environment—and the ways in which these characteristics form either a strong sense of place or a bland placelessness.

People perceive and navigate through places, and develop a sense of place, through mental maps, which are the way humans organize places in their minds. Most people can draw a map of their city, country, or the world, but in all likelihood, there will be more detail and precision in areas that they move through on a regular basis or are exposed to through various media. You probably have a detailed mental map of your neighborhood, which includes the location of local businesses, parks, the location of homes of people you like to see (and those whom you do not), places that are pleasant to travel through, and places that are dangerous or unpleasant.

Mental maps influence how we move through our cities and neighborhoods, as people tend to follow familiar routes and avoid unfamiliar routes. People also move on the basis of perceived characteristics

of areas in their mental maps, staying clear of areas seen as dangerous or staying in areas viewed as aesthetically pleasing.

Beyond our own neighborhoods and cities, we still form mental maps. Most Americans can draw a rough map of the United States as well as a partial map of the world with some countries and continents. Through the study of geography, our mental map becomes more detailed. Well-developed mental maps allow us to better understand our world and the events that take place in it. Knowing that Iraq's neighbors include countries such as Syria, Turkey, Iran, and Saudi Arabia makes it easier to understand the complex ethnic, religious, and political stresses tearing at that country: Islamic State fighters centered in Syria, ethnic Kurds that straddle Turkey and northern Iraq seeking autonomy, and sectarian rivals from Saudi Arabia and Iran that clash in Iraq.

Regions
Regions are locations with unifying characteristics that distinguish them from other locations. Unlike places, which are viewed in a more holistic way and are more often found on general reference maps (Paris, Yellowstone National Park, Manhattan, Venice Beach), regions are distinguished by a limited number of human and/or physical characteristics.

Placelessness

Many people view bland streetscapes and cookie cutter suburbs as placeless, from housing tracts in Phoenix, Arizona, to peripheral roadways in Moscow, Russia.

Figure 1.29. Placelessness. Homogenous landscapes with few distinguishing characteristics are often considered "placeless." Such places appear virtually the same in any location, with no visible ties to local culture and history. World Imagery basemap by Esri; data sources: Esri, DigitalGlobe, GeoEye, Earthstar Geographics, CNES/Airbus DS, USDA, USGS, AEX, Getmapping, Aerogrid, IGN, IGP, swisstopo, and the GIS User Community. Streetscape photo by Yuriy Stankevich. Royalty-free stock photo ID: 589563587. Shutterstock.

Regions are a useful way of categorizing the world for purposes of geographic research. Just as biologists categorize the living world into species and historians categorize time into eras, geographers categorize space into regions. By creating categories, biologists can compare wolves with dogs, historians can compare the Middle Ages with the Renaissance, and geographers can compare North America and Latin America.

There are three types of regions: formal, functional, and perceptual.

Formal regions can be identified by mapping one or more human or physical features. The Corn Belt in the United States can be identified by mapping acreage devoted to corn production, while Tornado Alley can be found by mapping tornado frequency (figure 1.30). The Bible Belt can be mapped by the number of people who state they attend church on a regular basis. A common cultural trait, such as language or religion, can be used to distinguish North America from Latin America.

Functional regions are delineated by a central place, or node, and a surrounding hinterland that interacts with the node. For example, regional shopping malls and other businesses collect customer address data that they map and use to determine their functional sales region, or market area. A functional metropolitan region can be identified by mapping commuting patterns of workers and consumers who travel to a city (figure 1.31). The advantage of mapping functional regions is that it avoids the use of artificial city or county or even state boundaries when determining the region surrounding a central place.

Perceptual regions, also called *vernacular* regions, are based on subjective criteria of individuals. Everyone knows that the South and Southern California exist in the United States, but where exactly are they? One

Formal Regions

Formal regions are identified by mapping **one or more variables.**

Tornado alley is a commonly used term to describe the US region heavily impacted by tornados.

However, there is no official definition of exactly where the region is.

One way of identifying Tornado Alley as a formal region is by mapping **clusters** of **high intensity tornados**

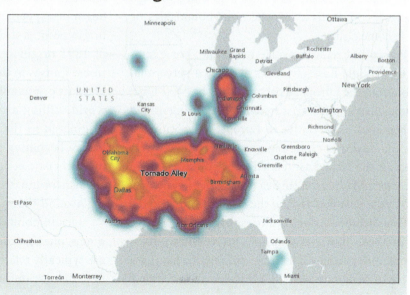

Figure 1.30. Formal regions: Tornado Alley. View the entire dataset of over 52,000 tornados from 1950 to 2008 at http://arcg.is/2lzJ4Rk. Map by author. Data source: National Weather Service, Storm Prediction Center. Publication Date: 200906. Title: United States Tornado Touchdown Points 1950-2008. Publication Information: Publication Place: Reston, VA. Publisher: National Atlas of the United States.

Functional Regions

Metropolitan areas can be seen as functional regions that consist of a central urban core and surrounding commuter neighborhoods.

Functional regions often reflect more realistic areas for studying cities than official city and county boundaries, in that they consider **spatial interaction** with surrounding areas.

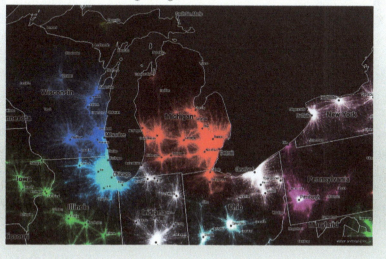

Figure 1.31. Functional regions: Commuter megaregions of the US. This map shows commuter flows between census tracts based on the 2006-2010 US Census, American Community Survey. Map by Rae, Alasdair; Garrett.G.D.Nelson@dartmouth.edu (2016): United States Commutes and Megaregions data for GIS. Figshare. https://doi.org/10.15131/shef.data.4110156.v4.

way of identifying this type of region is by having people draw boundaries on a map. With a large enough sample, a consensus as to where the boundaries lie will become clear.

Perceptual regions evolve over time. For example, the Middle East gained its name as a perceptual region from the European perspective. The Near East, Middle East, and Far East were historically identified by their locations relative to Europe. Perceptions of the boundaries of these regions have changed over time, but the Near East tended to be countries along the eastern Mediterranean Sea (those "near" Europe), the Middle East was around the Persian Gulf, and the Far East included Asian countries that face the Pacific Ocean (those "far" from Europe). Perceptions of these regions differ somewhat today, but again, having a sample of people draw "the Middle East" on a map would lead to a consensus on its boundaries.

It is important to keep in mind that the boundaries between all types of regions are typically fuzzy. Just as there is no specific day that divides the Middle Ages from the Renaissance, there is rarely a fixed line that separates one region from another.

Go to ArcGIS Online to complete exercise 1.5: "Places and regions."

Origin, spatial interaction, and spatial diffusion

Places, regions, and the spatial patterns of human features do not lie in isolation. Rather, patterns of human activity evolve through the movement and interaction of people and cultures from different locations. This movement and interaction helps explain why unique spatial patterns, places, and regions form.

Origin

Many spatial phenomena can be viewed in terms of origin and diffusion. The *origin* is a starting point, a location where something begins. It is often referred

to as a *culture hearth*. A disease outbreak, a new musical style, a new technology, or a new idea will begin in a specific part of the world. While its exact origin point is still debated, the deadly flu pandemic of 1891 originated in one of several possible locations: Kansas in the United States, China, or France. From its origin point, it then diffused throughout the world, killing around 50 million people. Hip-hop originated in the Bronx, New York, in the 1970s before becoming a global phenomenon. The major monotheistic religions of Christianity, Islam, and Judaism originated in the Middle East before spreading around the world.

Origin points must have the right conditions for a new phenomenon to form, and these conditions are typically related to human actions. For example, new disease outbreaks are more likely to occur in places with poor sanitation and weak health care systems (figure 1.32) than in places where sanitation and health care are adequate. New social and technological innovations are likely to form in societies that are open to new ideas and that already have the technological prerequisites for the innovation. For instance, some countries foment new ideas through the protection of free speech, whereas other countries stifle innovation through heavy censorship and limits on open debate. In addition to having an open society, innovations cannot materialize unless technological prerequisites are in place. The Wright brothers' airplane could not have been built without technical knowledge of structures and materials, motors, and basic physics.

Thus, it is important to remember that the origin of new phenomena comes from the combination of multiple influences. This combination is typically the result of spatial interaction, the movement of ideas and things between places; and spatial diffusion, the spreading of an idea or thing across space over time.

Spatial interaction

Spatial interaction takes place when two or more areas are linked by a network. Transportation and communications linkages tie places together and allow

Figure 1.32. Live poultry market in China. Geographers and others see parts of Asia as being the most likely origin point for an avian flu pandemic. High population densities, poor sanitation, and close interaction between chickens and humans create ripe conditions for disease. Photo by Fotokon. Stock photo ID: 186029792. Shutterstock.

for people, ideas, and things to move between them. The more spatial interaction one place has with other places, the more it will be exposed to new ideas and technologies. A place with air and seaports, road networks, and government policies that facilitate the movement of goods and people will tend to be more innovative and be the origin point for new ideas and technologies. The same holds true for communications network connections. Places that are linked by telephone and internet connections allow for the quick movement of information, which fosters creativity and the formation of new ideas (figure 1.33). Thus, spatial interaction results from connectivity and accessibility. Transportation and communications networks connect locations, allowing people, ideas, and things to have access to different places.

Figure 1.33. Internet café in Indonesia. People access the internet with cell phones and laptop computers at a café in Indonesia. Communications networks are an essential component of spatial interaction. When places are connected in this way, ideas and innovation spread. Photo by Lano Lan. Stock photo ID: 344025836. Shutterstock.

Spatial interaction is strongly influenced by distance. This influence is described by Tobler's first law of geography, which states that "everything is related to everything else, but near things are more related than distant things." The same concept can be described as distance decay, whereby there is more interaction between places that are close together than between places that are far apart (figure 1.34).

For instance, Mexican cities along the US border will be more like US cities, while Mexican cities in southern Mexico are less similar. Northern cities have more signage in English, people use more Spanglish terms, and the latest consumer goods from the US are more prevalent than in southern Mexican cities. This is because of greater spatial interaction—good connectivity and accessibility via roads and border crossings as well as TV and radio signals. This is also because of a greater diffusion of US culture and goods, especially through relocation and contagious diffusion. As one moves south from the US border, the influence of the English language and of American products and culture becomes less pronounced.

The same can be seen with Mexican influence in US cities along the southern border, where Mexican language, food, music, and other cultural features are more prevalent. Moving north, Mexican cultural influence declines as distance increases from the border.

Geographers also study spatial interaction in terms of core and periphery. Core areas include concentrations of things such as political power, economic activity, specific cultural characteristics, or population density. The periphery includes surrounding areas that have spatial interaction with the core. Often, the core is seen as having an advantageous position relative to the periphery. For instance, a political core makes laws that govern the periphery under its control, while an economic core creates wealth by using labor and natural resources from its periphery. At a global scale, core and periphery have been used to describe how wealthy countries (i.e., the core) exploit poor countries (i.e., the periphery) by extracting natural resources.

Core and periphery can also be seen in terms of regions, whereby the core of a region constitutes a heavy concentration of regional characteristics, while the periphery is the area where the characteristics gradually diminish. For instance, Louisiana may be considered part of the core of the South, but Texas can be seen as part of the periphery, as southern music, food, and dialects gradually fade as one moves west.

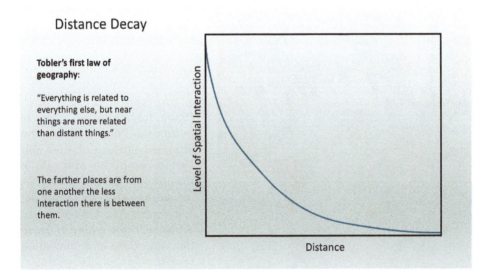

Distance Decay

Tobler's first law of geography:

"Everything is related to everything else, but near things are more related than distant things."

The farther places are from one another the less interaction there is between them.

Level of Spatial Interaction

Distance

Figure 1.34. Distance decay. Image by author.

With advances in communication and transportation technologies, some geographers refer to space-time compression. This is the idea that the world is "shrinking" as relative distance declines with changes in technology. New ideas now travel instantly to different places via communications technology, and people and goods move rapidly by car, ship, and airplane. As places become "closer" together, change brought about through spatial interaction happens more quickly. Global pop music stars now influence youth fashion around the world at the same time rather than just in smaller local or regional markets. Similarly, the 2008 financial crisis that began with banks in the United States quickly impacted the global economy. Whereas music, fashion, and economic crises were once local phenomena or spread slowly, they now impact people around the world in a short period of time.

Spatial diffusion

Spatial diffusion is another way that characteristics spread to new locations. From an origin point, an idea or thing can diffuse to new places, where it can then combine with other ideas or things and form something new.

Spatial diffusion can be broken down into two broad categories: relocation diffusion and expansion diffusion.

Relocation diffusion occurs when people move to a new location and take their ideas and possessions with them. With relocation diffusion, the number of people using a particular idea or possession does not change, but the place where they are used does change. When Spaniards first migrated to the Americas, they brought their religious ideas and weapons with them. Thus, via relocation diffusion, Christianity and European military technology spread to the New World. This diffusion process occurred even before the indigenous people of the Americas began practicing Christianity and using steel weapons.

Likewise, Latin American culture has diffused to many parts of North America as Latino immigrants relocate north (figure 1.35). The Spanish language, food, music, and other cultural elements can now be found in many American and Canadian cities, specifically because of relocation diffusion.

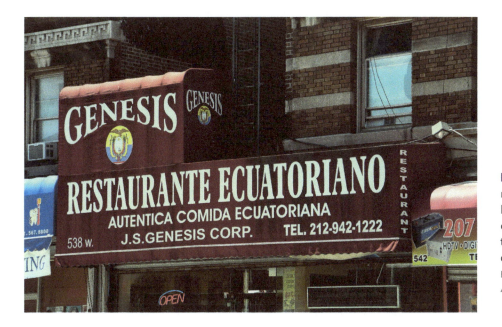

Figure 1.35. Ecuadorian restaurant in New York. Latino immigrants bring their culture with them through the process of relocation diffusion. Photo by Lee Snider Photo Images. Stock photo ID: 426459184. Shutterstock.

Expansion diffusion occurs when the number of people using an idea or item increases. This type of diffusion resulted as Spaniards in the Americas converted the indigenous population to Christianity and the indigenous people began using and making steel weapons. Likewise, many nonimmigrants in North America now eat foods from Latin America and know a few words or phrases in Spanish ("*Hasta la vista, baby*"). Via expansion diffusion, culture and technology spread to new people.

Expansion diffusion can be broken down into *contagious diffusion* and *hierarchical diffusion*. Contagious diffusion is when a characteristic spreads from person to person on the basis of proximity. In a sense, it can be visualized as when a pebble is thrown in a pond and the waves move outward in a circular pattern (figure 1.36). Places close to the origin point adapt the new idea or item before places that are farther away

do. Using the example of Spaniards in the Americas, the first indigenous people to be converted to Christianity and to use steel would be those who lived close to early Spanish settlements. Diffusion to indigenous people in more remote locations took much longer. Similarly, more people in the American Southwest than in other US areas eat dishes from Latin America due to their proximity to the southern border and Latino immigrant communities in southwestern cities. Over time however, Latino culture has spread to places beyond the border and immigrant-heavy cities. Guacamole is now a staple throughout the United States during the most all-American of events, the Super Bowl.

Hierarchical diffusion is when something spreads from a person or place of power and influence (figure 1.37). Geographers most often refer to hierarchical diffusion in terms of an urban hierarchy. If cities are ranked by population from large to small,

Contagious Diffusion of Walmart

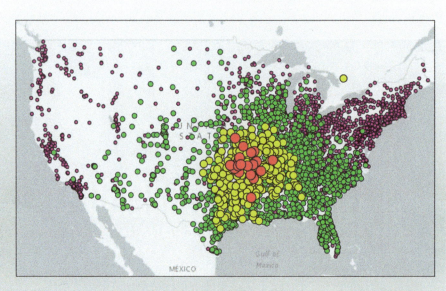

Walmart began in Arkansas and spread to nearby Missouri and Oklahoma in the **1960s**.

In the **1970s** it diffused to surrounding states in the South and Midwest.

It continued to spread outward in the **1980s**.

Ultimately, it reached the west coast and New England in the **1990s**.

Figure 1.36. Contagious diffusion of Walmart. Contagious diffusion represents the outward spread from a point of origin, like rings emanating from a pebble falling in a pond. Explore the Walmart store openings dataset (1962–2006) in ArcGIS Online at https://arcg.is/0Cfy5a. Map by author. Data sources: Thomas J. Holmes. University of Minnesota, Federal Reserve Bank of Minneapolis, and NBER.

it is more common for a new idea or item to originate in and diffuse to large cities first, then to medium-size cities, and later to small cities and towns. The latest musical trend or fashion typically begins in a large city, then diffuses to other large cities, even if they are far away—think New York, Los Angeles, London, and Paris. Medium-size cities will pick up on the trend a bit later, while diffusion to small towns will be later still.

In addition to urban hierarchies, diffusion can move along an income hierarchy, often from rich to poor. In most countries, the diffusion of things such as higher education, personal computers, and cars with airbags all began in households and neighborhoods with higher incomes. As time progressed, the cost of these innovations decreased, and they became more widely available in neighborhoods with lower income.

Stimulus diffusion is when a characteristic spreads to a new place, but rather than remaining in its original form, it stimulates a new innovation. For example, US fast-food restaurants have diffused around the world, but those that have been most successful have modified their menus to reflect local culture. For example, rather than a simple diffusion of burgers and fries to Japan, McDonald's offers crab croquette burgers and fries with white and dark chocolate sauce. McDonald's is a global company, but its diffusion worldwide has stimulated innovation in new menu items. In a similar sense, the idea behind ride-hailing services such as Uber has stimulated new services for hailing rickshaws in India.

Diffusion processes do not flow unimpeded across the landscape but rather face barriers to diffusion. Barriers can be physical or cultural. Mountains, oceans, rivers, dense forests, and deserts can act as physical barriers that slow or stop diffusion. Physical barriers can also include walls, trenches, and other human-built features. Cultural barriers can be just as powerful in stopping or slowing diffusion. Language, race and ethnicity, religion, income, and other cultural

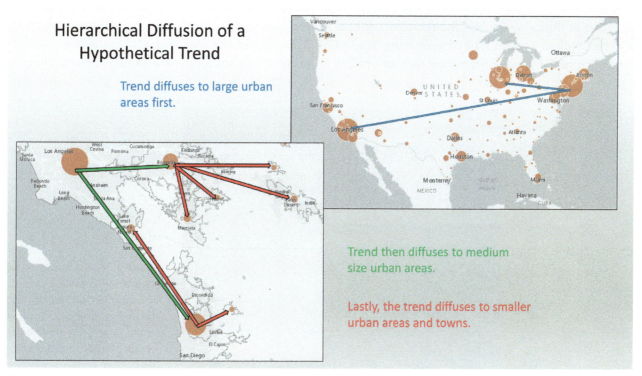

Figure 1.37. Hierarchical diffusion of a hypothetical trend. Geographers often study hierarchical diffusion in terms of urban hierarchies. New trends and products originate in large cities, then diffuse to medium and small cities over time. Maps by author. Data sources: Esri, HERE, Garmin, NGA, USGS.

differences can limit communication and interaction between groups of people, thus inhibiting the diffusion of ideas and items from one community to another, be it religion and philosophy, music, disease, or technology. Furthermore, cultures that are more conservative and tradition-based will resist the diffusion of new ideas and items from other people and places. The Taliban's prohibition of modernity in terms of music, technology, clothing, and much more limits diffusion into areas under its control.

Go to ArcGIS Online to complete exercise 1.6:
"Origin, spatial diffusion, and spatial interaction."

Human–environment interaction

The ecological perspective: Cultural ecology

As discussed so far, we have seen how spatial patterns are created from the interaction between places and the diffusion of things over space. But these patterns are also the result of another type of relationship studied by geographers: human-environment interaction. The interaction between humans and the environment is known as the *ecological perspective*, or *cultural ecology*—the interplay of human cultures with ecological patterns.

Humans impact the environment in many ways. They alter plant and wildlife distributions by converting natural habitats to farms and cities; they change the course of rivers through dams and canals; they alter the quality of air, water, and soils through pollution; and they remove hills and fill valleys for development projects (figure 1.38). Very few places are free of human impacts, and with human-induced climate change, the distributions of plants, animals, crops, and human settlements are likely to be further transformed at a global scale.

Similarly, the environment impacts humans. Human settlements tend to be less populous in areas that are too wet, such as the tropics; too dry, such as deserts; or too cold, such as the Far North and Far South (figure 1.39). These types of environments are poor for agriculture and make large-scale food production difficult, thus limiting human settlement. The natural environment also influences components of human culture. For example, due to climatic

Figure 1.38. Garbage-filled harbor in Malaysia. Humans' actions often have a profound negative impact on the environment, threatening the health of people and wildlife alike. Photo by Rich Carey. Stock photo ID: 214284142. Shutterstock.

differences, northern Europeans have a traditional diet that is high in fish, meats, and carbohydrates, while Mediterranean Europeans have a traditional diet that includes a wide variety of fruits and vegetables. The way people dress, obviously, is also a function of environment; just envision people in sun-soaked Rio de Janeiro, Brazil, and chilly Helsinki, Finland. How people build varies as well, with wood used in forests, adobe in deserts, steeply pitched roofs in areas of heavy snowfall, and flat roofs in arid regions.

Environmental determinism and possibilism

Environmental determinism is the idea that the natural environment determines much of the spatial patterns of human activity. As in the earlier examples, the environment is seen as determining where people live or do not live, what types of crops they grow and what foods they eat, how they dress, and how they build their houses. This theory has also been used to explain patterns of economic development. European thinkers once believed that the mid-latitudes, which were not too hot, too cold, too wet, or too dry, led to vigorous, hard-working, and productive societies. In contrast, tropical latitudes, with their heat and humidity, made hard work so unpleasant that societies remained primitive. Environmental constraints on hard work were said to hold true of hot desert regions and cold high-latitude regions as well.

Environmental determinism fell out of favor during the twentieth century. Historically, many successful societies have formed in areas once considered to have overly harsh environments, from the Mayans of Central America to the great ancient cities of Mesopotamia in the Middle East. More recently, tropical places such as Singapore and Hong Kong have become among the richest in the world, and major urban areas such as Phoenix and Las Vegas have grown in deserts. Irrigation allows for new crops in the desert, while fertilizers in the tropics can overcome poor soil quality. As is clear from these examples, environmental conditions do not directly determine the spatial patterns of human activity.

Instead, the concept of *possibilism* is more appropriately used when studying human-environment interaction. Possibilism is the notion that the natural environment creates possible outcomes for human activity but that humans can overcome many of the constraints imposed by nature. With human creativity,

Figure 1.39. Ittoqqortoormiit Village in Greenland. Arctic settlements such as this village tend to have low populations due to harsh environmental conditions. Photo by Adwo. Stock photo ID: 200898014. Shutterstock.

tropical Singapore used the natural conditions of a harbor located on trade routes between Asia, the Middle East, and Europe to become a wealthy state tied to international trade and services. Human taming of the Colorado River via construction of the Hoover Dam, combined with the invention of air conditioning, allowed for the massive urban areas of Las Vegas and Phoenix to grow (figure 1.40). Natural environments offer opportunities and constraints, but due to human ingenuity, they do not determine spatial patterns.

Environmental perception and hazards

Geographers are also interested in how people perceive their environments and how these perceptions influence cultural ecology. *Environmental perception* relates to the way in which people view the environment and how this view influences their interpretation and use of the natural landscape. One group of people may view the forest as a place for recreational hiking and will want to preserve it in a natural state, while another may view it as a source of economic development and will want to harvest wood for sale. Different perceptions lead to very different uses of the land.

Humans can perceive natural landscapes in terms of exploitation, preservation, or sustainability. The exploitative approach is to use natural resources and modify natural landscapes in the unlimited pursuit of economic growth. Preservation is aimed at leaving natural resources and landscapes untouched by humans, with use geared, at most, toward limited-impact recreation. In contrast to both, sustainability is the idea that natural resources and landscapes can be used by humans for economic growth, but they must be used in a manner that is sustainable in the long run. Lumber can be harvested from forests, but only at a rate that allows for regeneration of trees and protection of wildlife and other flora; fishing can be sustainable by restricting the amount of fish taken per season; natural landscapes can be converted to urban

Figure 1.40. Phoenix, Arizona. While the natural environment influences human settlement, people overcome many of its challenges with technology. Without air conditioning, the growth of Phoenix would be highly constrained. Photo by Tim Roberts Photography. Stock photo ID: 91397345. Shutterstock.

uses or farming as long as waterways are protected and wildlife preserves are incorporated. Sustainability can be accomplished either by cultural norms in a society or by government regulations that restrict overuse.

Environmental perception is useful when studying how people react to *natural hazard risk*. Many residents of modern urban societies view natural hazards as something that humans can control. People build homes in areas prone to flooding along rivers and coastlines, in wooded and canyon areas at risk for fire, and in areas of landfill subject to earthquake liquefaction. As cities in much of the western United States expand, more and more homes are being built in scenic yet fire-prone hillside areas. Residents assume that firefighters and forest crews will keep their homes safe, which in many cases is true, but at a great cost in terms of economic resources (figure 1.41). Similarly, residents of the southeastern United States increasingly live along coastal areas subject to hurricanes and other

flooding. Again, the perception is that these scenic areas are safe to live in, resulting in a great deal of housing and urban infrastructure that is exposed to serious environmental risk.

In some societies, people may not believe that humans can control natural hazard risk but rather must leave their fate in the hand of gods. Still others may avoid living in areas of risk, avoiding flood zones or canyons at risk for landslide or fire. In some cases, perceptions of natural hazard risk can change over time, altering how people inhabit the land. In Chile, a major earthquake in 2010 sent tsunami waves rushing over many coastal towns, destroying tens of thousands of homes and killing over five hundred people. As a result, people's perceptions of coastal risk changed. Reconstruction aimed to mitigate future risk, with some coastal areas being reserved for parkland and housing in high-risk areas being built to withstand future inundation (figure 1.42).

Figure 1.41. Homes destroyed by Hurricane Sandy in Far Rockaway, New York. Many people perceive natural hazard risk as minimal, and they continue to build in places prone to hurricane wind and flood damage. Photo by Leonard Zhukovsky. Stock photo ID: 130759928. Shutterstock.

Figure 1.42. Tsunami warning sign in Castro, Chile. After a deadly earthquake and tsunami in 2010, some towns began establishing buffer zones between housing and the coast. Photo by Matyas Rehak. Stock photo ID: 568647175. Shutterstock.

Go to ArcGIS Online to complete exercise 1.7: "Environmental perception: Flood risk in Miami."

References

Bolstad, P. 2016. *GIS Fundamentals: A First Text on Geographic Information Systems.* Acton, MA: XanEdu.

Buchanan, Robert A. n.d. "History of Technology." *Encyclopædia Britannica.* Accessed April 20, 2017. https://www.britannica.com/technology/history-of-technology#toc10382.

Castree, N., R. Kitchin, and A. Rogers. 2013. *A Dictionary of Human Geography.* Oxford, England: Oxford University Press.

Dunbar, B. 2014. "Global Positioning System." NASA. Accessed April 20, 2017. https://www.nasa.gov/directorates/heo/scan/communications/policy/GPS.html.

Graham, S. 1999. "Remote Sensing." NASA Earth Observatory. Accessed April 20, 2017. https://earthobservatory.nasa.gov/Features/RemoteSensing.

Kimerling, A. J., A. R. Buckley, P. C. Muehrcke, and J. O. Muehrcke. 2016. *Map Use: Reading, Analysis, Interpretation.* Redlands, CA: Esri Press.

Light, Alan, and Greg Tate. n.d. "Hip-hop." *Encyclopædia Britannica.* Accessed April 20, 2017. https://www.britannica.com/topic/hip-hop.

Mitchell, A. 2012. *The ESRI Guide to GIS Analysis.* Redlands, CA: Esri Press.

National Geographic Society. "Looking at the World in Multiple Ways." Accessed April 20, 2017. https://www.nationalgeographic.org/education/about/national-geography-standards/geographic-perspectives/.

National Geography Standard Index. n.d. Retrieved April 20, 2017. http://www.nationalgeographic.org/standards/national-geography-standards.

Vergano, Dan. 2014. "1918 Flu Pandemic That Killed 50 Million Originated in China, Historians Say." *National Geographic.* Accessed April 20, 2017. http://news.nationalgeographic.com/news/2014/01/140123-spanish-flu-1918-china-origins-pandemic-science-health.

Chapter 2
Population

A natural starting point for a book on human geography is to examine where humans live and how these patterns impact places and regions. As we move through busy cities, gridlocked with traffic, a brown haze of smog in the air, concrete in all directions, we may get the feeling that the world has reached a breaking point and cannot support any more people. But then we can visit vast open spaces, from forests and deserts to savannas and steppes, where humans have left a much more minimal imprint on the landscape. Clearly, the distribution of human population is uneven. As we will see, human populations are clustered and thus impact some locations more than others. In addition, rates of population growth vary substantially from place to place, so that the spatial distribution of clusters can shift. Population centers move over time, as people are born in particular places and move to new locations.

Population patterns and growth rates have a profound impact on places, as seen in their spatial relationships with economic and social conditions. As you can imagine, some countries face fast-growing populations, where women have many children and there is a sensation of overpopulation. In parts of the world, governments struggle to grow their economies and provide enough jobs, housing, and food for rapidly expanding populations. Shortages in these areas can sometimes lead to political instability, as peoples' anger over unmet needs erupts into street protests, coups and revolutions, and crime.

Less obviously, some places are seeing very slow population growth or even negative growth. This can create a completely different set of problems, as governments face conditions where there may not be enough people to work and grow (or even maintain) an economy. Low population growth can lead to a situation where places have too many elderly residents relative to young, active workers.

There is a strong spatial relationship between economic and urban development, gender roles, culture, and differing rates of population growth. As people move away from farms and into cities, economic and social change has a profound impact on population dynamics. As the reader of this book, you are most likely a resident of an affluent country and live in a city or town. Given your life experience, how many children do you plan to have? Why did you choose that number? If you lived in a poor rural setting, your life experience would be very different and would have a significant impact on the number of children you would want. Where you live has a strong relationship with how much you will contribute to population growth.

Population issues also lead to important discussions on spatial interaction. When population growth is high in some places and low in others, forces of migration can come into play. Migration is covered in more detail in chapter 3, but let it suffice that population pressures on economies and environments can push people to leave some places and move to new countries or cities.

Spatial distribution of population

World population numbers

For much of human history, populations grew slowly, but once humans moved from hunting and gathering to agriculture, improved food supplies allowed population growth to speed up. Later, as more consistent food supplies were complemented by improvements in human health through medical innovation and sanitary conditions, population growth rates increased even more as in Tokyo, Japan (figure 2.1). The global human population did not reach one billion until around 1800 (figure 2.2). After that, however, the population growth rate increased quickly. In just 123 years, by 1927, the population reached two billion. Then, in just thirty-three years, by 1960, world population reached three billion. Four billion was reached by 1974, and five billion by 1987. In 1999, there were six billion, and by 2011, there were seven billion people. The human population as of 2017 was over 7.3 billion, and growing at over eighty million people per year.

These growth patterns appear to be following the S-shaped curve (figure 2.2), whereby a population grows slowly during an initial lag period and then sees a rapid increase of exponential growth.

But before getting too worried about the fact that over eighty million people are added to our planet each year, it is important to see how the growth rate is changing over time (figure 2.3). Globally, peak growth was reached between 1965 and 1970. However, since that time, the growth trend has typically sloped downward. So, while the earth's population is still growing, it is growing less quickly than in the past. Population geographers and others believe that this downward trend will continue, with population reaching 11.2 billion by the year 2100. After that time, projections become more uncertain. The global population could continue to gradually increase, or it could ultimately reach zero growth, where the human population returns to a relatively stable equilibrium. This pattern would represent the later stages of the S-shaped curve, where population growth slows and possibly reaches a more stable plateau.

While the world's population is over 7.3 billion people, these people are not evenly distributed over the

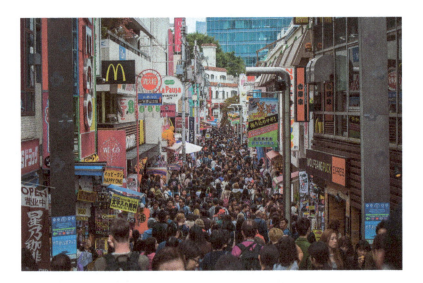

Figure 2.1. Tokyo, Japan. Understanding human population patterns is a first step in understanding human geography. Photo by aon168, Stock photo ID: 519265849. Shutterstock.

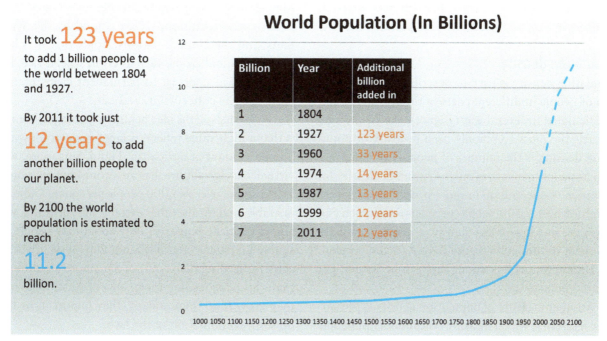

It took **123 years** to add 1 billion people to the world between 1804 and 1927.

By 2011 it took just **12 years** to add another billion people to our planet.

By 2100 the world population is estimated to reach **11.2** billion.

World Population (In Billions)

Billion	Year	Additional billion added in
1	1804	
2	1927	123 years
3	1960	33 years
4	1974	14 years
5	1987	13 years
6	1999	12 years
7	2011	12 years

Figure 2.2. World population over time. Human population size remained relatively flat until the past one-hundred years or so. Data sources: United Nations, 1999; United Nations, 2015.

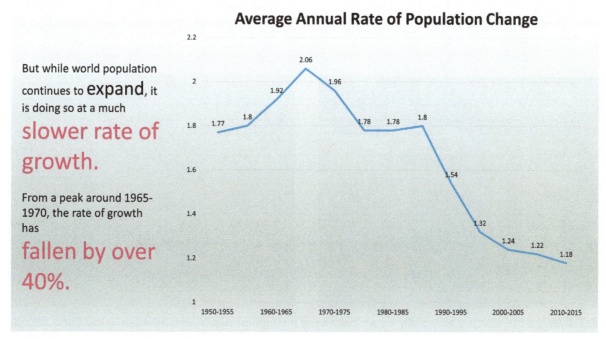

But while world population continues to **expand**, it is doing so at a much **slower rate of growth.**

From a peak around 1965-1970, the rate of growth has **fallen by over 40%.**

Average Annual Rate of Population Change

Figure 2.3. Population growth rate. The human population continues to grow but at a slower rate than in the past. Date source: United Nations, 2015.

planet. As pointed out earlier, humans are clustered, resulting in very distinct levels of population density.

Population density

Population density is used to describe the concentration of people in different parts of the world and can be calculated in several ways (figure 2.5). *Arithmetic density*, the most commonly used measure of population density, is the number of people per unit area, such as people per square mile. Arithmetic density (in countries of one million or more people) ranges from a high of over 7,500 people per square kilometer in Singapore to a low of less than two people per square kilometer in Mongolia (figure 2.4). As a comparison, the United States has about thirty-four people per square kilometer.

Another measure of population density is *physiological density*, which measures the number of people per unit of arable land and is intended to compare the number of people in an area with the amount of land available to feed them. Arable land is defined in a couple of different ways and thus can be confusing when mapping and analyzing physiological density. One definition holds that arable land is land that is suitable for cultivation. It includes land with the proper

soils, elevations, slopes, and climates for growing crops. A narrower definition, which is used by the World Bank and the United Nations, holds that it is land used for annual crops, such as corn, wheat, rice, and vegetables, in contrast to permanent crops planted once, such as coffee, fruit, and nuts.

Nevertheless, the idea behind physiological density is that a high value reflects a large population with a limited amount of agricultural land. Singapore, with virtually no agriculture, has an astonishing physiological density of over 945,000 people per square kilometer of arable land. At the other extreme lies Australia, with a physiological density of only about fifty people per square kilometer of arable land. Obviously, the strategies for feeding the people of Singapore are different than those for feeding the people of Australia: either they grow food domestically, they import it, or they do some combination of the two.

A third measurement is *agricultural density*, the number of rural residents per unit of arable land, which indicates how many people are involved with agricultural production. This measure helps illuminate which countries are efficient at growing food and which are not. For instance, Egypt's agricultural density is over 1,800, whereas it is 39 for the United States. This means

Figure 2.4. Singapore, which is essentially urban, has one of the highest population densities in the world. Mongolia, a sparsely populated country, has one of the lowest population densities in the world. Singapore photo by Martin Ho Smart. Stock photo ID: 559395556. Shutterstock. Mongolia photo by Jan Peeters. Stock photo ID: 225794263. Shutterstock.

Measures of Population Density

Agricultural Density:

High: Egypt. Bangladesh.

> Poorer. Inefficient agriculture. Many rural laborers needed to work arable land.

Low: Australia. United States. Japan.

> Richer. Efficient agriculture. Few rural laborers needed to work arable land.

Physiological Density:

Problem: Egypt. Bangladesh.

> Often struggle to pay for imported food.

No Problem: Singapore. Japan.

> Easily purchase imported food.

Arithmetic Density:

No relation to a country's standard of living.

High density Singapore. Low density Australia. Both wealthy places.

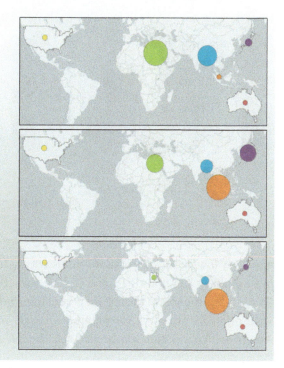

Figure 2.5. Measures of population density. Common measures of population density include agricultural density, physiological density, and arithmetic density. Explore population densities in ArcGIS Online at https://arcg.is/1LaKue. Data source: World Bank.

that there are forty-six times more people in rural areas of Egypt relative to arable land compared to the United States. Presumably, these people are involved directly or indirectly in the agricultural economy. A lower proportion of Americans in rural areas relative to arable land implies that they are much more efficient than the Egyptians in agricultural production, possibly because of the availability of agricultural technology such as farm machinery, agricultural chemicals, and precision agricultural mapping and monitoring with geographic information systems.

When physiological and agricultural densities are high, pressure to expand arable land into new areas may increase. This expansion can cause negative environmental impacts as natural landscapes are converted to agricultural uses, displacing wild plant and animal life. One way to reduce this pressure is to import food. For example, Egypt was the world's largest importer of wheat in 2016. If a country such as Egypt, with high physiological and agricultural densities, has a weak economy and is unable to earn foreign currency through exports, then it will be difficult to import food, and hunger can ensue. With Egypt's economy weakened from political instability and violence in recent years, foreign currency shortages and a declining exchange rate have strained the country's ability to import food (figure 2.6). Prices have increased for consumers, forcing the state to spend more on subsidizing bread. In contrast, countries with high physiological and agricultural densities and relatively strong economies, such as South Korea and Japan, can import food more easily and face virtually no risk of hunger.

Population clusters

As can be seen, the spatial distribution of people around the world varies greatly. Some places have high densities, known as *population clusters*, whereas others

Figure 2.6. Waiting to buy bread in Aswan, Egypt. Egyptians rely heavily on imported wheat for their bread. With a weakened economy, prices rise, and many people struggle to purchase this staple food. Photo by Olga Vasilyeva. Stock photo ID: 418291645. Shutterstock.

have very low densities. Population clusters are found in several key locations (figure 2.7). East Asia, South Asia, and Europe form the largest population clusters, with other localized clusters found in parts of the Americas, Africa, and Southeast Asia. In East Asia, China alone has over 1.3 billion people, making it the largest country in the world in terms of population size. The region also includes large populations in South and North Korea as well as Japan. South Asia is dominated by India, with a population of just under

Major World Population Clusters

The East Asian population cluster is dominated by **China**, which has the world's largest population:

Over 1.3 billion.

South Asia is dominated by the second largest country, **India**. Its population is

over 1.2 billion.

The European cluster sweeps from the Iberian Peninsula eastward into Russia and Turkey.

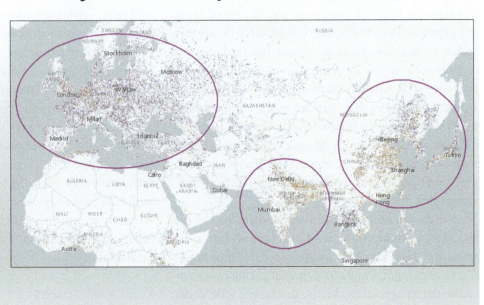

Figure 2.7. Major world population clusters in Europe, South Asia, and East Asia. Sign in to your ArcGIS Online account and explore this map at http://arcg.is/2lDz8WW. Data sources: World Population Estimated Density 2015, Esri.

1.3 billion, the second largest in the world. Bangladesh, Pakistan, and Sri Lanka also have substantial populations in South Asia. Dense populations can be found in Europe as well, stretching from the Iberian Peninsula (Portugal and Spain) into the western portions of Russia and northwest Turkey. In the United States, dense populations can be found along the northeast seaboard, stretching from Boston to Washington, DC. Smaller population clusters can be found throughout the world in the form of large urban agglomerations, the result of ongoing urbanization of human society.

The distribution of population can be partially explained by the natural environment. As humans migrated out of Africa millions of years ago and spread to the far reaches of the earth's surface, some environments proved more suitable than others for supporting large populations. Temperate climates (those that are not too hot or too cold, too wet or too dry) tend to form soils well suited for agriculture, a prerequisite for large populations. Environments with more extreme climates, such as deserts and subarctic regions, are suited only for small populations, which tend to be nomadic. Larger human populations also tend to be located at lower elevations. Exceptions to this rule are the mountain valleys of tropical regions, such as Central America, which have milder climates and better soils than the lowland tropics. Other environmental characteristics, such as natural resources or waterways for trade, can influence population distributions. For instance, arid locations with large mineral deposits or tropical rainforests with timber resources can attract people to small local clusters. Likewise, populations can cluster along coastal and river locations that facilitate trade with other areas.

While the natural environment has been an important force shaping where human populations clustered for much of human history, it is no longer the most important determinant. The differing rates of fertility, mortality, and migration that determine world population distributions now depend on a wider range of cultural, environmental, political, technological, and economic forces. For instance, since food can be easily imported from far away, the importance of agricultural potential in the growth of population clusters is diminished. Also, population growth in trade centers can be just as fast along human-built routes such as railroads and highways as along natural rivers and harbors. Airports can now replace natural seaports. With fewer restrictions tied to the natural environment in terms of food production and trade, population clusters can form in many more locations than in the past.

There are several good examples of places where populations are growing despite the natural environment, not because of it. Phoenix, Arizona, with its arid landscape and over 100 days per year when temperatures reach 100 degrees or higher, is now a thriving metropolis. The same goes for Las Vegas, Nevada (figure 2.8). In both cases, economic development and technology have allowed for populations to grow in places with difficult natural environments. Air conditioning and complex systems to deliver food and water from far away allow these places to grow in ways that were unthinkable in the past (figure 2.9).

On the opposite end of the temperature spectrum is the growing population of Astana, Kazakhstan. It lies in the flat, open steppe of Central Asia, where temperatures never rise above freezing for over 100 days per year. Its barren and remote location made it the ideal

Figure 2.8. Las Vegas, Nevada. Technology such as air conditioning and water delivery infrastructure have allowed cities such as this to grow in relatively inhospitable climates. Photo by littleny. Stock photo ID: 105916268. Shutterstock.

Figure 2.9. Astana, Kazakhstan. Like Las Vegas, this city has a growing population in an inhospitable climate. Coincidentally, both Las Vegas and Astana have fanciful urban landscapes that reject not only their natural environments but also historical and cultural traditions. Photo by ppl. Stock photo ID: 216420808. Shutterstock.

site for Soviet gulag prison camps when it was part of the Soviet Union. With Kazakhstan's post-Soviet independence, its capital city was relocated to Astana for political reasons, given that city's central location in the country. Again, modern transportation and food-delivery systems mean that populations can thrive in large numbers despite the inhospitable environment.

As population clusters shift within countries, such as in movement to a new capital city or into territories previously viewed as undesirable, the population centroid changes location. This point represents the center of the country weighted by the location of the population. Imagine a flat map of the US, with weights representing people stacked up for every city, town, and rural area. The population centroid would be the point where the map would balance perfectly. Figure 2.10 shows how the US population has shifted westward and slightly southward since 1790. This shift is in response to westward agricultural settlement, development of the industrial economy of the Great Lakes region, migration to the West Coast with the railways, and later growth of the Sun Belt states.

Go to ArcGIS Online to complete exercise 2.1: "Spatial distribution of population."

Population Centroid of the United States

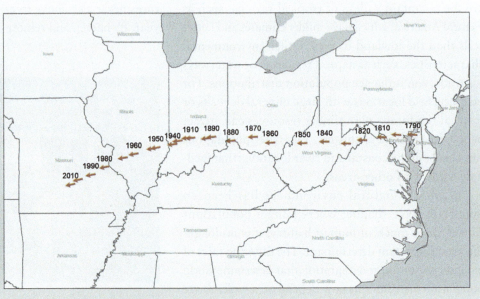

Each point represents the center of the country weighted by the location of the population for a given decade.

The centroid has shifted over time in response to population growth

west and

south from the original colonies.

Figure 2.10. Population centroid of the United States. The centroid has shifted over time in response to population growth west and south from the original colonies. Data source: US Census.

The components of population

As seen so far in this chapter, the spatial distribution of population varies substantially from place to place. Environmental and economic forces play an important role in these patterns, as people cluster in areas with land suitable for agriculture or along important trade routes. But these forces tell only part of the story. As you can imagine, birth and death rates greatly vary from place to place as well. In some countries, women have many children, while in others, women have very few. Likewise, death rates can be high in some places and low in others. Naturally, the relationship between birth and death rates plays a role in population distributions. When more people are born than die, populations increase, while populations decrease when death rates are greater than birth rates. The reasons for the variation in birth and death rates include additional economic and political forces as well as cultural attitudes and beliefs.

Population growth or decline in a specific place can be calculated with a simple demographic equation:

$$\text{Population change} = \text{Births} - \text{Deaths} + \text{Immigration} - \text{Emigration}$$

This intuitive equation simply states that the population of a place changes as babies are born, people die, immigrants move in, and emigrants move out.

This chapter deals with the birth and death portion of the equation, and chapter 3 covers immigration and emigration.

Births

The *crude birth rate* (CBR) is the number of births per 1,000 people in a given year. The lowest CBR in 2015 was in Monaco with 6.65 births per 1,000, and the highest was in Niger with 45.45 (table 2.1). The CBR for the United States falls within the lower third of national rankings, at 12.49 births per 1,000. This measure is useful for calculating how quickly countries'

Fertility, 2015

COUNTRY	CRUDE BIRTH RATE 2015	TOTAL FERTILITY RATE 2015
MONACO	6.65	1.52
NIGER	45.45	6.76
SINGAPORE	8.27	0.81
UNITED STATES	12.49	1.87

Table 2.1. Measurements of fertility vary greatly from place to place. Some countries face rapidly growing populations while in others births are not sufficient to replace previous generations. Data source: World Bank.

populations are growing, but its weakness is that it can be affected by the proportion of women in the population (those who may give birth) and by the age structure of the population (the proportion of young people to old people). These factors are discussed in more detail later in the chapter.

A somewhat more intuitive way of understanding population change, and one that accounts for differences in age structures and the proportion of sexes, is the *total fertility rate* (TFR) (table 2.1, figure 2.11). The TFR represents the average total number of children a woman will have during her lifetime. TFRs vary significantly. In Singapore, women in 2015 were having an average of just 0.81 children—less than one child per woman! The United States' 2015 TFR fell in a moderate range of 1.87 children per woman. At the high end, Niger's TFR was 6.76. As seen in the map, clusters of low TFRs can be found in Europe, and clusters of high TFRs are seen in sub-Saharan Africa.

Replacement fertility is the TFR necessary for a population to replace itself from one generation to the next without growing or shrinking. Replacement fertility of 2.0 would be the theoretical value, since 2.0 children would replace their two parents, resulting in no net gain or loss in population. However, because some women will die prior to reaching their reproductive years, replacement fertility is slightly over 2.0. In general, a TFR of 2.1 is used for replacement fertility. However, in developing countries with high mortality rates for infants and young adults, the rate can be a few decimal points higher. If a country has a TFR above replacement fertility, its long-term trend is toward an expanding population. Conversely, if a country's TFR is below replacement fertility, its long-term trend is toward a shrinking population.

Look at figures 2.11 and 2.12. The countries represented in white have TFRs close to 2.1 and are in the ballpark range of replacement fertility.

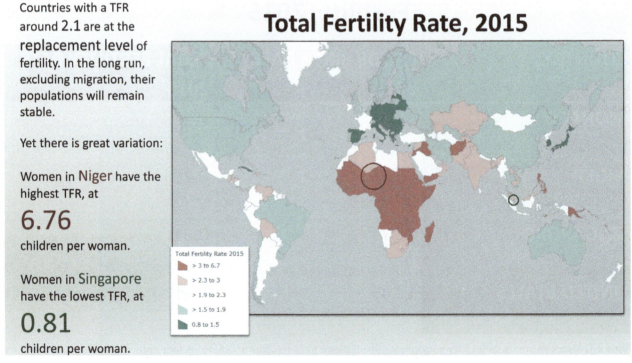

Countries with a TFR around 2.1 are at the replacement level of fertility. In the long run, excluding migration, their populations will remain stable.

Yet there is great variation:

Women in Niger have the highest TFR, at

6.76

children per woman.

Women in Singapore have the lowest TFR, at

0.81

children per woman.

Total Fertility Rate, 2015

Total Fertility Rate 2015
- > 3 to 6.7
- > 2.3 to 3
- > 1.9 to 2.3
- > 1.5 to 1.9
- 0.8 to 1.5

Figure 2.11. Total fertility rate, 2015. Explore this map at http://arcg.is/2lDAgd5. Data source: World Bank.

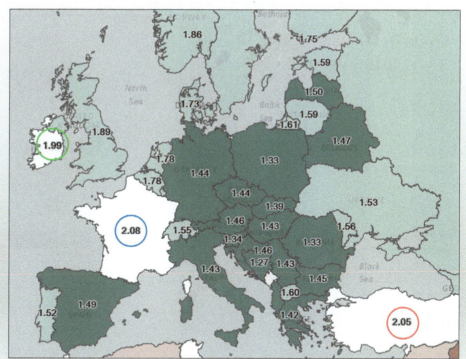

Total Fertility Rate in Europe, 2015

Nearly all of Europe has total fertility rates below replacement level.

Generally, northern Europe has higher rates, while southern and eastern Europe have lower rates.

Only three countries are close to replacement:

Ireland: 1.99
France: 2.08
Turkey: 2.05

Figure 2.12. Total fertility rates in Europe. Europe has some of the lowest fertility rates in the world, with most falling well below replacement levels. This is especially acute in Southern and Eastern Europe. Explore this map at http://arcg .is/2IDHFJs. Data source: World Bank.

Those in the highest category have very high TFRs and should have rapidly growing populations. Most interesting, though, is the lowest category, those with TFRs substantially below 2.1. All those countries are facing the prospect of shrinking populations, since women are having fewer children than required for replacement. It is very rare for a species to voluntarily reproduce below the replacement level, but humans appear to be doing exactly that in many parts of the world.

The rate of births in a country results from a complex mixture of variables. Obviously, access to contraception is an important variable, but other factors include the spatial relationship of where a woman lives, economic conditions, political and social stability, gender equality, education, and levels of urbanization. These are discussed in more detail later in the chapter.

Go to ArcGIS Online to complete exercise 2.2: "Fertility rates."

Deaths

Just as populations grow when babies are born, they shrink when people die. For this reason, the second essential component of population to understand is death and the different measurements used for quantifying it.

The *crude death rate* (CDR) is the number of deaths per 1,000 people in a given year (figure 2.13). As with the CBR, there is great spatial variation in the CDR. At the low end, Qatar had a 2015 rate of 1.53 deaths per 1,000, while Lesotho had a rate of 14.89. The United States falls close to the top third, with 8.15 deaths per 1,000.

When looking at figure 2.13, you can see that the CDR can lead to some surprising results. As you may

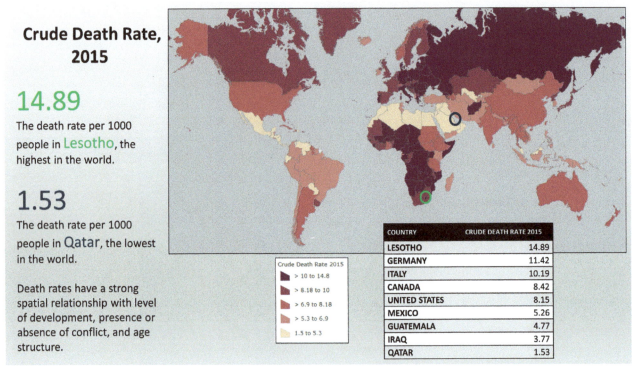

Crude Death Rate, 2015

14.89

The death rate per 1000 people in Lesotho, the highest in the world.

1.53

The death rate per 1000 people in Qatar, the lowest in the world.

Death rates have a strong spatial relationship with level of development, presence or absence of conflict, and age structure.

Crude Death Rate 2015
- \> 10 to 14.8
- \> 8.18 to 10
- \> 6.9 to 8.18
- \> 5.3 to 6.9
- 1.5 to 5.3

COUNTRY	CRUDE DEATH RATE 2015
LESOTHO	14.89
GERMANY	11.42
ITALY	10.19
CANADA	8.42
UNITED STATES	8.15
MEXICO	5.26
GUATEMALA	4.77
IRAQ	3.77
QATAR	1.53

Figure 2.13. Crude death rate 2015. Explore this map at http://arcg.is/2lDNfM0. Data source: World Bank.

expect, sub-Saharan Africa includes a large cluster of countries with high CDRs. Poverty, hunger, disease, and lack of medical care are prevalent in many parts of this region, so most people will not find it surprising that there are many deaths. But from there, the map illustrates death rates that many would not expect. Mexico and other Latin American countries have lower CDRs than the United States and Canada, for example. It is also apparent that Germany and Italy have CDRs higher than Iraq's (figure 2.14). In fact, nearly all developed countries, such as those of North America and Europe, have higher CDRs than less developed countries in Latin America, Asia, and the Middle East.

At first glance, it seems strange that rich, peaceful countries can have higher CDRs than poor, war-stricken countries. The reason is that although the CDR can be influenced by obvious factors such as war and violence, hunger and malnutrition, and disease and lack of adequate health care, another factor has an even greater

impact on the CDR: the age structure of the population. When a baby boom is followed years later by low birth rates, the proportion of elderly in a society increases. For example, many people were born in Europe after World War II ended in 1945. By 2015, that large group was reaching seventy years of age. But younger generations had low fertility rates, which declined through the 1960s, 1970s, and 1980s, meaning there were fewer children. With a large number of elderly and few children in a country, the proportion of elderly is higher. Since the elderly die at a higher rate than the young, the CDR is high (figure 2.15). In contrast, countries with high birth rates, such as Iraq, have a growing number of young people in relation to the number of elderly people. Since the young die at lower rates than the old, these countries can have a low CDR (figure 2.16).

Another measure of mortality is the *infant mortality rate*, which is a measure of the death rate of infants from birth through their first year per 1,000 live births. This

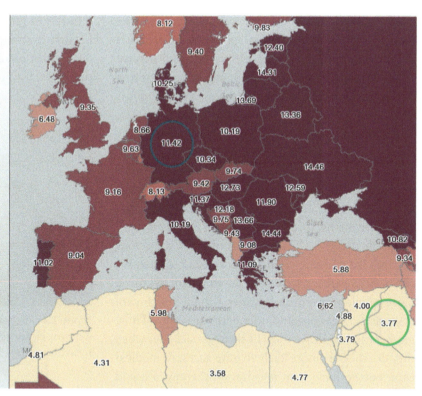

Crude Death Rate in Europe, North Africa, and the Middle East, 2015

11.42

The death rate per 1000 people in affluent and safe Germany.

3.77

The death rate per 1000 people in conflict ridden Iraq.

This seeming incongruity is due to differences in age structure. Low birth-rate countries in Europe have fewer children and more elderly. Higher birth-rate countries in North Africa and the Middle East have many children and fewer elderly.

Figure 2.14. The crude death rate in Europe, North Africa, and the Middle East, 2015. Unexpectedly to many, much of affluent Europe has higher CDRs than poorer and politically unstable countries in the Middle East and North Africa. This disparity relates to the differing age structure of the populations in each country. Explore this map at http://arcg .is/2kUSGcj. Data source: World Bank.

Figure 2.15. Dusseldorf, Germany. Countries such as Germany with a large number of elderly and a smaller number of young people have a higher crude death rate. Photo by Isarescheewin. Stock photo ID: 477151594. Shutterstock.

Figure 2.16. Kirkuk, Iraq. Countries such as Iraq with a large number of children and a smaller number of elderly have a lower crude death rate. Photo by Serkan Senturk. Stock photo ID: 509928916. Shutterstock.

rate has come down significantly over time but still varies considerably (figure 2.17). The highest rate in 2015 was in Afghanistan, at 115. That tells us that 11.5 percent of all children born in Afghanistan die before their first birthday, an astonishingly high number. At the low end is Monaco, with a rate of 1.82 per 1,000, or 0.182 percent. The United States has a rate of 5.87 (0.587 percent), which is higher than many other developed countries.

Declines in infant mortality are among the most important achievements in global health in recent decades. In 1960, there were roughly 122 deaths per 1,000 live births, meaning that over 12 percent of all babies died within their first year of life. Continuous declines led to a rate of under 32 by 2015, a decrease of about 73 percent. Economic development has given more people access to clean water, formal health-care systems, sanitation, and adequate nutrition. Thus, mothers and infants are less likely to get sick, but when they do, they are more likely to have access to medical care. For example, oral rehydration therapy, a liquid

mixture of glucose and electrolytes, is a simple cure for dehydration, one of the leading killers of infants.

Education, especially for women, has also helped reduce infant mortality rates (figure 2.18). Educated women are more likely to know how to access and

Figure 2.18. Schoolgirls in Skardu, Pakistan. Educating women is an important strategy for reducing infant mortality rates. Photo by Khlong Wang Chao. Stock photo ID: 426040138. Shutterstock.

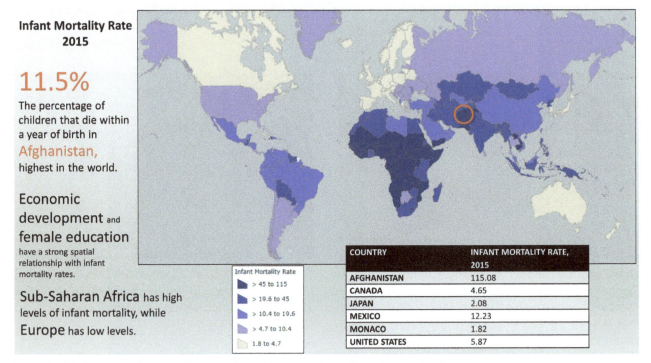

Infant Mortality Rate 2015

11.5%

The percentage of children that die within a year of birth in Afghanistan, highest in the world.

Economic development and female education

have a strong spatial relationship with infant mortality rates.

Sub-Saharan Africa has high

levels of infant mortality, while

Europe has low levels.

Infant Mortality Rate
- > 45 to 115
- > 19.6 to 45
- > 10.4 to 19.6
- > 4.7 to 10.4
- 1.8 to 4.7

COUNTRY	INFANT MORTALITY RATE, 2015
AFGHANISTAN	115.08
CANADA	4.65
JAPAN	2.08
MEXICO	12.23
MONACO	1.82
UNITED STATES	5.87

Figure 2.17. Infant mortality rate, 2015. Explore this map at http://arcg.is/2mg6FKf. Data source: World Bank.

navigate public health systems to obtain health care for themselves and their babies. They are also more likely to have a better understanding of nutrition and sanitation to properly feed their children and avoid waterborne and foodborne contaminants. Additionally, educated women are more likely to work and therefore have fewer children, leaving more resources for a smaller number of children.

Afghanistan, which consistently has among the highest infant mortality rates in the world, illustrates how low levels of economic development and low levels of education for women contributes to a high rate of infant deaths. Most babies are born with midwives at home rather than in medical clinics that can more quickly and safely deal with complications. Poverty and high levels of malnutrition mean that many pregnant women give birth to low-weight babies, increasing their babies' susceptibility to disease. When babies do get sick, medical care is often lacking, so common infections, respiratory illnesses, and diarrhea frequently lead to death. Lack of education, with its corresponding

lack of knowledge about health and nutrition, further contributes to a high infant mortality rate. Only about 50 percent of babies under six months of age are exclusively breast-fed. Instead, mothers often feed their babies tea and biscuits, leading to an increased risk of consuming contaminated food or water in a country with limited potable water and inadequate food storage systems.

The United States also offers a good case study on infant mortality. Although the infant mortality rate in the US is low by global standards, it is significantly higher than in other affluent countries (figure 2.19). For instance, a baby born in the US is 2.8 times more likely to die than a baby in Japan. The explanation for this difference is not clearly understood, but what is known is that more babies are born prematurely in the US than in the countries with lower infant mortality rates. Premature birth, which is associated with cigarette smoking, drinking alcohol during pregnancy, diabetes, and high blood pressure, puts infants at greater risk of mortality.

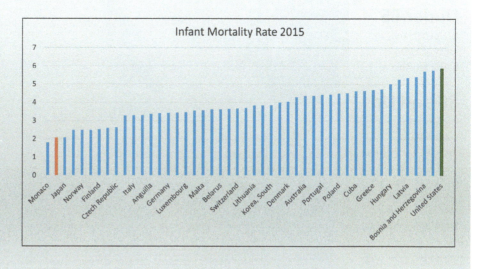

Infant Mortality Rates Lower than the United States'

Nearly 6 children per 1000 live births die within their first year in the United States.

This is **2.8 times greater than** in Japan

Premature births associated with pregnant mothers' cigarette smoking, drinking alcohol, diabetes, high blood pressure, and fragmented healthcare services have been attributed to higher rates in the United States.

Infant Mortality Rate 2015

Monaco, Japan, Norway, Finland, Czech Republic, Italy, Anguilla, Germany, Luxembourg, Malta, Belarus, Switzerland, Lithuania, Korea, South, Denmark, Australia, Portugal, Poland, Cuba, Greece, Hungary, Latvia, Bosnia and Herzegovina, United States

Figure 2.19. Infant mortality rates lower than the United States' rate. Data source: World Bank.

This accounts for some, but not all, of the mortality difference. It is possible that the fragmented health care that some young women receive, especially those of lower incomes, prevents clinicians from detecting and monitoring risk factors.

Life expectancy is the average number of years a person is expected to live. As with all demographic variables we have seen so far in this chapter, life expectancy varies widely around the world (figure 2.20). At the high end in 2015 was Monaco, at 89.52 years, and at the low end was Chad at 49.81 years. The United States has a life expectancy of 79.68 years.

Life expectancy figures can sometimes paint a misleading picture. For instance, life expectancy in the Ancient Roman Empire was 22 years. More recently, Chad had a 2015 life expectancy of 49.81 years. One may get the impression that Ancient Rome had few people beyond their mid-20s, while Chad has few over 50. This is not true, however, since life expectancy is calculated as an average of the age of death of all people in a year. If a country has a high infant mortality rate, those infant deaths pull down the average. Just imagine a place where five people died in the same year at the ages of 1, 19, 56, 79, and 95. The life expectancy for this group would be (1 + 19 + 56 + 79 + 95)/5 = 50. In this case, three people out of five lived beyond the average of 50 years, with one well beyond. What pulls down the average very quickly is when there are many 1s in the equation from infant deaths. Typically, there is a strong spatial relationship between high rates of infant mortality and low life expectancies. Note the similarities between the two in figures 2.17 and 2.20.

Life expectancy has a strong spatial relationship with socioeconomic and lifestyle variables. As with infant mortality, places with stronger economies are more likely to have the infrastructure necessary for longer lives. In more affluent places, clean water and contaminate-free food, vaccinations, medical care, health and safety standards, and proper housing are more common. Thus, people are less likely to get sick

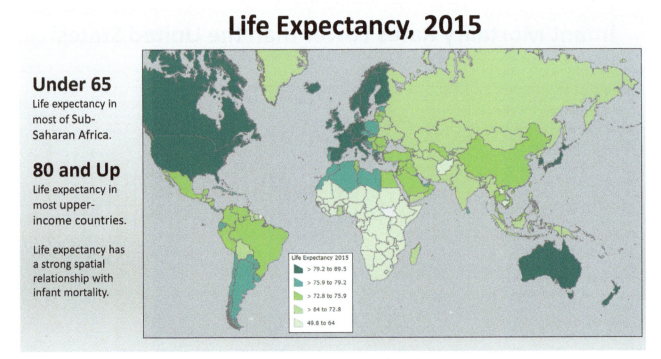

Figure 2.20. Life expectancy, 2015. Explore this map at http://arcg.is/2kUSEkK. Data source: World Bank.

or injured, but when they do, they are more likely to get proper medical attention. Less developed places lack many of these features, resulting in higher mortality and lower life expectancies. But lifestyle also plays an important role in life expectancy. Populations with higher rates of smoking, alcohol consumption, drug use, sedentary lifestyles, and poor diet have lower life expectancies. For instance, the US states with the lowest life expectancies also rank below average in terms of exercise and other healthy lifestyle factors (figure 2.21).

Globally, life expectancy increased by five years between 2000 and 2015. This was largely due to improvements in Africa. During this time, public health improvements have lowered infant mortality rates, but death from diseases such as malaria and AIDS has also been reduced (figure 2.22). Gains in Eastern Europe and Russia have also contributed to an increasing global life expectancy. After the fall of the Soviet Union in 1991, a collapse of public health systems, combined with stress-related increases in alcohol consumption, suicide, and other factors, led to a sharp decline in life expectancy. As the region has adjusted to new economic and political systems, health has recovered.

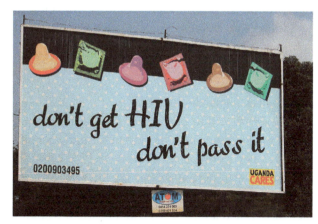

Figure 2.22. Kampala, Uganda. Improved public health in Africa has helped increase life expectancy in the region. Photo by Robin Nieuwenkamp. Stock photo ID: 189950711. Shutterstock.

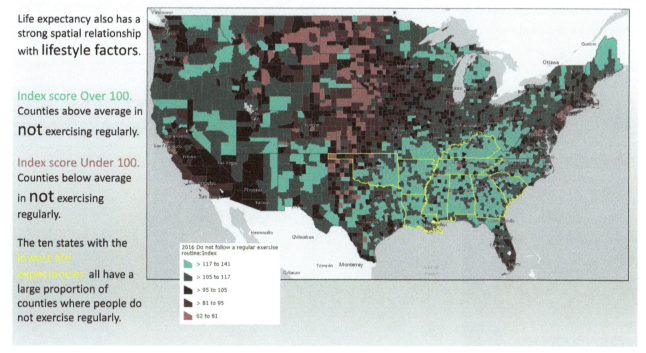

Life expectancy also has a strong spatial relationship with **lifestyle factors**.

Index score Over 100. Counties above average in **not** exercising regularly.

Index score Under 100. Counties below average in **not** exercising regularly.

The ten states with the lowest life expectancies all have a large proportion of counties where people do not exercise regularly.

2016 Do not follow a regular exercise routine:Index
> 117 to 141
> 105 to 117
> 95 to 105
> 81 to 95
62 to 81

Figure 2.21. Lack of regular exercise by US county. Lifestyle factors, such as exercise or lack of it, can have a significant impact on life expectancy. Explore this map at http://arcg.is/2eBVJjl. Data sources: 2016 USA Adults That Exercise Regularly. Esri and GfK US, LLC, the GfK MRI division.

Natural increase

Returning to the idea of the demographic equation from earlier in this chapter, we know that births and deaths are key components of population change. The difference between these two gives the rate of *natural increase*. Simply stated, Natural Increase is calculated by adding in the number of people born each year and subtracting the number who die.

Natural increase = Crude birth rate – Crude death rate

As an example, the 2015 CBR for the United States was 12.49 per 1,000 people, the CDR was 8.15 per 1,000 people, and thus the rate of natural increase was 4.34 per 1,000 people (12.49 – 8.15 = 4.34). To see the result in percentage terms, divide by 10, for a natural increase rate of 0.434 percent. The highest current estimated rate of natural increase is Malawi, at 3.31 percent (a CBR of 41.56 and a CDR of 8.41). At the low end is Bulgaria, at –0.552 (a CBR of 8.92 and a CDR of 14.44) (table 2.2).

It may be difficult to visualize what these natural increase percentages mean for countries. Nevertheless, we can begin to get a feel if we look at the percentages relative to the US. Malawi's natural increase rate of 3.31 percent, divided by the United States' natural increase of 0.43 percent, shows that Malawi's population is growing at about 7.6 times the rate of that of the US! For a low-income African country, that rate represents significant challenges in terms of growing jobs, housing, and food supply at the same rate or more. At the low end of the natural increase rankings are negative numbers. These result when death rates are higher than birth rates and mean that populations are shrinking unless offset by immigration. Just as a fast-growing population can present challenges, a shrinking population presents a whole different set of challenges, which are discussed in the next section of this chapter.

Another way natural increase can be put into context is by calculating *doubling time*, the number of years it will take for a population to double in size. The rule of 70 is an easy tool for estimating doubling time by dividing 70 by the natural increase rate. Using the preceding data, the doubling time for the US population

Natural Increase, 2015

	COUNTRY	RATE/1000
HIGHEST NATURAL INCREASE	Malawi	33.15
	Uganda	33.1
	Niger	33.03
	...	
	United States	4.34
	Canada	1.86
	...	
LOWEST NATURAL INCREASE	Latvia	-4.31
	Serbia	-4.58
	Bulgaria	-5.52

Table 2.2. Highest and lowest natural increase rates, 2015. Data source: World Bank.

is 70/0.434 = 161 years. Likewise, the doubling time for Malawi is 70/3.32 = 21 years. It must be remembered that when using natural increase to calculate doubling time, we include only births and deaths; migration is not included in the calculation. Nevertheless, these doubling time calculations illustrate that Malawi is facing much more rapid population growth from high rates of births and low rates of deaths than is the United States. This implies that Malawi will need to produce more schools, housing, and jobs in a relatively short time, while the United States should have the luxury of a much longer timeframe.

When natural increase is zero, meaning crude birth rates and crude death rates are the same, a place is said to be in zero population growth. In 2015, a handful of European countries, including Denmark, Slovakia, and Austria, were very close to zero population growth. Again, it must be recalled that migration is not included in these calculations. Some people consider zero population growth to be a desirable goal, since both growing and shrinking populations can lead to problems, as discussed in the next section.

Go to ArcGIS Online to complete exercise 2.3: "Death rates and natural increase."

Population structure

As patterns of births and deaths change over time in a society, they create different population structures. *Population structure* refers to the age and sex distribution of people in a society in terms of the proportion of men and women, young and old in a place. This structure can have profound impacts on a society, determining whether limited economic resources go to the young or the old and influencing opportunities for economic development.

Population pyramids

Population structure can be illustrated as a *population pyramid*, which shows men and women by five-year age-sex cohorts (figure 2.23). Traditionally, population structures have a pyramid shape, with many young people at the bottom and fewer old people at the top.

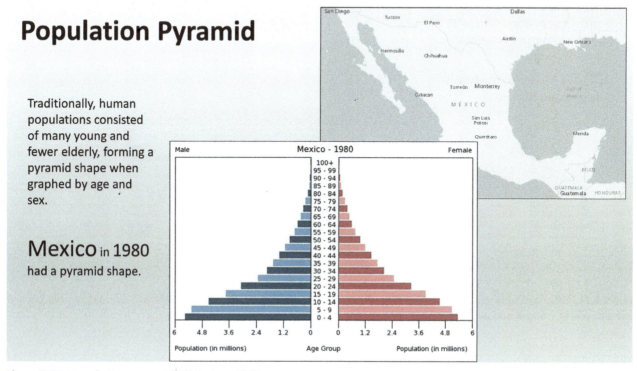

Figure 2.23. Population pyramid of Mexico, 1980. Data source: US Census.

This form is traditional in that, for most of human history, many babies would be born, creating a wide base to the pyramid, while people would die as they aged, creating a progressive narrowing toward the top. However, as discussed earlier in this chapter, birth rates have declined in many countries, causing population structures to narrow at the bottom. As age-sex cohorts from previous higher-fertility generations age, a bulge of people moves up the pyramid. Eventually the population structure comes to resemble more of an inverted pyramid, with fewer children and more elderly people.

In figure 2.24, Japan's population pyramid shows this change. Japan, by 1990, was already moving toward low fertility rates. Two bulges are apparent in 1990. The first is of people ages 40 to 44 who were born as part of the "baby boom" just after World War

II. Another smaller baby boom can be seen among 15- to 19-year-olds, who would have been born between 1971 and 1975. By 2017, these two bulges had aged and can be seen in the 65- to 69-year and 40- to 44-year age cohorts. Japan's ongoing decline in fertility meant that no new bulges in young cohorts formed, resulting in a narrowing base. Projections to 2050 show that the population will continue to age, as previous cohorts get older and low fertility rates result in smaller cohorts of young.

The dependency ratio

Changing population structures present distinct challenges to societies, which can be illustrated by the *dependency ratio*. The dependency ratio measures the proportion of people ages 15–64, compared to those under 15 and over 64. The idea behind this ratio is

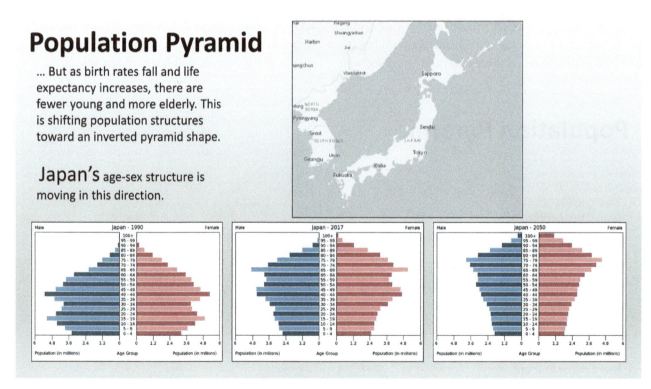

Figure 2.24. Population pyramid for Japan, 1990, 2016, 2050. Low fertility rates are resulting in an increasingly narrow base of young people and wider top of elderly people. Data source: US Census.

to calculate the proportion that are of working age in relation to those that are typically considered too young or too old to work.

When birth rates are high (resulting in a wide-based pyramid), there is a large proportion of people under the age of 15 (figure 2.25). With a large, dependent population of young people, a country must make large investments in things such as child care and education. High birth rates and a young population also typically mean that a population is growing quickly and will continue to do so as young people enter reproductive ages and have their own children. This means that food supplies, housing, and jobs must increase at least as fast as population to maintain standards of living.

The Middle East and North Africa had the highest youth unemployment of all regions in 2014, in part from relatively high birth rates combined with weak economies and low levels of job creation. The unemployed, combined with many more who are underemployed in the informal economic sector with unstable work selling goods on the street and working for cash under the table, often become disenchanted with their governments. Some researchers argue that this large group of young people with limited opportunities to better their future is more susceptible to radical ideologies, including terrorism. Demographic trends and economic conditions can thus create volatile situations.

Conversely, when birth rates are low, the eventual result will be a higher proportion of dependent elderly people (a smaller base of the pyramid and larger top), which requires greater investment in health-care and

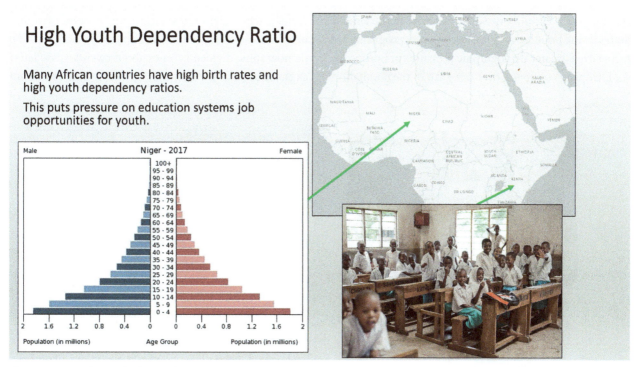

Figure 2.25. High youth dependency ratio. Countries with a high dependency ratio due to many children face costs associated with education. Once these children are a little older, the economy will have to provide sufficient jobs as well. Photo: Schoolchildren in Mombasa, Kenya. Data source: US Census. Photo by Juliya Shangarey. Stock photo ID: 619096334. Shutterstock.

retirement systems (figure 2.26). But with fewer young people entering the workforce, paying for these systems becomes a challenge that can lead to cuts in health-care and retirement benefits, higher taxes on the shrinking working population, or increases in the age of retirement. Obviously, all of these potential solutions face resistance from some segments of the population.

This type of population structure represents a declining population, with few children being born and a larger elderly population dying. Maintaining economic strength with a declining population can be difficult. Continuous increases in worker productivity or immigration are necessary to maintain strong economies.

Many European countries are facing the consequences of rapidly aging populations. In some towns, schools are closing for lack of pupils, and nursing homes are opening at an increasing rate as the elderly population grows. You may have heard concerns about the long-term viability of Social Security in the United States as the population ages, but pension programs are facing even greater demographic pressures in Europe. In six European countries, over one-fifth of the population

is already over 65 years of age. Furthermore, many people retire well before 65. The effective age for men to retire in France is 59.4, while in Italy and Greece, it is just over 61. Only in Switzerland and Portugal is the effective age of retirement over 65. Lower retirement age puts immense pressure on pension plans. Again, demographic trends can have profound impacts on societies.

The demographic dividend

While countries with high dependency ratios face a series of challenges, what about those with low dependency ratios? When the bulge in the population pyramid is largest among the working-age cohorts, countries have the possibility of taking advantage of a *demographic dividend*. The demographic dividend occurs when a country with previously high birth and death rates transitions to one with low birth and death rates. As explained previously, when birth rates fall, the proportion of young people (those at the bottom of the pyramid) declines. The demographic dividend comes about twenty years later, as those born in the last high birth-rate cohort mature

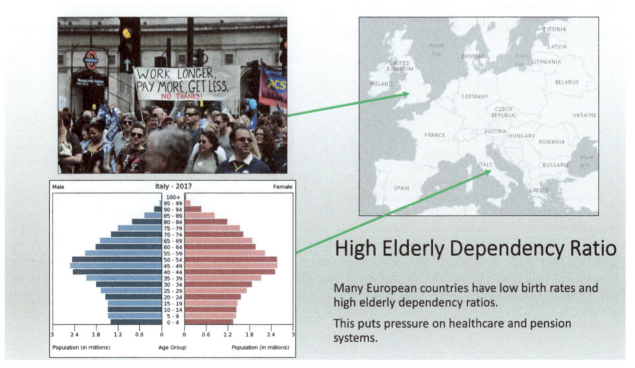

High Elderly Dependency Ratio

Many European countries have low birth rates and high elderly dependency ratios.

This puts pressure on healthcare and pension systems.

Figure 2.26. High elderly dependency ratio. Photo: Unions protest proposed pension reforms in London. Data source: US Census. Photo by Matt Gibson. Stock photo ID: 80226988. Shutterstock.

to the point where they enter the workforce. When this occurs, the population structure has a low dependency ratio: there are fewer young people, not yet too many elderly, and lots of working-age people. Thus, costs associated with supporting the young and old are minimal. At the same time, a large number of workers can help increase economic output. Essentially, the demographic dividend is a several-decade window of opportunity that countries can use to grow their economies. Taxes can be kept low, allowing for more private investment, or tax revenue can be used for roads, ports, and power plants rather than schools and pension plans.

China is an example of a country that took great advantage of its demographic dividend. In 2000, when China entered the World Trade Organization and became fully integrated into the world economy, its population structure was nearly perfectly placed (figure 2.27). A large cohort of people born in the mid-1960s through 1971 were roughly in their 30s by that time. In turn, this group had given birth to a second cohort that was entering adolescence. By opening up to the world economy, China was able to move large numbers of these people from low-productivity farm work to higher-productivity factory jobs, fueling its rapid economic ascent.

Of course, a demographic dividend cannot last forever. Eventually, the large working-age cohorts reach the age of retirement and leave the workforce. Due to ongoing low birth rates, no new large cohort is able to replace them, resulting in a higher dependency ratio with a large number of elderly, as with Italy (figure 2.26). By 2017, China's population was reaching this point, which will have major impacts on the economic model that has been driving its growth. The days of placing large numbers of low-skilled workers into factories is coming to an end.

A big question remains as to how well other developing countries will take advantage of their demographic dividend. As birth rates fall in Africa and the

The Demographic Dividend: China

When China joined the world economy in 2000 it had a bulge of working and nearly working-age people.

Many of these people were effectively moved from low productivity agriculture to higher productivity manufacturing, boosting economic growth.

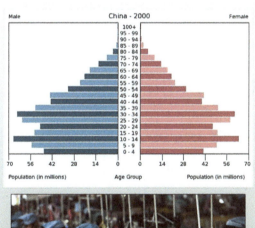

Figure 2.27. The demographic dividend: China. Large working-age cohorts helped drive growth of its global industrial export sector. China took advantage of its demographic dividend by moving large numbers of workers from inefficient farm work to more productive factory work. Photos: Women in Jiujiang, China, working in tea fields. Women in a clothing factory in Huaibei, China. Data source: US Census. Farm photo by Humphery. Stock photo ID: 611411201. Shutterstock. Factory photo by Frame China. Stock photo ID: 390471148. Shutterstock.

Middle East, they will face great opportunities and challenges. If the bulge of working-age people can find employment, these countries can make a big push toward development. If they cannot find jobs, unemployment and discontent can rise, and the demographic dividend will be squandered.

The sex ratio

Population structures can also vary in terms of the proportion of men and women. Generally, the proportions are similar, although in older cohorts, women tend to outnumber men. The *sex ratio* is the ratio of males to females in a population. Due to higher rates of health problems in male infants and the propensity of males to die younger, nature attempts to even out the sex ratio at birth by making a slight preference for males. Under normal circumstances, the sex ratio at birth is 105 males for every 100 females. However, some societies have a strong preference for male babies, resulting in an even greater difference between male and female births. For instance, in India, the sex ratio

at birth in 2015 was 111 males for every 100 females, while in China it was 116 to 100.

The "excess" number of males is the result of sex-selective abortion and female infanticide. Pregnant women sometimes choose to abort after ultrasound examination reveals that the fetus is a female, or less commonly, they kill or abandon newborn girl babies prior to registering their births. The ratio can become more skewed toward males even after birth. A preference for male children means that girls suffer malnutrition more than boys and are given less medical care. This results in higher child death rates for girls than for boys. This imbalance can create social stresses due to a surplus of men unable to marry and form families. In some cases, this leads to human trafficking of women, as male "demand" outstrips female "supply" of spouses. There is also evidence that rates of crime, including rape, are disproportionately committed by unmarried men between the ages of 15 and mid-30s.

Figure 2.28 illustrates how sex selection has become more prevalent in recent decades in three highly skewed

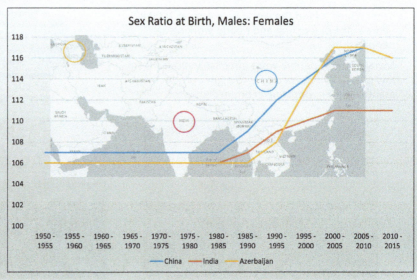

Sex Ratio at Birth

116:100

The sex ratio of men to women in China, 2010-15.

Countries with cultural and legal systems that favor males have a disproportionate number of men.

As ultrasound technology becomes more accessible, more parents elect sex-selective abortion in favor of sons.

Figure 2.28. Sex ratio at birth. Data for China, India, and Azerbaijan show a rapid rise in the sex ratio after the diffusion of ultrasound technology. Data source: United Nations.

countries, China, India, and Azerbaijan. Prior to 1980, the sex ratio at birth in each country was just slightly above the natural rate of 105. However, around the 1980s, the rates rose substantially. This was due to better access to ultrasound technology that allowed the sex of fetuses to be determined. Furthermore, as incomes rose, more families could afford to pay for ultrasound scans. Increased access and affordability of ultrasound scans allowed for greater sex-selective abortion.

In countries with highly skewed sex ratios, cultural norms and laws favor males over females, resulting in a preference for sons. For instance, in India, the families of daughters must pay a dowry when she gets married. In addition, once she is married, she often moves in with her in-laws, caring for them rather than her own parents. Women generally cannot inherit property and are not equal guardians of their children. Furthermore, girls often cannot choose who to marry, and families sometimes feel obliged to commit "honor killings" of their daughters who disobey family wishes. Because of these cultural and legal traditions, parents prefer to have sons.

In some cases, skewed sex ratios come not from a preference for males but from immigration. The most highly skewed sex ratios in 2016 were found in the Middle Eastern countries of Qatar and the United Arab Emirates (figures 2.29 and 2.30). In both cases, the population pyramid shows a dramatic increase

Figure 2.30. Foreign construction workers from South Asia in the United Arab Emirates. Photo by Rob Crandall. Stock photo ID: 584348938. Shutterstock.

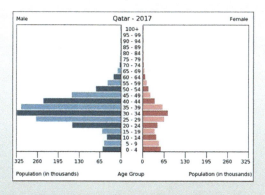

Sex Ratio and Immigration

The majority of the population in Qatar and the United Arab Emirates consists of foreign workers.

The creates a population structure heavily skewed toward working-age males.

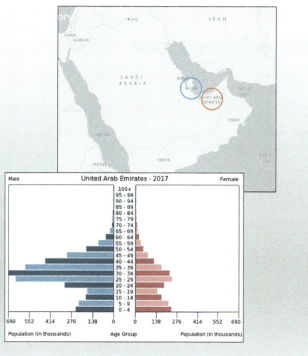

Figure 2.29. Sex ratio and immigration. Data source: US Census.

in the male population ages approximately 20 to 49, which reflects a large number of male foreign workers in each country.

Go to ArcGIS Online to complete exercise 2.4: "Population structure."

Population theories

Demographic transition

After observing patterns of fertility and mortality, natural increase, population structure, and more, we want to know why these patterns vary over time and place. The *demographic transition* model offers a useful explanation by describing how places move through four demographic stages (figure 2.31). In each stage, there are changes in fertility and mortality, which result in differing population growth rates and age structures.

Stage 1: Preindustrial society

Preindustrial societies are in stage 1 of the demographic transition. From hunter and gatherer societies through agricultural societies, birth rates and death rates tend to be high. Birth rates are high because children are economic assets, helping with hunting, gathering, and farming from an early age. Likewise, children serve as insurance for their parents. They are their "pension

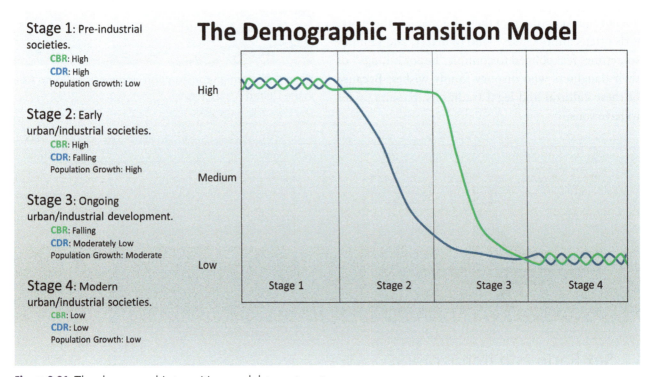

Figure 2.31. The demographic transition model. Image by author.

plan," caring for them in their later years, and can also mitigate risk by working for others if output on the family farm is limited. A large number of children are also desirable due to high infant mortality rates. Families must have enough children who survive into adulthood and can, in turn, have their own children and ensure survival of the group. For these reasons, numerous cultures developed that bestow status and prestige on parents with many children and shame those with few children. Gender roles also have a significant impact on birth rates. Stage 1 societies tend to be culturally conservative, where women's roles are restricted to that of mother and helping with local village tasks. As part of this role, women marry and become mothers at a young age. Given the restricted role that women play in these societies, there is often a preference for sons. Having at least two sons is often the goal of parents to ensure that at least one survives to care for and support the family. Thus, women need, on average, 4.1 children to get two boys. With a reality of high infant mortality rates, an even larger number of births is often required.

Death rates also tend to be high in stage 1 societies due to disease and unstable food supplies. A lack of proper nutrition, along with limited calorie intake in hard times, results in many deaths, especially among infants and young children. At the same time, communicable diseases kill many children due to a lack of medical knowledge. Diseases that are spread by other humans, animals and insects, and unsanitary water—such as influenza, cholera, malaria, and plague—ensure that death rates remain high.

In stage 1, population growth (natural increase) remains flat, or close to zero, given that both birth rates and death rates are high. This was the situation for most of human history.

Stage 2: Early urban-industrial societies

Societies enter stage 2 as they begin to industrialize and urbanize. In this stage, birth rates remain high, but death rates begin to fall substantially for several reasons. With industrial and urban development, agricultural technology leads to more stable food supplies and thus better nutrition and calorie intake. Furthermore, knowledge about germs improves, and the development of public health programs expands. Clean water and sewer systems are gradually built, which reduces waterborne disease. Meanwhile, rodent and insect controls are put in place, and the use of soaps and disinfectants is promoted. In addition, medical technology improves and accessibility diffuses to a larger proportion of the population.

For instance, the smallpox vaccine was widely used by the early 1800s, resulting in many fewer deaths from that deadly disease. Also in the 1800s, English physician John Snow linked a water well to an outbreak of cholera, using spatial analysis to show that the disease was waterborne and not airborne (figure 2.32). His discovery contributed to the eventual development of urban water and sewer infrastructure.

However, during stage 2, birth rates remain high as people maintain the cultural tradition of having large families both in rural areas and in expanding urban areas. Cultural traditions that limit the role of women and esteem large families and deride small ones still influence family size. Furthermore, a preference for sons pushes parents to keep having children until at least two boys are born.

With a large difference between births and deaths, natural increase rises and population growth increases significantly in this stage. Countries in stage 2 see a boom in population growth.

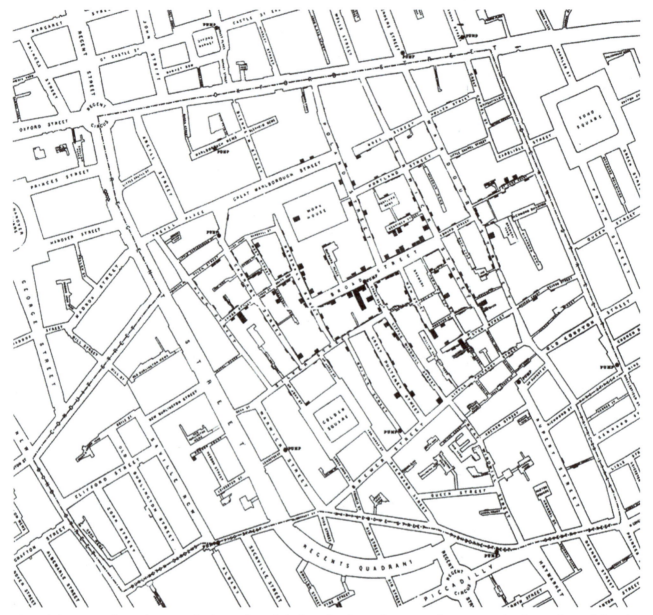

Figure 2.32. By studying the spatial relationship of cholera deaths and water-well pumps, John Snow showed that the disease was waterborne. Map by John Snow is in the public domain.

Stage 3: Ongoing urban-industrial development

With further industrialization and urbanization, societies enter stage 3. In this stage, births fall significantly (figure 2.33). This occurs for several important reasons. First, lower infant mortality rates mean that fewer children are needed to ensure that some survive to adulthood. Second, industrialization and modernization make contraceptives accessible to a larger share of the population. But most important, women want to use contraceptives by stage 3. In contrast with agricultural societies, children in industrial societies are more of an economic liability. They cannot contribute to the household until they are much older, as urban industrial jobs require more physical strength or education. Also, housing is more crowded and expensive in the city than on the farm. Lastly, women have more

Figure 2.33. Economic and social change leads to smaller family size as societies transition from rural-agricultural to urban-industrial. A large rural family in Nannilam, India. A small urban family in Delhi, India. Photo of large rural family by Ashok India. Stock photo ID: 490150441. Shutterstock. Photo of small urban family by Paul Prescott. Stock photo ID: 52607926.

options and opportunities as societies industrialize and modernize. Whereas a woman on the farm in stage 1 or 2 is largely limited to having children and caring for the house, she has many more options in the city. Traditional gender roles typically loosen as time passes in urban areas and preferences for sons weaken. Women attend school for more years, and their role in the paid workforce grows. As women study and work more, they delay marriage and childbirth until they are older, thus reducing the total number of children they have.

Death rates in stage 3 continue to decline, although at slower rates than in stage 2, as medical care and nutrition improve. Better treatments for communicable diseases are developed—for example, antibiotics became widely used around the 1940s—while new treatments for degenerative diseases such as heart disease also become more widely available.

As the difference between births and deaths narrows, natural increase slows and population growth becomes moderate.

Stage 4: Modern urban society

Stage 4 consists of modern urban societies. In this stage, both birth rates and death rates are low. Births are low for the same reasons as in stage 3 but by stage 4 have diffused to a larger proportion of the population. Low infant mortality rates guarantee that nearly all

children will survive to adulthood, and contraceptives are widely available. Child-rearing is focused more on quality than quantity, as parents have few children but invest great amounts of time and money in their upbringing. Opportunities for women other than motherhood abound. Education levels for women are high, with most finishing high school and many continuing with higher education. This translates into generally high levels of female workforce participation. As women study and work more, they further delay marriage and childbirth to later ages and limit the number of children they have. This can be seen in the median age at first marriage in the United States. While women married at age 20.3 and men married at 22.8 in 1960, by 2010, the age had risen to 26.5 for women and 28.7 for men. Japan and other stage 4 countries similarly have average ages of first marriage around 30 years old.

Likewise, death rates are low due to the availability of modern medicine and ample food supplies. Deaths from communicable disease become less common, while the focus for new medical treatments shifts to degenerative diseases such as heart disease and cancer that are associated with old age.

In stage 4, as in stage 1, births and deaths are about equal, and natural increase approaches zero and population growth is small. However, in this case, both rates are low rather than high.

Stage 5: A new stage of population decline?

Some geographers now discuss a fifth stage to the demographic transition as a handful of countries move toward birth rates that are lower than death rates, with ensuing negative natural increase. Interestingly, while birth rates are universally low in modern, urban stage 4 countries, societies moving into a stage 5, with extremely low birth rates, may have some unique characteristics. In very-low-birth-rate countries, it is more common for women to have little help in raising children. Southern European countries such as Italy, as well as wealthy Asian countries such as Japan, tend to have limited state-run programs, including low-cost day-care and after-school programs that allow women to have children and pursue their careers. At the same time, men in these societies tend to help less with child-rearing, leaving the burden on mothers. It seems that when women get little help from the state and little help from fathers, they tend to have well under an average of two children. In stage 4 countries with family-friendly government programs and more engaged fathers, women tend to have slightly more children, since they can more easily be mothers and pursue their careers. France and the countries of Scandinavia fit this description.

Go to ArcGIS Online to complete exercise 2.5: "The demographic transition in Latin America."

Malthusian Theory

Thomas Malthus presented a theory in the eighteenth century linking resources and population growth. *Malthusian theory* holds that populations will grow at a faster rate than food supply. Once population outstrips food, famine and disease will cause populations to collapse (figure 2.34). This theory contrasts significantly with the demographic transition model, which states that population growth declines as urbanization and modernization cause people to choose to have smaller families.

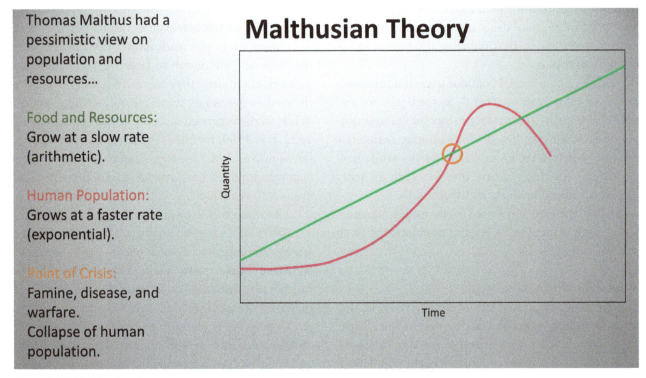

Figure 2.34. Malthusian theory. Image by author.

Whereas Malthus's theory was developed in the eighteenth century, his predictions have yet to be fulfilled by the twenty-first century. According to the World Health Organization, calories consumed per person have increased, not decreased, over time (figure 2.35). Malthus did not account for increases in agricultural productivity that allow production of more food per acre. Nor did he predict the expansion of agricultural land into arid regions and regions with poor soils through the use of irrigation and fertilizers (figure 2.36).

However, this does not mean that global hunger has been conquered. Whereas food supply has grown at a more rapid rate than population, hunger exists due to natural disasters and drought, disruption of food production and distribution from war, poverty, limited distribution due to poor agricultural infrastructure, and overexploitation of the environment. More on food and hunger is discussed in chapter 6.

Figure 2.36. Center pivot irrigation in Saudi Arabia. Malthus did not foresee such advances in agricultural technology. Explore this image at http://arcg.is/2ioJJ63. Data sources: World Imagery–Esri, DigitalGlobe, GeoEye, Earthstar Geographics, CNES/Airbus DS, USDA, USGS, AeroGRID, IGN, and the GIS User Community.

Neo-Malthusians concede that the world has not yet run out of food but say that it is still a possibility. Human populations continue to grow, especially in

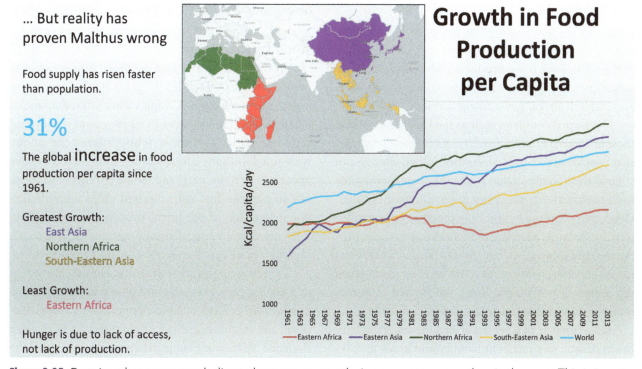

Figure 2.35. Despite what many may believe, there are more calories per person now than in the past. This is in spite of ongoing population growth. In essence, our ability to produce food has grown faster than human populations. Data source: Food and Agriculture Organization of the United Nations.

developing regions such as Africa, the Middle East, and South Asia. This growth will continue to put strains on food supplies, which will have to grow at least at the same rate. At the same time, more people will be using more resources, such as energy that fuels the agricultural sector. Not only that, but as societies grow richer, people's diets change as they demand more meat. Producing meat uses more resources in terms of land and energy than does farming plant-based food. For these reasons, some people say that Malthus ultimately will be correct.

Government population policy

The demographic transition model and Malthusian theory do not explicitly discuss the role of governments in population growth. Nevertheless, in some cases, governments try to manipulate population change within their borders, typically through interventions in fertility.

Fertility control

Governments use a variety of means to help women control their fertility. This is often seen as an interest of the state, since uncontrolled births and rapid population growth can pose a threat to economic prospects and put a drain on government services.

Developed countries such as the United States and those of western Europe often have government programs that provide access to birth control. When women can control their fertility more easily, they can devote time to education and careers prior to starting their families. Developed countries benefit from the effect of controlled fertility because an educated workforce is more productive and innovative than an uneducated workforce. Lower-income countries may also want to give women the option of postponing birth, but limited resources make effective birth control less widely available.

Governments in some lower-income countries have taken drastic steps to limit fertility and in some cases have been accused of violating human rights in the process. In the case of India, sterilization programs, beginning in the 1970s but still widespread today, have been an important component of government-sponsored family planning. Critics of this program accuse officials of coercion by using cash incentives and pressure from government family-planning workers to get poor men and women to be sterilized. The process involves health risks as well. In 2014, eighty women in a rural village were sterilized by one doctor and his assistants during a three-hour period. Thirteen of the women died and dozens more were sickened due to insufficient cleaning of surgical instruments. In many cases, women do want to limit family size, but a lack of alternatives makes sterilization the only option. The Indian Ministry of Health and Family Welfare reported in 2016 that contraceptive use among married women ages 15 to 49 years was 56.3 percent, which leaves a substantial portion of women without contraceptives.

Interestingly, male vasectomies are a much quicker and safer means of sterilization but are less commonly promoted in many less-developed countries. Many men see family planning as a woman's issue and are less likely to voluntarily be sterilized. There are efforts to promote vasectomies, however, including DKT International, the nonprofit organization that organizes World Vasectomy Day.

In China, a state-mandated one-child policy was implemented in 1980. This program also led to coercive policies in some cases. Couples were offered financial incentives and preferences in employment if they had only one child, and those who violated the policy were fined. In some cases, women were forced to have abortions or to be sterilized. Some parents who chose not to pay the fine for a second child were denied documentation of the birth, thus making their child "invisible" from a legal standpoint and unable to get identification, enroll in school, or find employment.

Despite being a rather draconian policy, according to a 2010 report, the one-child policy likely applied to less than 40 percent of China's population. Rural residents and ethnic minorities could typically have more than one child, while some urban residents ignored the ban, choosing to pay penalties and fines.

As pointed out previously in this chapter, with fewer children, China's population structure has been aging. China's ability to be the factory floor of the world economy depends on a large workforce—a workforce that will shrink in coming decades. And remember, an aging population also puts pressure on health-care systems and pension funds. China may well become an "old" country before it becomes rich enough to cover age-related expenses. For these reasons, in 2016, China began to allow families to have two children.

It should also be mentioned that in some countries, there is tension between state family-planning programs and conservative cultural forces. For instance, in the Philippines, a country with a very conservative and influential religious establishment, a family-planning law passed in 2012 was held up in court for two years due to lawsuits despite that the country is poor with a high CBR and high maternal mortality rates. With 72 percent of the population in support of the family-planning law, it was eventually upheld by the Philippine Supreme Court, allowing for free contraceptives at government health clinics, sex education in schools, and medical care for women who have had illegal abortions.

Pronatal policies

As seen in the TFR map (figure 2.11), many countries now have rates below the replacement level of 2.1. In some cases, the number is well below replacement level. As stated previously, with low birth rates, population structures skew toward the elderly, causing a shortage of working-age people and financial strains on pension and health-care systems. Consequently, some countries now promote childbirth through *pronatalist* policies.

Pronatal policies come in many forms but most often are designed to reduce the costs of child-rearing and promote family-friendly labor laws. In Singapore, cash "baby bonuses" help offset the cost of giving birth; in the United States, child tax-credits help offset the costs of raising children; in France, families are paid a cash child benefit for every child after the first.

While the cost of raising children keeps some couples from having more, negative impacts on parents' careers also contribute to lower fertility rates. This is especially important for women, who tend to bear the greatest burden of caring for children. Paid maternity leave and, in some cases, paternity leave allow parents to care for newborns without having to sacrifice income (figure 2.37). Both rich and poor countries frequently offer paid leave for mothers. For instance, Kenya and Afghanistan offer thirteen weeks at 100 percent pay, while Germany offers fourteen weeks and Greece offers seventeen. Paternity leave is less common or typically of shorter duration. Kenya offers two weeks at 100 percent pay, while Afghanistan, Germany, and Greece offer none. It must be noted when viewing these benefits that not all mothers and/or fathers qualify. In lower-income countries, few people have formal employment contracts with legal benefits. Therefore, although Kenya and Afghanistan offer generous maternity leave, few women take advantage of it. In more affluent countries, such as those of Europe, most people do have formal employment contracts, so parental leave can be used by most workers (figure 2.38).

Perhaps more important than one-off baby bonuses and parental leave, benefits such as free or subsidized child care, government incentives for companies to provide on-site child care, and low-cost after-school programs can reduce the ongoing burden of balancing careers with raising children.

Pronatalist policies can help increase the TFR in some cases, but regardless of the effort of governments, culture plays a role as well. TFRs in France, Norway, and Sweden are higher than those in Italy, Greece, and Japan. This is partially because fathers help more with child-rearing in the first set of countries than in the second set. Even with government benefits, when fathers help less with child-rearing, women can feel they are left to choose between being stay-at-home moms and pursuing their career. In many cases, they choose the latter, which means having one or possibly no children.

Overpopulation and carrying capacity

Discussions of population growth and population distributions often turn to the idea of *overpopulation*. Overpopulation is a commonly used term, but the

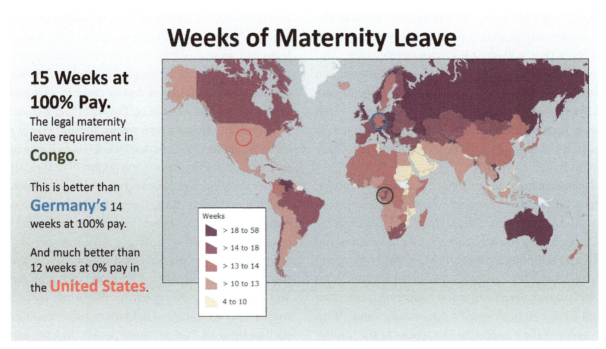

Figure 2.37. Weeks of maternity leave. Many countries offer paid maternity leave to help women balance having children with remaining in the paid workforce. Explore this map at http://arcg.is/2laPTHp. Laura Addati, Naomi Cassirer and Katherine Gilchrist. 2014.

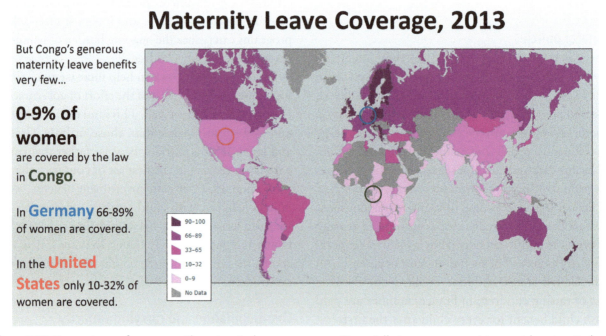

Figure 2.38. Percentage of women with maternity leave coverage. Especially in poorer countries, paid maternity leave covers only a small fraction of women, since few have formal employment contracts with benefits. Explore this map at http://arcg.is/2laTAwH. Laura Addati, Naomi Cassirer, and Katherine Gilchrist, 2014.

concept can quickly become contentious on closer inspection. For most people, it means there are too many people in a given place. Sometimes the concept is used in terms of a city or country, while other times it is seen in global terms, focusing on the question of how many people planet Earth can support. The concept can become contentious because defining "too many" people is very subjective. When stuck in traffic on the freeway, people may say their city is overpopulated. If housing is too expensive or jobs are scarce, people may say the same thing. But in these cases, it can also be argued that the problem is not too many people but too little public transit, too little housing, or not enough economic growth (figure 2.39).

The definition of overpopulation can be limited to the concept of *carrying capacity*: the number of people that can be supported by the land they live on. This can be seen in terms of land for food, clean air for breathing, and clean water for drinking. But this definition also leads to problems. First, as discussed earlier in this chapter, human populations can be very dense without being overpopulated and, via trade, can exceed their local carrying capacity. Food and water can be brought in from far away to support millions of people. For instance, Los Angeles, California, gets the majority of its water via canals from the northern parts

of the state and the Colorado River, and most food is imported from other parts of California, the United States, and the world. Thus, the concepts of overpopulation and carrying capacity depend a great deal on infrastructure, technology, and economic development. Places with sufficient infrastructure, technology, and economic development can use resources from around the world and support massive, dense populations. Places without these advantages can more easily be considered overpopulated because their populations face shortages of food, water, and other resources. There is no fixed population density that can be considered an overpopulation threshold.

Although overpopulation is difficult to define at a local scale, many argue that at a global-scale population will eventually outstrip the earth's carrying capacity. A 2012 report by the United Nations Environmental Program states that most estimates of the earth's carrying capacity fall within a fairly wide range of eight to sixteen billion people. As of 2015, the United Nations median estimate for total world population by the year 2100 was 11.2 billion people, a point where population growth may level out.

Earth's carrying capacity will depend on several factors. One is how much technology can help humans use resources more efficiently. If goods and services can be produced with less energy and material inputs, then

Figure 2.39. Is the world overpopulated? Defining the concept can be difficult to agree on. Crowds in Times Square, Manhattan. A lone farmhouse in Tuscany, Italy. Times Square photo by Andrew F. Kazmierski. Stock photo ID: 36538936 Shutterstock. Tuscany photo by Shaiith. Stock photo ID: 141591835. Shutterstock.

the world will be able to support a larger number of people. One reason Malthus has not yet been proven correct is that technological innovations allow more food to be produced with less land, labor, and capital resources. Another factor is the global standard of living. If all 11.2 billion people in the year 2100 live at the same standard of living that Americans do today, then there will be a much greater strain on resources. Consumption of energy and other resources per capita is much higher in the United States and other rich countries than in less developed countries. Technological innovation will have to improve efficiency dramatically to support increased levels of consumption as people get richer.

References

Addati, Laura, Naomi Cassirer, and Katherine Gilchrist. 2014. *Maternity and Paternity at Work: Law and Practice Across the World.* Geneva, Switzerland: International Labour Organization.

Carberry, S. 2014. "An Afghan Success Story: Fewer Child Deaths." NPR. http://www.npr.org/2014/02/04/269551459/an-afghan-success-story-fewer-child-deaths. February 4.

Centre des Liaisons Européennes et Internationales de Sécurité Sociale. 2018. "The French Social Security System IV—Family Benefits." Centre des Liaisons Européennes et Internationales de Sécurité Sociale. http://www.cleiss.fr/docs/regimes/regime_france/an_4.html.

"The Child in Time." 2010. *The Economist,* August 19, 2010. http://www.economist.com/node/16846390?story_id=16846390&CFID=145777375&CFTOKEN=91195822.

Cohn, D'Vera, Jeffrey S. Passel, Wendy Wang, and Gretchen Livingston. 2011. "Barely Half of US Adults Are Married—A Record Low." Pew Research Center's Social & Demographic Trends Project. http://www.pewsocialtrends.org/2011/12/14/barely-half-of-u-s-adults-are-married-a-record-low.

DeSilver, D. 2015. "Refugee Surge Brings Youth to an Aging Europe." *Factank: News in the Numbers,* October 8, 2015. Pew Research Center. http://www.pewresearch.org/fact-tank/2015/10/08/refugee-surge-brings-youth-to-an-aging-europe.

De Soto, H. 2018. "The Capitalist Cure for Terrorism." *Wall Street Journal,* October 5, 2018. https://www.wsj.com/articles/the-capitalist-cure-for-terrorism-1412973796.

Food and Agriculture Organization of the United Nations (FAO). 2017. "Country Brief on Egypt." GIEWS—Global Information and Early Warning System. http://www.fao.org/giews/countrybrief/country.jsp?code=EGY.

International Labour Organization. 2015. *Global Employment Trends for Youth 2015: Scaling Up Investments in Decent Jobs for Youth.* https://www.ilo.org/wcmsp5/groups/public/—dgreports/—dcomm/—publ/documents/publication/wcms_412015.pdf.

Kimpton, D. 2014. "75 Million Catholics Just Got Free Birth Control." *VICE News,* April 9, 2014. https://news.vice.com/article/75-million-catholics-just-got-free-birth-control.

McFalls, J. A. 2007. *Population: A Lively Introduction,* 5th ed. Washington, DC: Population Reference Bureau.

Omran, A. R. 2005. "The Epidemiologic Transition: A Theory of the Epidemiology of Population Change." *The Milbank Quarterly* 83, no. 4: 731–57. doi: http://doi.org/10.1111/j.1468-0009.2005.00398.x.

Organisation for Economic Co-operation and Development (OECD). 2014. *Ageing and Employment Policies—Statistics on Average Effective Age of Retirement.* Paris: OECD Publishing.

Pletcher, K. 2016. "One-Child Policy." *Encyclopædia Britannica.* https://www.britannica.com/topic/one-child-policy.

Population Reference Bureau. 2015. "2015 World Population Data Sheet." Population Reference Bureau Data Sheet. https://assets.prb.org/

pdf15/2015-world-population-data-sheet_eng
.pdf.

Samman, E., E. Presler-Marshall, and N. Jones.
2016. *Women's Work: Mothers, Children and
the Global Childcare Crisis*. London: Overseas
Development Institute. https://www.odi.org/sites/
odi.org.uk/files/odi-assets/publications-opinion-
files/10333.pdf.

Singh, K. 2013. *Laws and Son Preference in
India—A Reality Check*. New Delhi : United
Nations Fund for Population Activities.
https://www.unfpa.org/sites/default/files/
jahia-news/documents/publications/2013/
LawsandSonPreferenceinIndia.pdf.

Timmons, A. T. H. 2013. "India's Man Problem."
New York Times, January 16, 2013. https://
india.blogs.nytimes.com/2013/01/16/
indias-man-problem/?_r=2.

United Nations. 1999. *The World at Six Billion*. New
York: United Nations. https://www.un.org/esa/
population/publications/sixbillion/sixbillion.htm.

———. 2015. *The World Population Prospects: 2015
Revision*. New York: United Nations.
https://www.un.org/en/development/desa/
publications/world-population-prospects-2015-
revision.html.

———. 2017. "Sex Ratio at Birth."
UN Data. http://data.un.org/Data
.aspx?q=Sex+Ratio+at+Birth&d=PopDiv&
f=variableID%3a52.

United Nations Children's Fund (UNICEF).
n.d. "Health: Ending Preventable Maternal,
Newborn, and Child Deaths." UNICEF
Afghanistan. https://www.unicef.org/
afghanistan/health_nutrition.html.

United Nations Environment Programme. 2012.
"One Planet, How Many People? A Review
of Earth's Carrying Capacity." UNEP Global
Environmental Alert Service (GEAS). https://
na.unep.net/geas/archive/pdfs/geas_jun_12
_carrying_capacity.pdf.

Walker, J. 2014. "Instruments 'Rusty' at Indian
Sterilization Camp Where 13 Women Died." *The
Huffington Post,* November 12, 2014. http://
www.huffingtonpost.com/2014/11/12/india-
sterilization-deaths_n_6143894.html.

Weeks, J. R. 2016. *Population: An Introduction
to Concepts and Issues*. Boston, MA: Cengage
Learning.

Williams, S. C. P. 2013. "Gone Too Soon: What's
Behind the High U.S. Infant Mortality Rate."
Stanford Medicine 30, no. 3: 12–15. http://
sm.stanford.edu/archive/stanmed/2013fall/
article2.html.

Wolchover, N. 2011. "Why Are More Boys Born
than Girls?" *Live Science,* September 9, 2011.
http://www.livescience.com/33491-male-female-
sex-ratio.html.

World Food Programme (WFP). n.d. "Zero Hunger."
http://www1.wfp.org/zero-hunger.

World Health Organization (WHO). n.d. "Global
and Regional Food Consumption Patterns
and Trends." http://www.who.int/nutrition/
topics/3_foodconsumption/en.

———. 2016. "Life Expectancy Increases by 5 Years,
But Inequalities Persist." Press release. http://
www.who.int/mediacentre/news/releases/2016/
health-inequalities-persist/en.

Worldometers. n.d. Current World Population.
Accessed April 25, 2017. http://www
.worldometers.info/world-population.

Yüksel-Kaptanoğlu, I., M. A. Eryurt, I. Koç, and
A. Çavlin. 2014. *Mechanisms Behind the Skewed
Sex Ratio at Birth in Azerbaijan: Qualitative
and Quantitative Analyses*. Baku, Azerbaijan:
United Nations Population Fund.

Chapter 3
Migration

Since humans first left Africa sixty thousand years ago, ours has been a story of migration. From our origin on the African continent, we have settled all corners of the world, from within the Arctic Circle in the north to Tierra del Fuego in the south, from the most remote parts of Siberia to the jungles of Brazil. Only Antarctica has been spared human migrations, although even there a small number of scientists live on a temporary basis.

But humans have not been content to stay in place since settling the far reaches of our planet. We continue to move in search of better opportunities for our families or, all too often, as others force us to go elsewhere. Migration affects economies, environments, and cultures in profound ways. Migrants can help boost economic growth, as new workers provide needed labor in agriculture, manufacturing, and service enterprises. They can also send money back home, helping to improve the places they came from. But at the same time, migration can threaten the jobs of native workers, leading to societal tension and political battles, or lead to broken families and a lack of workers in the places they left. The movement of people and the mixing of languages, religions, customs, and beliefs can add vibrancy to places but can also strain social cohesion. Consequently, migration is one of the most important aspects of human geography (figure 3.1).

In Chapter 2, we were introduced to the demographic equation, which was stated as:

$$\text{Population change} = \text{Births} - \text{Deaths} + \text{Immigration} - \text{Emigration}$$

Obviously, birth and death rates help determine population size and population density on the surface of the earth. But that is only half of the equation. To make our explanation of these patterns complete, we must add one more component: *migration*. Migration is defined as the change in location of one's permanent or semi-permanent residence. Thus, the population of a particular place is determined by the number of births and the number of deaths, as well as the number of people who migrate in and the number of people who migrate out. Immigration refers to in-migration; thus, we talk about immigrants who come from other countries to the United States. Emigration refers to out-migration, so those leaving, say, Mexico, for work abroad are called emigrants.

Migration (both immigration and emigration) can be analyzed at a variety of scales. International migration patterns are those between countries, such as between Mexico and the United States. But we also discuss national migration patterns, such as from the Northeastern United States to the Southwest. We can also examine local migration patterns, such as from rural areas to cities, known as rural-to-urban

Figure 3.1. US Immigrants, 1910 and 2016. Human migration is an ongoing phenomenon. Immigrants to Ellis Island c.a. 1910 and a naturalization ceremony in Portland, Oregon, in 2016. Photo by Bain, George Grantham. Immigrants, Ellis Island. ca. 1910. Library of Congress Prints and Photographs Division Washington, DC. 20540 USA. Photo by Diego G. Diaz, Royalty-free stock photo ID: 483311986. Shutterstock.

migration, and other moves within states or even within counties or cities.

Spatial distributions

Migrant stock and flow

Before we discuss why people migrate, it is useful to look at the spatial distribution of migrants. This distribution can be viewed in terms of stock and flows. *Migrant stock* is the number of people who reside in a place where they were not born. When viewed as a percentage, it can show the proportion of the population that is not native to a place. While measures of migrant stock reflect the number of newcomers to a place, they do not tell when the people migrated there. Thus, migrant stock includes recent arrivals as well as migrants who may have arrived decades ago as children.

Migrant flows refer to migration within a certain timeframe. A count of the number of people moving into and out of a place within the past year represents a migration flow. Flows can be viewed in different time frames, so one geographer may be interested in how migration flows during the past fifty years have impacted growth of US Sun Belt cities, while another geographer may be interested in how flows during the last month affect support for nativist political parties in Europe.

At a global scale, 71 percent of immigrants live in high-income countries, while the vast majority of emigrants (65 percent) come from middle-income countries. As seen in figure 3.2, there are higher concentrations of immigrants as a percentage of population in high-income places such as North America, much of Europe, Australia, and New Zealand. In the United States, 14.5 percent of the population was born abroad, substantially fewer than in Canada with 21.8 percent and Australia with 28.2 percent (figure 3.3). A cluster of countries with high immigrant stocks can also be seen in the Middle East, where over 70 percent of residents of the United Arab Emirates, Qatar, and Kuwait are foreign born.

Less-developed countries tend to have many fewer immigrants. The so-called global south, consisting largely of Latin America, Africa, and much of Asia,

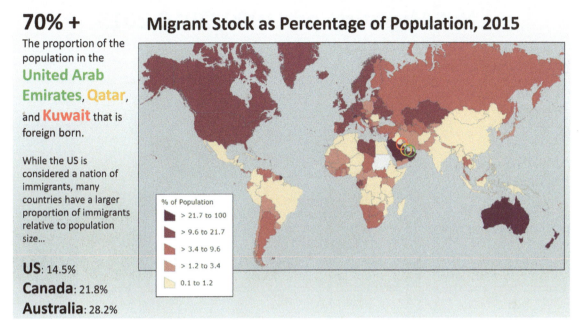

70% +

The proportion of the population in the **United Arab Emirates**, **Qatar**, and **Kuwait** that is foreign born.

While the US is considered a nation of immigrants, many countries have a larger proportion of immigrants relative to population size...

US: 14.5%
Canada: 21.8%
Australia: 28.2%

Migrant Stock as Percentage of Population, 2015

% of Population
- > 21.7 to 100
- > 9.6 to 21.7
- > 3.4 to 9.6
- > 1.2 to 3.4
- 0.1 to 1.2

Figure 3.2. Migrant stock as a percentage of population. 2015. High-income countries tend to have a larger migrant stock than lower-income countries. Explore this map at http://arcg.is/2dDt4bf. Data source: United Nations.

Figure 3.3. Chinatown in Melbourne, Australia. The foreign-born population of Australia is double the proportion of that of the United States. Photo by ChameleonsEye. Stock photo ID: 188715929. Shutterstock.

stands out in figure 3.2 as having low proportions of immigrants. Among the countries with the lowest percentage of immigrants are China, Vietnam, Indonesia, and Cuba, all of which have a foreign-born population of only 0.1 percent.

In 2015 raw numbers, the largest diasporas, or people who have left their homelands, were from India, Mexico, Russia, and China. All four of these countries had at least ten million emigrants living in other countries, many of which have formed ethnic enclaves (more in chapter 4) in cities around the world.

Figure 3.4 illustrates international migrant flows by showing net migration per 1,000 people between 2010 and 2015. Larger negative numbers mean that more people left a country than arrived during this five-year period. Larger positive numbers mean that more people arrived than left. The Middle East shows the greatest movement of people during this time. The war-torn states of Syria and Libya show the largest proportional outflow of migrants, while the greatest proportional inflow of migrants was to Oman, Lebanon, Qatar, and Kuwait.

Net Migration per 1000, 2010-2015

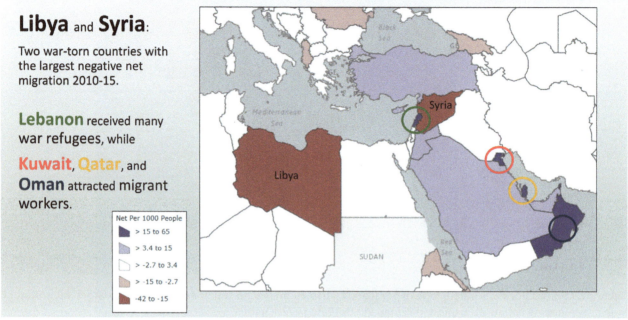

Libya and **Syria**:

Two war-torn countries with the largest negative net migration 2010-15.

Lebanon received many war refugees, while

Kuwait, **Qatar**, and **Oman** attracted migrant workers.

Net Per 1000 People

- > 15 to 65
- > 3.4 to 15
- > -2.7 to 3.4
- > -15 to -2.7
- -42 to -15

Figure 3.4. Net migration 2010–2015 per 1,000 people. The Middle East has seen massive migration flows during this time. Explore this map at http://arcg.is/2dDt4bf. Data source: United Nations.

While immigration is an ongoing topic of political discussion in the United States, net migration between 2010 and 2015 was 3.2 per 1,000, substantially less than in some other rich democracies. For instance, US net immigration was less than half that of Canada (6.7) and nearly one-third that of Australia (8.9) and Norway (9.3).

Migration stock and flow in the United States

Even though the United States has a smaller migrant stock and lower migration flows than some other countries, immigrants have still played an important role in shaping American society. Early American history consisted primarily of immigration from Europe (figure 3.5). The nineteenth century was dominated by immigration from the United Kingdom, Ireland, and Germany, while the early twentieth century saw many Italians, Russians, and Austria-Hungarians.

After World War II, immigration from the Americas picked up, specifically from Mexico, which by the 1960s outpaced European immigration. Asian immigration also picked up around the 1960s.

Annual trends since 1980 show how immigration from Latin America dominated the late twentieth century and early twenty-first century (figure 3.6). A huge spike in immigration from Latin America occurred in the late 1980s, in large part due to military conflict in Central America. While Latin American immigration outpaced Asian immigration for most of the time-period, by around 2011, new Asian immigrants began to outnumber new immigrants from Latin America, especially from countries such as China, India, and the Philippines.

International immigrants made up 14 percent of the US population in 2015, and immigrants and their children have been the main driver of population

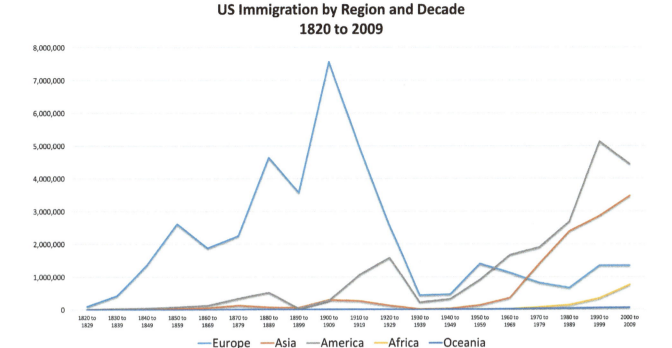

Figure 3.5. US Immigration by region and decade, 1820–2009. Immigration was dominated first by Europeans and later by Latin Americans and Asians. Data source: US Department of Homeland Security.

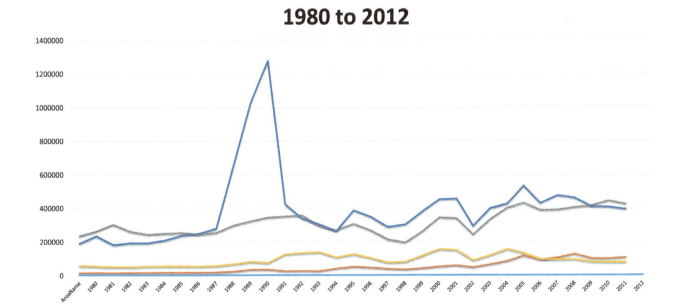

Figure 3.6. Immigration levels by year, 1980–2012. Around 2011, Asian immigrants began to outnumber Latino immigrants. Data source: United Nations.

growth since the 1960s (figure 3.7). Without immigrants and their descendants, it is estimated that the US population in 2015 would have been 252 million as compared to the actual number of 324 million. By 2065, immigrants and their children will make up 36 percent of the US population. This change will have significant impacts on the human geography of the United States in terms of culture, politics, and economics.

At the state level within the United States, migration stock and flows can be observed in terms of international migration and of intrastate migration.

As the United States becomes increasingly influenced by immigrants and their descendants, the country will look more like the states of California, New York, Texas, and Florida. These are the states with the largest international migrant population counts, all with over 3.5 million international immigrants in 2015. These states are all important coastal and border gateway states that also have overall large populations—of both the native born and immigrants. When mapping international immigrants as a proportion of the total population, the states of Nevada, New Jersey, and Massachusetts also stand out, with over 15 percent of their populations being foreign born (figures 3.8 and 3.9).

Native-born migrants within the United States continue to move to the Sun Belt states. The largest native-born migrant stock is found, starting with the highest proportion, in Florida, California, and Texas, followed by Georgia, North Carolina, and Arizona.

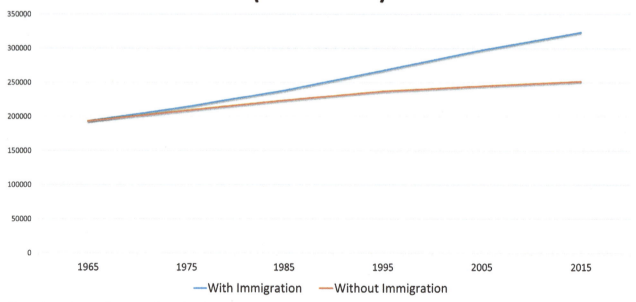

Figure 3.7. US population with and without immigration. Without immigrants and their children, it is estimated that the US population would have had 72 million fewer people in 2015. Data source: Pew Research Center, 2015.

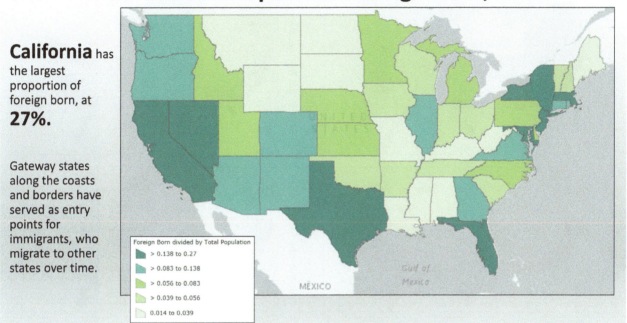

US States: Proportion Foreign Born, 2014

California has the largest proportion of foreign born, at **27%.**

Gateway states along the coasts and borders have served as entry points for immigrants, who migrate to other states over time.

Foreign Born divided by Total Population
- > 0.138 to 0.27
- > 0.083 to 0.138
- > 0.056 to 0.083
- > 0.039 to 0.056
- 0.014 to 0.039

Figure 3.8. Proportion of foreign-born residents, 2014. States with more immigrants represent demographic changes that will be faced by the country as a whole. Explore the map at http://arcg.is/2dDtVJc. Data source: US Census.

Figure 3.9. Tet Lunar New Year celebration in Little Saigon, Westminster, California. Gateway states such as California, Texas, New York, and Florida have large immigrant populations. The United States as a whole is becoming more diverse due to immigration. Photo by Joseph Sohm. Stock photo ID: 297524915. Shutterstock.

Figure 3.10 shows the movement of native-born Americans by region. Generally, people stay within the region where they were born—a clear sign of distance decay—so those born in the West tend to move to other western states, while those born in the Midwest stay within their own region as well. The same holds true for people born in the South and Northeast. However, it can also be seen that when people leave their region

Migration by US State of Origin

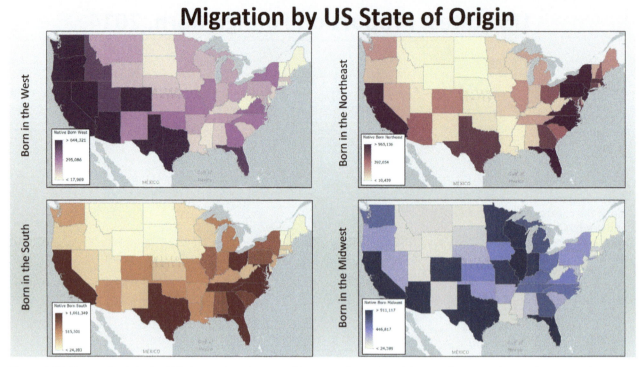

Figure 3.10. State of origin for US domestic migrants. Distance decay is evident in that people tend to move between states within the same region. Explore the data in ArcGIS Online at https://arcg.is/1ymjTa. Data source: US Census.

of birth, there are clear migration flows to the Sun Belt—California, Texas, and Florida stand out as having large numbers of immigrants from all four major US regions.

Go to ArcGIS Online to complete exercise 3.1: "Spatial distributions."

Push and pull forces

As seen in the previous section, some places clearly attract immigrants, while other places induce emigration. To understand the driving forces behind these patterns, it is essential to look at the place-specific characteristics of sending and receiving locations as well as the spatial interaction that ties places together.

Different theories have been given to help explain how, where, and why people migrate. One of the first and most influential thinkers on migration was

E. G. Ravenstein, who in 1885 wrote *On the Laws of Migration*. He presented a set of "laws" based on his research in Great Britain in the nineteenth century. While some of his laws do not apply to all contemporary migration trends, many still inform modern migration theory. Ravenstein's laws of migration are as follows:

- Migrants move short distances along fixed currents toward "great centers of commerce and industry."
- Rapidly growing cities absorb migrants from nearby places. The places vacated by these migrants are then filled by migrants from places farther away, in a step-by-step fashion. Communication flows between places can countervail the disadvantages of distance.
- Some places see a dispersion of their populations, as they migrate to growing urban areas.
- All migration currents have a compensating countercurrent.

- Long-distance migrants generally move to one of the great centers of commerce or industry.
- Rural residents migrate more than urban residents.
- Females migrate more than males.

Implicit in Ravenstein's work is the idea that certain places attract immigrants, as in his "great centers of commerce and industry." The idea that some places attract people led other migration researchers to expand on the idea of push and pull factors. These theories are based on the fact that places of origin have factors that "push" people to leave, while places of destination have factors that "pull" people to come. These forces can be viewed in terms of the economic, social and cultural, political, and environmental conditions of places.

Ravenstein also noted that migration is not random but rather is tied to currents of migration—flows between specific places that people follow in regular patterns. This spatial interaction between places is fed by communication, as people share information on the opportunities and challenges involved with making the move. Furthermore, he noted that the characteristics of migrants take a specific form. Migrants are not randomly selected from the general population but tend to have specific characteristics in terms of sex, age, income, and other variables.

Economic push and pull

As you may expect, some of the most common reasons to migrate are economic. One way of understanding how economic forces influence migration is through neoclassical *migration theory*. This theory focuses on wage differentials between origin and destination. If wages are low in one place relative to another place, then individuals will be pushed from the low-wage to the high-wage location. This theory is useful in understanding much migration, in that historically and globally, economic forces are the key drivers of migration.

For example, most migration from Mexico to the United States has been driven by significant wage differences between the two countries. The average Mexican income in 2014 was US $12,850, adjusted for cost of living differences with the United States. This compares to average US earnings of $57,000. But neither most Mexicans nor most Americans earn the average income. In 2014, 14.8 percent of Mexican workers earned less than two-thirds of the median wage, the equivalent of US $707 or less per month, again accounting for differences in cost of living. Those making only the minimum wage earned less than US $160 per month. By migrating to the US and earning the 2014 federal minimum wage of $7.25 per hour, a full-time worker would earn about $1,160 per month. That is a 60 to 600 percent increase in wages—certainly enough of an incentive for many to migrate. The gains are even greater for immigrants who arrive in states with minimum wages above the federal level. For instance, the 2014 minimum wage in California was $9 per hour, while in Illinois it was $8.25 and in New York, $8.

But wages alone are not enough to explain all migration patterns. Economic forces can also include unemployment rates, working conditions, and opportunities for career advancement. Places with high unemployment rates are likely to push people away, while places with low unemployment will pull people in. Similarly, a place with a cluster of large corporations in a specific industry may offer attractive benefits packages, interesting coworkers, and opportunities to move into positions of greater responsibility. The technology cluster in Silicon Valley, for instance, pulls in people from all over the US and the world for good wages but also for fun and interesting work and myriad opportunities for professional growth. Likewise, people migrate from around the world to New York and London because of ample career opportunities in their financial sectors.

The level of emigration from a place can also be related to level of economic development (figure 3.11). This relationship is described by the migration transition theory, which is closely related to the demographic transition theory discussed in chapter 2. In preindustrial societies, emigration is low. High, yet roughly equal, rates of births and deaths mean that population growth is minimal, so there is no demographic pressure

Migration Transition Theory

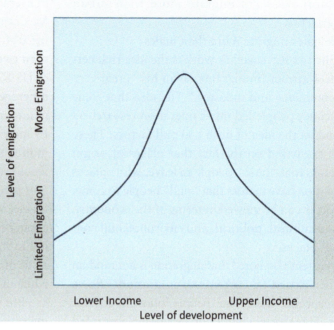

Emigration is **low** when places are very **poor and rural**.

Emigration **increases** as places **industrialize** and **urbanize**.

Emigration **falls** once places are **modern urban societies**.

Figure 3.11. Migration transition theory. Emigration increases as countries industrialize and urbanize, but as they get richer, emigration falls. Image by author.

to emigrate. Furthermore, because of limited income, people do not have the resources to travel to new places.

As countries begin to industrialize and urbanize, as in stage 2 of the demographic transition, high birth rates and falling death rates result in a population boom. Incomes also increase during this time as economies shift from basic agriculture to mechanization on farms and in factories. With population pressure and higher incomes, migration increases significantly. During this stage, there is massive rural to urban migration as people shift from agricultural work in rural areas to manufacturing work in cities. There is also emigration to other countries, as people seek new opportunities farther from home.

As countries continue to develop through urbanization and industrialization, they enter stage 3. During this stage, birth rates fall, resulting in a slowing of population growth. As population pressures decline, rural to urban migration and emigration rates slow somewhat.

Finally, modern, urban stage 4 countries with low birth and death rates, have low levels of emigration. Migration tends to be city to city, while net international migration is positive as more migrants arrive than leave.

Given that Mexico has been one of the most important sources of immigrants to the US in recent decades, it is useful to view Mexican immigration in terms of the migration transition. Figure 3.12 shows how income and fertility have changed in Mexico since 1990. Whereas in 1990 the per capita income was $5,820, by 2014 it had risen in real dollars to $17,200. At the same time, the total fertility rate had fallen from nearly 3.5 children to about 2.2. As Mexico transitions to a higher income and lower fertility country, the migration transition theory predicts that emigration will decline.

As pointed out previously, Asian immigration to the US is now higher than that from Latin America. As the economies of Asia grow, a larger number of

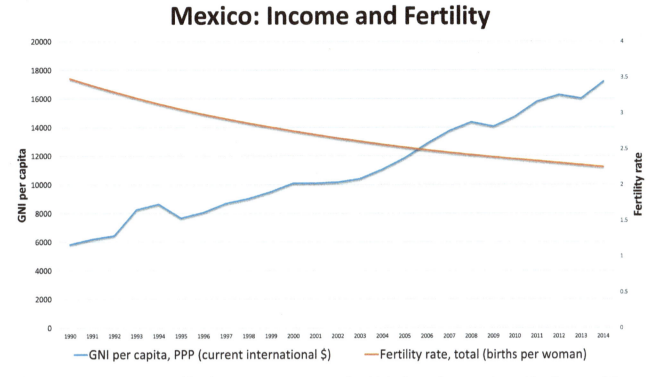

Figure 3.12. Mexico: Income and fertility. Mexican emigration should decline as incomes rise and fertility rates fall. Data source: World Bank.

people have the financial resources to immigrate to the United States. While birth rates in much of Asia are relatively low, the sheer number of people in the region means that a substantial number are available and have the financial resources to emigrate.

Social and cultural push and pull

Social and cultural factors, including social connections, language, and religion, also drive migration. In terms of personal social connections, many people are pulled to places where friends and family already live, while in other cases, people may be pushed away by troublesome friends or abusive family relationships.

Language can also play a role in migration. For instance, the English-speaking population of Quebec, Canada, decreased from the late 1960s through the late 1980s as a number of laws promoting the French language were put into effect (figure 3.13), which pushed to other Canadian provinces some businesses and residents who spoke only English. At the same

time, French-speaking migrants from former French colonies in the Caribbean and Africa, while probably migrating primarily for economic reasons, were pulled to Quebec because of a common language.

Especially in the case of migration pull forces, language plays a similar role in many locations. France attracts migrants from former North African colonies such as Algeria and Tunisia, while the United Kingdom and the United States attract more English-speaking Nigerians. Likewise, many emigrants from Latin America have moved to Spain because of their common language, which allows them to more quickly adapt to and integrate with the host country.

Along with language, one of the most powerful facets of identity for some people is religion. For this reason, it can play a significant role in migration. Mass movements of Hindus and Muslims in South Asia came with the region's independence from Great Britain in the late 1940s and the formation of Hindu-dominated India and Muslim-dominated Pakistan

Figure 3.13. French language signage in Quebec, Canada. Laws promoting the French language led to emigration of some English-speaking Canadians and immigration from some French speaking countries. Photo by Rob Crandall. Stock photo ID: 248614414. Shutterstock.

and Bangladesh. Religion was also a major force during the formation of Israel in shaping who lives there and in the Palestinian Territories. In more recent years, the Middle East has seen a dramatic shifting of religious populations. Sunni and Shiite Muslims increasingly live in separate parts of Iraq, Syria, and Saudi Arabia, while small Christian, Jewish, and other minority populations in places such as Egypt and Iraq have been pushed out of the region in large numbers.

The Yazidi people in Iraq constitute a recent example of how religion can be used as an excuse to push people away from their homeland. In 2014, as the Islamic State terrorist organization was gaining territory in Iraq and Syria, the Yazidi people were forced to flee their land (figure 3.14). In the eyes of the Islamic State, the Yazidis, whose religion is one of the most ancient of the Middle East, were seen as devil worshipers who had to be eliminated. At least 5,000 Yazidi men were killed by the Islamic State, and

Figure 3.14. Religion and migration. A refugee camp in Dohuk, Kurdistan, Iraq, and a Yazidi grandmother with two children in the Kanke refugee camp, Kurdistan, Iraq. Many Yazidis forced to flee persecution by the Islamic State sought safety in refugee camps in Iraqi Kurdistan. Dohuk photo by Paskee. Stock photo ID: 176894732. Shutterstock. Kanke photo by Owen_Holdaway. Stock photo ID: 294848531. Shutterstock.

thousands of women and children were taken captive, some for sexual enslavement and forced marriage to the organization's fighters. Those who were not killed or captured fled their villages to survive. Only the eventual defeat of the Islamic State will allow these people to return to their homes.

Political push and pull

Political push and pull forces can also be significant determinants of migration. This includes issues such as conflict and peace or persecution and human rights. Places with high levels of violent crime or war will frequently push people away. Young migrants from Honduras, Guatemala, and El Salvador have been emigrating to escape violence in some of the most crime-ridden places in the world, where youth are often recruited by force into street gangs (figure 3.15). In 2015, El Salvador had the highest homicide rate of any country not at war. Extortion by gangs is rampant in these countries, where bus drivers, small business owners, and residents of poor neighborhoods are forced to pay protection money under threat of violence. For these reasons, emigrants continue to flow out of these Central American countries, many of whom attempt to immigrate to the United States.

Likewise, there has been mass migration as people have been pushed away from war-torn countries such as Syria. Between 2011 and 2016, roughly nine million Syrians had fled their homes, migrating to nearby Jordan, Iraq, Turkey, Lebanon, and various European countries as well as to other locations within Syria (figures 3.16 and 3.18).

As people are pushed away from violence, be it crime or war, people are pulled to places with relative peace and security, be it nearby Costa Rica in Central America, Jordan in the Middle East, or further away in the United States and Europe.

Figure 3.15. Crime and migration. Soyapango, El Salvador. Special Forces officers from the Grupo Reacción Policial guard an alley during a gang raid. Disrupting criminal gangs in Central America is essential for reducing violence and resulting pressure to emigrate. Photo by ES James. Royalty-free stock photo ID: 30996073

Figure 3.16. War and migration. A destroyed tank in Azaz and war damage in Serekaniye, Syria. Civil war in Syria has led to massive destruction and the death of hundreds of thousands of people, pushing millions to emigrate to safer places. Azaz photo by Christiaan Triebert. Stock photo ID: 161912165. Shutterstock. Syria photo by fpolat69. Stock photo ID: 138083423. Shutterstock.

People are also pushed away from places where there is political persecution and pulled to places where human rights and the rule of law apply equally to all. Political push factors can be distinguished from social and cultural push factors by the fact that persecution involves punitive sanctions by a government. Thus, people often flee repressive and authoritarian regimes for fear of arbitrary seizures of financial assets, detention, or arrest and move to stable democracies.

North Korea is probably the most extreme case of political persecution and violation of human rights. Each year, around 2,000 North Koreans flee the country, often migrating to China with hopes of getting to democratic South Korea. Legal emigration is prohibited, so emigration typically involves illegally crossing the border, individually or with the help of human smugglers. In North Korea, no political opposition is allowed, cell phones and other technology that can communicate with the outside world are illegal, and criticism of the leadership is prohibited. Punishments include torture, execution, and lifetime imprisonment with hard labor. In some cases, entire extended families are sent to prison camps for the actions of one family member. An estimated 80,000 to 120,000 people are in these camps, which are plagued with high death rates from poor nutrition, lack of medical care, mistreatment by guards, executions, and backbreaking work (figure 3.17).

As another example, Russian emigres have cited anti-LGBT laws and political persecution as reasons for being pushed away from their country. A 2013 law banned "propaganda of nontraditional sexual relations" toward minors, which many interpreted as a means of persecuting Russia's LGBT community. In other cases, members of opposition political parties have faced what they see as trumped-up criminal charges and harassment by the police in response to

North Korea: Kwanliso 16 Prison Camp

20,000
The estimated number of prisoners in 2011.

3 times the size of Washington, DC.
Area encompassed by the camp.

Places of **brutal and deadly forced labor in mining, logging, and agriculture**.
Identified with satellite imagery.

Figure 3.17. Kwanliso 16 prison camp, North Korea. This image shows a small section of the camp, the largest in North Korea. Explore this map at http://arcg.is/2dDroii. Map based on Amnesty International, 2013.

their criticism of government policies. In some cases, opposition figures have been found shot dead.

In recent years, journalists' and opposition political activists' fear of detention has also spurred emigration from other authoritarian-leaning countries, such as Venezuela, Turkey, and Cuba. Typically, people who are pushed away from countries because of political conditions choose to migrate to countries with strong pull forces of democracy, protection of individual rights, and the rule of law, such as Western Europe, Canada, and the United States.

Refugees are a specific type of social or political migrant. The United Nations 1951 Refugee Convention defines a refugee as a person who, "owing to a well-founded fear of being persecuted for reasons of race, religion, nationality, membership of a particular social group or political opinion, is outside the country of his nationality, and is unable to, or owing to such fear, is unwilling to avail himself of the protection of that country." Per international law, the legal rights of refugees are different from the rights of those who are migrating for economic or other reasons. Therefore, most countries have systems for evaluating refugee claims and admitting those deemed to fit the definition.

In the United States, for instance, refugee status has been given to Cubans facing political persecution under the communist Castro regime, Christians and Jews fleeing the Islamic Republic of Iran, and Iraqis and Afghans who assisted the US military.

Around 2014 to 2016, Europe faced one of the largest refugee migration flows since World War II. Because of conflict in Syria, Iraq, and Afghanistan, large numbers of people started migrating in search of asylum (figure 3.18). Over one million immigrants crossed the Mediterranean Sea, with the majority arriving in Greece, followed by Italy. The journey proved fatal for thousands, as smugglers sent them in overcrowded boats of questionable quality. In one single incident, over 800 people drowned when a boat sank while crossing from Libya on its way to Italy. Hundreds of thousands more have attempted to take land routes through Turkey and Eastern Europe in attempts to reach countries such as Germany and Sweden. While the majority of the refugees were from Syria, Afghanistan, and Iraq, people from other parts of Southwest Asia and Africa also immigrated. Syrians were given priority for refugee status because of a brutal civil war, while most others faced rejection as economic migrants.

But while news reports in the West often focus on refugees immigrating to Europe and North America, the vast majority of refugees, 86 percent, reside in

Figure 3.18. Refugee flows to Europe. Refugees on an overcrowded raft arriving at Lesvos, Greece; refugees in Hungary making their way to Germany. Greece photo by Anjo Kan. Stock photo ID: 390513937. Shutterstock. Hungary photo by Istvan Csak. Stock photo ID: 385009378. Shutterstock.

developing countries (figure 3.19), largely due to the concept of distance decay discussed in chapter 1. Given that there is more spatial interaction between two places that are close together, it is most likely that refugees facing push forces in developing countries will migrate to another developing country nearby. Life for refugees in developing countries can be difficult. Developing countries do not have large sources of revenue to assist those in need and strong economies to absorb many immigrants. Sometimes, the United Nations assists with funding refugee camps, but often there are many more refugees than space in the camps. Too often, adults lack jobs and children lack schools. In 2016, one estimate showed that 80 percent of Syrian children in Turkey did not attend school. Often, children must work to support their families, with some working twelve-hour days for US $60 in weekly wages. Unable to fully integrate into host countries and unable to return to their home countries, many refugees remain in precarious positions for years, leaving children with limited education and limited opportunities for the future.

Environmental push and pull

In addition to the economic, social and cultural, and political push and pull forces directly caused by humans, environmental conditions drive migration. In the most benign form of environmental push and pull, people can be pushed away from areas with uncomfortable climates and pulled to places with pleasant climates. For instance, since the 1970s, population growth in the US southern and western Sun Belt states has increased at a faster rate than in the Snow Belt states of New England and the Midwest.

But the environment can also cause migration for more serious reasons. Natural disasters can drive

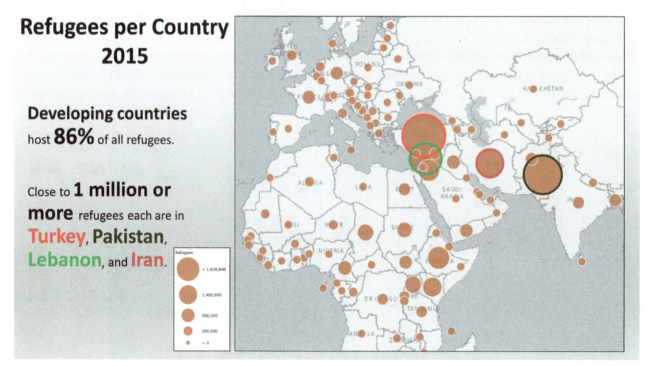

Figure 3.19. Number of refugees per country in 2015. Despite news reports in the United States and Europe that show resistance to accepting refugees, the vast majority are found in developing countries. Data source: UNHCR.

people away, as in the case of New Orleans after Hurricane Katrina in 2005 (figure 3.20). Because of that natural disaster, over 200,000 people moved out of New Orleans after 80 percent of the city was flooded. Many people who left never returned. By 2014, the city's population had nearly recovered, as some emigres returned and new residents moved in, but it remained somewhat below pre-hurricane levels nevertheless.

Droughts and resultant crop failure and famine can also push people to migrate. For example, during nearly a decade in the 1930s, drought conditions in the Plains states created the Dust Bowl, which forced millions of people to move to other parts of the United States.

While refugees fall under the social and political migrant categories, internally displaced migrants are more broadly defined as including those displaced by armed conflict, human rights violations, and natural disasters. Those who are internally displaced have not crossed an international border, yet have been pushed away from their regular place of residence.

The severity of internal displacement caused by natural disasters relates to a society's degree of vulnerability. Vulnerability is the level of susceptibility to harm people have when exposed to a disaster. At the individual scale, this relates to characteristics such as income, age, mobility, and gender. For example, the poor and elderly may be more vulnerable to harm when displaced by an earthquake, as they struggle more than others to obtain food and shelter.

Vulnerability also varies at smaller scales, such as by country. Countries differ on a range of factors, such as political stability, level of economic development, and more. A society will be more vulnerable to natural disasters, such as drought, when it is intertwined with poverty, political instability, and armed conflict. For instance, both Afghanistan and South Sudan had over one million internally displaced people in 2015 due to both armed conflict and drought conditions, while Ethiopia had nearly 300,000 (figure 3.21). All of these countries are poor, with governments that lack resources to end conflict and improve agricultural infrastructure.

When armed conflict and natural disasters occur in the same place, internal displacement can be much more severe. In some cases, drought worsens or causes armed conflict. In the case of Ethiopia in 2015, drought conditions led pastoralists to move livestock onto lands traditionally use by pastoralists from different tribes.

Figure 3.20. Natural disaster and migration. Hurricane Katrina damage in the 9th Ward of New Orleans. Massive flooding from Hurricane Katrina pushed many to emigrate from New Orleans. Photo by Patricia Marroquin. Stock photo ID: 3638784. Shutterstock.

Figure 3.21. Internal displacement. Families in South Sudan gather to collect water. South Sudan's internally displaced population was well over one million in 2015. Photo by Paskee. Stock photo ID: 176894750. Shutterstock.

This led to armed conflict between tribes over increasingly limited pasture land for their livestock.

In other cases, armed conflict can worsen impacts from drought, leading to or exasperating internal displacement. During conflict, food and agricultural aid may not be able to reach people in conflict zones. This lack of aid can force many of them to move to new areas.

Residents of richer countries, such as the United States, are less vulnerable overall when disaster strikes. More people have insurance that can help cover losses from floods and fires, while the US government is better equipped to assist with food and shelter. Not only can richer countries respond more effectively to disasters, but disasters are less likely to occur in the first place. For instance, richer countries have better irrigation systems to reduce vulnerability to drought, buildings are better built and thus resist earthquakes, and fire departments are better equipped to control wildfires. For these reasons, internal displacement from disasters tends to be less severe and for a shorter time in more developed countries. However, those who are more likely to be internally displaced and remain that way for a longer period tend to be individuals who are more vulnerable because of poverty, old age, and other socioeconomic characteristics.

Forced and voluntary migration

Push and pull forces are often discussed in terms of migration that is forced or voluntary. Generally, economic migration is seen as voluntary. While some people may feel forced to move due to unemployment or a lack of economic opportunity, for the most part, people are free to choose whether or not to migrate. For this reason, those deemed economic migrants are rarely given refugee status or counted as internally displaced persons.

People who migrate because of armed conflict, persecution, or environmental disasters, however, are more often seen as part of forced migration. People in this category are more likely to receive refugee status or be counted as internally displaced people. During the European migrant crisis of 2015–16, countries of the European Union struggled to separate migrants who were forced by warfare or persecution to flee to Europe from those who arrived as voluntary economic migrants. In general, Syrians fleeing civil war were regarded as forced migrants and given priority for refugee status. Others, such as those from Afghanistan and countries of North Africa, were classified as economic migrants and denied legal entry.

In reality, the difference between forced and voluntary migration can be blurred, and it can be difficult to fit migrants into just one category. Migrants search for security and opportunity for themselves and their families. Poor economies and insecurity often go hand in hand, so teasing out who migrated primarily because of risk to their safety and who migrated primarily for economic opportunity is difficult in most cases.

Go to ArcGIS Online to complete exercise 3.2: "Push and pull forces."

Spatial interaction and migration

Understanding how some places push migrants to leave and other places pull migrants in is an important first step in understanding migration patterns. But once migrants decide to leave a place, they typically have more than one choice of where to move to. Thus, is it important to return to the idea of spatial interaction between places.

Distance decay and intervening obstacles

While differences between places in terms of economic, social, political, and environmental characteristics are powerful determinants of migration, intervening obstacles lie between places. These obstacles can direct the flow of migration between different places. One important obstacle is distance. As you recall, distance decay refers to the tendency for spatial interaction to occur more between places that are close together than between places that are far apart. Interaction decays, or declines, as distance increases between two places. Thus, the farther two places are from each other, the more difficult it is to migrate between them. With distance come increases in costs and the potential for risk to life and limb. Economic costs for transportation will be higher if one wants to migrate a greater distance. Similarly, cost in time will increase. Risk can also increase with distance, with greater odds of becoming the victim of crime, being stopped by government authorities, or facing inclement environmental conditions, such as dangerous extremes of heat or cold.

To use the example of Syrian emigrants again, many more migrated to nearby Jordan, Lebanon, or Turkey than to Europe. As these countries lie adjacent to Syria, costs in terms of money and time were much less than the costs of migrating to Europe. Furthermore, short moves by land to these countries were much safer than attempting dangerous crossing of the Mediterranean Sea.

The dangers involved with migrating longer distances are also evidenced by those who make their way from sub-Saharan Africa north to Libya as they attempt to reach Europe. Migrants are frequently abducted by criminal gangs to be held for ransom and beaten until family members pay for their release. Those who cannot pay are forced to work as slaves, and sexual abuse and forced prostitution is common. Upon arrival in Libya, they are held for weeks or months in squalid housing with limited food and water until a boat is ready to take them across the Mediterranean Sea. Even after embarking for Europe, the risks continue. Too often, overcrowded boats sink, killing dozens or hundreds of people (figure 3.22). In other cases, the Libya coast guard detains migrants and sends them to detention centers that are little better than where they were held by smugglers. People risk their lives to migrate when push and pull forces are strong enough, but long distances can present incredible risk.

Because of the power of distance decay, most international immigration is intraregional. For instance, 52 percent of African migrants live within Africa, 60 percent of Asian migrants live within Asia, and 66 percent of European migrants live within Europe. The

Figure 3.22. Survivors of a Mediterranean Sea crossing arrive in Italy. Forty-nine bodies were found in the overcrowded hold of the smuggler's boat. Photo by Wead. Stock photo ID: 328040444. Shutterstock.

only exception to the pattern is with Latin American and Caribbean migrants, 70 percent of whom migrate to North America.

Physical barriers, both natural and human-built, can also act as intervening obstacles. Natural barriers, such as mountains, rivers, deserts, and oceans, limit the movement of people between places. Likewise, human-built physical barriers, such as walls, can also restrict migration. No doubt there would be more Cuban immigrants in the United States if the countries were not separated by ninety miles of ocean. Likewise, fences, cameras, motion sensors, and other security barriers along the US-Mexico border severely limit the number of people crossing there (figure 3.23).

In addition, government immigration policies can create intervening obstacles. Restrictive policies in destination countries and in-transit countries can limit the flow of migrants. Conversely, open immigration policies, or policies that actively recruit migrants, will increase flows. More on immigration policies is discussed later in this chapter.

The concept of *cultural distance* can also play a role in migration patterns, as people tend to move between places with similar cultures. As discussed in the section on push and pull forces, flows of migrants tend to be greater between places with the

Figure 3.23. Physical barriers to migration. Photo of the border fence and US Border and Customs Protection patrol. The border wall separating San Diego, California (left), and Tijuana, Mexico, seen in ArcScene. Explore this scene at https://arcg.is/1PWeSW. Border fence photo by mdurson. Stock photo ID: 595401863. Shutterstock. Border wall image from ArcScene, Esri.

Figure 3.24. Cultural distance and migration. Places with similar cultural characteristics, such as language, will have more spatial interaction, "shortening" the distance between them. A Lebanese-Moroccan food restaurant in Paris. Photo by Hadrian. Stock photo ID: 275980151. Shutterstock.

same language and religion (figure 3.24). English speakers tend to migrate more between the United States, Great Britain, Australia, New Zealand, and Canada, while Arabic speakers move more between the countries of North Africa and the Middle East. This also holds true for former European colonies. A significant number of migrants from the former French colonies of Algeria and Vietnam migrated to France, while the same occurred with Great Britain and its former colonies, such as India, Pakistan, Hong Kong, and Nigeria.

The gravity model

The *gravity model* adds another component to the analysis of migration flows: size. Based on the ideas of Isaac Newton, the gravity model states that both distance and size will influence the flow of migrants between two places. The degree of migration between two places is proportional to the population size of each place and inversely proportional to distance. In essence, places with large populations that are close to each other will have large migration flows between them, while places with small populations that are far from each other will have small migration flows between them. Mexico and the Philippines are both developing countries and have populations around 100 million, yet Mexico sends many more migrants to the US. This is because Mexico is much closer, so migration is less costly in terms of time and money. By contrast, China sends more migrants to the US than does Nicaragua. Nicaragua is much closer to the United States, but the large population of China means that there is more spatial interaction with the

US in terms of migration. Out of a population of nearly 1.4 billion people, even a small emigration rate represents many migrants.

Evidence of the gravity model at work can be seen in interstate migration within the United States. Using the example of Nebraska, located near the center of the country, we can see that more immigrants came from nearby Iowa and Missouri, two states that also have populations larger than many other surrounding states (figure 3.25). These migration flows would be predicted by the model, based on both distance and population size. Being nearby, people from these states will have more information on the economic opportunities in Nebraska, and the costs associated with moving and with return visits to friends and family will be lower. At the same time, we see evidence of population size playing a large role in where immigrants to Nebraska came from. Both California and Texas were the origins of many of these immigrants. With their large populations, these states have a larger pool of potential emigres. Inevitably, some of the immigrants will relocate to other states, some nearby but others farther away.

Social networks and information flows

Distance, intervening obstacles and population size influence migration patterns, but these flows are further complicated by social networks and information flows from individuals and the media that connect specific places. When migrants make the decision to leave a place because of push forces, they do not consider all potential destinations. Rather, they base their decisions on available information. This information is inherently incomplete and is dependent on social connections and media sources. Information flows are enhanced when places have economic or political ties. For example, linkages formed through trade and investment, colonization and political influence, and cultural connections can facilitate information flows and lead to migration.

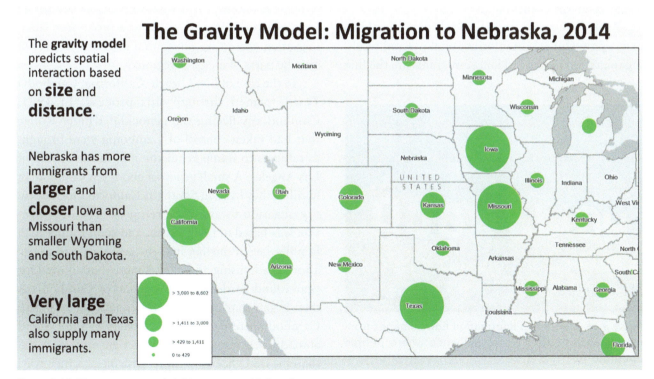

Figure 3.25. The gravity model: Migration to Nebraska. Data source: US Census.

Media information can come from advertisement campaigns that promote the economic benefits of a place as well as images in newspapers, magazines, television, and online (figure 3.26). Many people around the world have seen media images of the United States and Western Europe and their high levels of economic development. Partially due to these images, there are always large numbers of migrants who desire to move to these places.

Media images also shape migration patterns at more regional and local scales. For example, many Zimbabweans have migrated to South Africa on the basis of information that the economic situation there, while difficult, was more stable than in Zimbabwe. Likewise, within the United States, media information flows impact migration. Many Californians have migrated to Texas, partially because of Texan media campaigns promoting the state as an affordable destination with a strong pro-business government. On the other hand, Texans also move to California, no doubt in part because of media images of beaches and a year-round pleasant climate.

Aside from media images, social connections also play a powerful role in information flows and resulting migration patterns. Pioneer migrants are the first migrants to arrive in a new place. These trailblazers may arrive for any number of reasons: for university studies, a chance job opening, a romantic partner, or a search for novelty and adventure. They tend to be young and single, without family and financial responsibilities that tie them to their home place. These pioneer migrants then provide information to friends and family. Through letters, phone calls, email, and social media, they can share information on their lives in a new location, including job and wage information and quality of life. If those friends and relatives then feel push forces in their place of origin that are strong enough, they are more likely to be pulled to the place about which they have positive information.

Once social networks between pioneer migrants and their place of origin take hold, chain migration can begin. Friends, family, and neighbors begin to migrate to where pioneer migrants have established a foothold. Early migrants smooth the process by helping new migrants find work and housing. New migrants then pass information on to their own friends and family back home, further reinforcing and deepening social network connections. As time progresses, chain migration can create substantial migrant enclaves in specific parts of a city or region.

In the case of Long Beach, California, Cambodia Town formed through this process. In 1975, Cambodians fleeing the genocidal Khmer Rouge that had taken control of Cambodia were brought as refugees to Camp Pendleton marine base south of Los Angeles. A small group of about ten Cambodian families that lived in Southern California formed an organization to assist the refugees. The organization was based in Long Beach, and soon Cambodian families began moving into the city for its relatively affordable housing and mild climate that allowed for year-round gardening. This group of Cambodians then began sharing information about the city with Cambodian refugees in other parts of the US and abroad. These informational flows were more powerful than the US Office of Refugee Settlement, which tried to disperse Cambodian refugees in smaller clusters

Figure 3.26. Media and information flows. Images from media, such as Hollywood, influence where people emigrate to. Photo by bannosuke. Stock photo ID: 85556125. Shutterstock.

around the US. As word spread about the benefits of living in Long Beach and the presence of a growing Cambodian cluster, more Cambodians migrated to the city. Long Beach now has an officially designated Cambodia Town, with a thriving community of stores and services that serve the Cambodian population. Immigrant communities throughout the US and around the world form in a similar way, as a small group of pioneer migrants share information and assist the next wave of migrants, ultimately forming strong migration chains between two specific places.

Through cumulative causation, migration flows increase in strength over time. This is the process whereby migration becomes self-sustaining from ongoing positive feedback between immigrants and their place of origin. Early migrants transform places, making them more inviting for new immigrants as businesses open that cater to them and social networks for jobs and housing develop. At the same time, remittances sent to family in the place of origin provide financial resources for more people to emigrate. At a certain point, emigration can become the cultural norm, where all of those who are capable are expected to migrate.

Social and media networks have only become stronger in recent decades. For much of human history, information flows between places were slow, traveling by word of mouth or by handwritten letters. With the invention of the telegraph and later the telephone, information could travel more quickly, but costs and accessibility to these technologies were limited. In recent years, however, communication technology has reached the point that costs have fallen dramatically and accessibility has increased. Low-cost cell phones and internet cafés around the world link people in ways that were never before possible. Migrants can maintain contact with friends and family on a regular basis and share information about job opportunities and quality of life in their new home at limited costs.

Go to ArcGIS Online to complete exercise 3.3: "Refugees."

Characteristics of migrants

Rubenstein's pioneering work on migration in the 1800s included the observation that females migrate more than males. While that was true for Great Britain in the nineteenth century, it is not a fixed rule that applies to all times and places. Nevertheless, Rubenstein was correct in that the demographic profile of migrants is not representative of their place of origin. Except in the case of forced migration, where entire populations are forced to move, the characteristics of migrants will differ from the general population they come from in terms of age, sex, education, skills, marital status, and housing situation. This migrant selectivity reflects the fact that migrants are not randomly selected from their place of origin. Rather, certain subgroups are more likely than others to migrate.

There are few fixed patterns in the characteristics of migrants, except for age. People in their late teens through mid-30s are disproportionately likely to migrate (figure 3.27). There are many reasons why young people are more likely to move. They are at an age for seeking new job or educational opportunities, they may marry and follow a spouse's opportunities, they are less likely to be tied to a home mortgage, they are in better physical condition, and they often have a more adventurous mindset.

Looking at global migrant stock, 72 percent are age 20 to 64, whereas only 58 percent of the total world population falls within this age range. This reflects young adults who migrate and stay abroad during their working years. Upon retirement, some return to their place of origin.

Other than age, the characteristics of migrants can vary based on the place of origin and place of destination. While the sex ratio of migrants to the US is relatively even, in other cases the sex ratio of migrants can be highly skewed. Of those seeking asylum in Europe during 2014–15, fully 73 percent were men. Likewise, a disproportionate number of men migrate to the Middle East to work in the oil and construction industries. In 2015, 75 percent or more

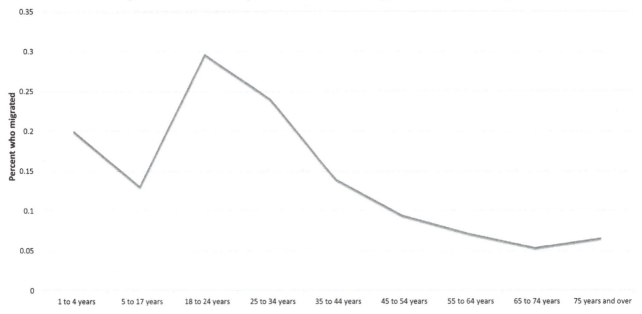

US Domestic Migration by Age, 2014
(Intra-county, Inter-county, Inter-state)

Figure 3.27. US domestic migration by age, 2014. Young adults migrate much more than other age groups in the US. Data source: US Census.

of immigrants living in the United Arab Emirates, Oman, and Qatar were men (figure 3.28). On the other end of the spectrum, women constitute a larger proportion of immigrants in parts of the former Soviet Union. Women represented close to 60 percent or more of the migrant population in Moldova, Latvia, Estonia, Kyrgyzstan, and Armenia, and Montenegro in 2015. This higher percentage of female immigrants

Figure 3.28. Construction workers in Dubai, United Arab Emirates. Not only does the UAE have many immigrant workers relative to the native population (as discussed in chapter 2), but over 75 percent of them are male. Photo by Draw. Stock photo ID: 416791933. Shutterstock.

occurred because the region's growing middle class sought domestic help, such as nannies, nurses, and cleaners.

The education level of migrants also varies from place to place, depending on market needs and government immigration policies. Domestically, those with less than a high school education in the United States between 2010 and 2015 were the least likely to move from state to state. This is likely due to a lack of opportunity for those with limited education. Migrating away from one's social network of friends and family with limited job prospects means that this group tends to stay close to home.

Foreign immigrants to the United States between 2010 and 2015 differ from the overall US population in terms of education levels. As seen in figure 3.29, a disproportionate number had less than a high school education compared to the entire US population, and a disproportionate number had bachelor's and graduate degrees. This reflects a demand for low skilled workers

in sectors such as agriculture, construction, and domestic work as well as a demand for skilled workers in technology, health care, and other knowledge-based industries. The distribution of education varies by place of origin, with about 52 percent of recent Mexican immigrants to the United States in 2013 having less than a high school education and over 57 percent of Asians and 61 percent of Europeans having at least a bachelor's degree.

Immigration control and government policy

Who migrates and where they migrate to is heavily influenced by government immigration policies. Even though a place may have strong pull forces, governments typically try to regulate the flow of immigrants for political and economic reasons.

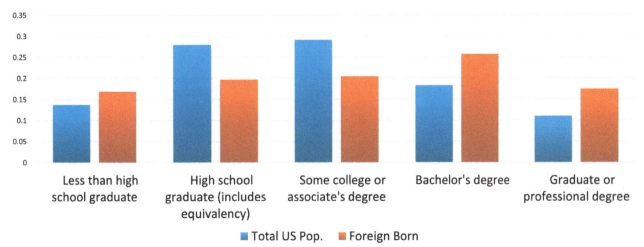

Figure 3.29. Education levels for the US population and foreign-born immigrants. Immigrants are found disproportionately at the high and low ends of the education spectrum. Data source: US Census.

US immigration policy

For most of the history of the United States, the government has enacted policies to regulate the number and/or characteristics of immigrants. Early legislation did not restrict the number of immigrants but rather focused on who could become a US citizen. The 1790 Naturalization Act limited citizenship to "free white persons" of "good moral character." This was the first legislation that attempted to control the racial or ethnic makeup of the country by limiting who could become a citizen. Citizenship rules were changed with the Naturalization Act of 1870, in which African Americans were given citizenship rights along with whites. However, Asians remained excluded.

The Immigration Act of 1875 placed the first restrictions on immigration, not just citizenship, by prohibiting criminals and forced Asian laborers. In 1882, the Chinese Exclusion Act was passed, and in 1917, immigration was banned from most other Asian countries. During this period, restrictions other than those based on race and ethnicity were put in place as well, including bans of anarchists, beggars, polygamists, "lunatics," illiterates, prostitutes, and those with contagious diseases.

From around 1880, immigration shifted away from Northern and Western Europe toward Southern and Eastern Europe (figure 3.30), prompting fears of a changing ethnic profile among many Americans. This resulted in the 1921 Emergency Quota Act, which placed numerical limits on immigration by creating quotas equal to 3 percent of the foreign-born population based on the 1910 census. It also placed a cap on the total number of immigrants per year. Asian immigration was not allowed. Even more restrictive quotas were enacted with the Immigration Act of 1924, which limited European immigration to 2 percent of the foreign-born population based on the 1890 census. It reduced the total number of

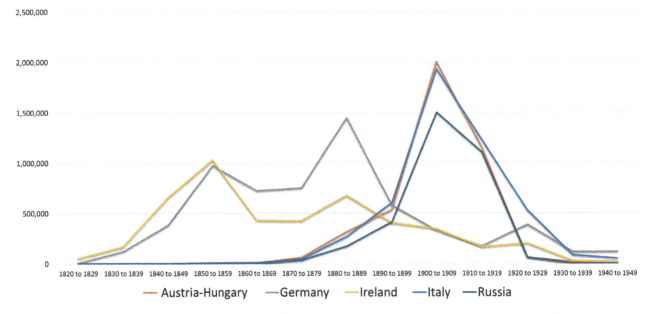

Figure 3.30. Major legal European immigration, 1820–1949. Around 1880, immigration from Western Europe began to be eclipsed by immigration from Southern and Eastern Europe. Data source: US Department of Homeland Security.

annual immigrants and continued to bar Asian immigrants. These quotas were strongly biased in favor of Northern and Western European immigrants, who dominated the US population in the 1890 and 1910 censuses.

It was not until 1943 that some Asians were allowed to legally immigrate and become citizens. During that time, a small number of Chinese immigrants were allowed (105 per year), and Chinese residents could become naturalized US citizens. This change in immigration law occurred, in part, because anti-Chinese sentiment subsided somewhat with World War II, as the United States fought alongside Nationalist forces against the Japanese (figure 3.31). Finally, in 1953, with the Immigration and Nationality Act, all race

restrictions were removed from immigration policy. Quotas remained but were open to all countries and were based on each nationality's population from the 1920 census. Naturally, this policy continued to favor Northern and Western Europeans, whose numbers were greater in the 1920 census.

From the 1950s onward, the national quota system was loosened and disconnected from nationality numbers in the census. Immigration policy gave priority to family reunification and skilled workers. With the Immigration Act of 1990, a global annual quota was established, excluding immediate family members of current US citizens. The quota is divided into categories, including family members, such as adult children and siblings of US citizens, and skilled workers

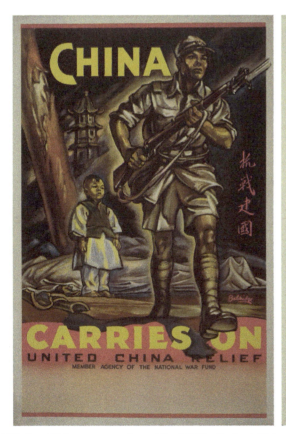

Figure 3.31. Pro-China posters from World War II. Anti-Chinese sentiment declined somewhat during this time as the United States fought alongside the Chinese against Japan. China Carries On image by Everett Historical. Stock illustration ID: 251930245. Shutterstock. China Fights On image by Everett Historical. Stock illustration ID: 249573538.

and investors. Refugee quotas are set separately by the president and Congress each year. Lastly, about 50,000 annual visas are granted as part of the Diversity Program. This program allows residents of countries that send lower numbers of immigrants to the US to apply for a visa lottery.

In line with the concepts of distance decay and the gravity model, changes away from immigration policies that favored Europeans led to a steady increase in legal immigration from Latin America, with a sharp rise in the 1980s and 1990s (figure 3.32). Likewise, Asian immigration, largely from China, India, and the Philippines, increased significantly from the 1960s and 1970s on (figure 3.33).

Since the 1980s, several laws have focused specifically on illegal immigration and enforcement. In 1986, the Immigration Control and Reform Act granted "amnesty" to 2.7 million people living without legal

residency in the US. In an effort to discourage further illegal immigration, sanctions were established for employers that knowingly hired illegal immigrants. The program largely failed to stem illegal immigration, however, as the estimated five million illegal immigrants living in the United States at that time continued to increase. The law's ineffectiveness was attributed to a number of factors. First, business groups watered down requirements for employers to confirm legal residency. Second, border enforcement funding was slower to be appropriated than expected. And third, there was no change to the quantity of legal employment visas granted. Thus, the push and pull forces driving immigration continued, but without a legal mechanism for the flow of people to be regulated.

Due to these failures, additional legislation in 1996 increased worksite enforcement and limited social welfare benefits for illegal immigrants. It

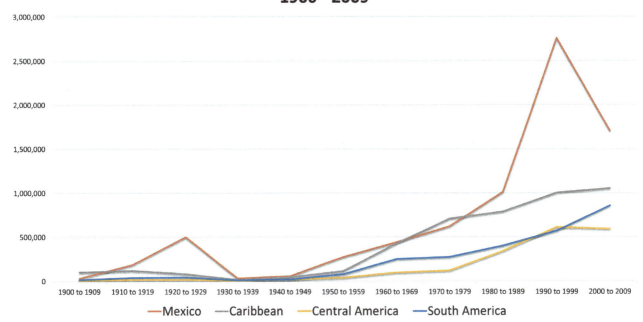

Figure 3.32. Legal immigration from Mexico and Regions of Latin America, 1900–2009. Migration to the US from Latin America rose sharply in the 1980s and 1990s. Data source: US Department of Homeland Security.

Major Asian Immigration
1940-2009

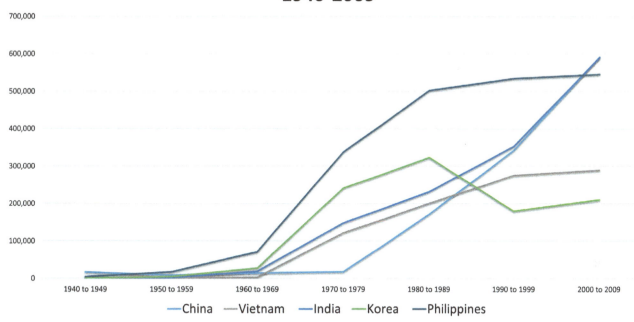

Figure 3.33. Major Asian immigration, 1940–2009. Asian migration to the US began to rise substantially in the 1960s and 1970s. Data source: US Department of Homeland Security.

also enhanced border enforcement and authorized fences along sections of the US border. After 2001, immigration legislation focused on border security. Immigration authority was brought under control of the Department of Homeland Security, and authorization for additional border fencing was established (figure 3.34).

It is estimated that the illegal immigrant population in the United States peaked in 2007 at about 12.2 million people. Since that time, the number has decreased to just over 11 million. Increased border enforcement, along with a sluggish US economy after the Great Recession of 2007 to 2009 and greater economic and political stability in much of Latin America, have all contributed to this decline.

As Congress has blocked comprehensive immigration reform under both the Bush and Obama administrations, the controversy over illegal immigration continues. In 2012 and 2014, executive orders by

Figure 3.34. US Customs and Border Protection agents at work along the US-Mexico border. Photo by Sherry V. Smith. Royalty-free stock photo ID: 753014410. Shutterstock.com.

President Barack Obama aimed to temporarily fix a small part of the problem. These orders, known as Deferred Action for Childhood Arrivals, established temporary deportation relief for young adults brought illegally to the US as minors. They also granted temporary work permits, allowing participants to legally work. However, the policy does not include a path to permanent residency or citizenship.

Immigration policies in other countries

Nearly all countries in the world have government policies to regulate the flow of immigrants, both in terms of numbers and in terms of migrant characteristics. Often, it is countries with the strongest pull forces that have more stringent policies aimed at controlling immigration.

Australia, like the United States, tried to limit nonwhite immigration for much of the twentieth century. After large numbers of Asians immigrated in the mid-1800s in search of gold and to work in agriculture, the Australia government feared that the country would lose its British roots. In 1901, it enacted the Immigration Restriction Act, commonly known as the White Australia Policy, which remained in place until 1958. Today, migrants are admitted first by skills and second by family ties. In addition, Australia has taken a very strict stance on illegal immigration, turning away boats with those trying to enter illegally and using immigration detention centers in other Pacific countries until they can be repatriated (figure 3.35).

Canada also had a "white only" immigration policy from the mid-1800s until 1962 that restricted immigration from Asia. Today, as with Australia, immigrants are admitted on the basis of skills and family reunification. Because of the focus on skills, immigrants to Canada have higher levels of education than the native-born Canadian population.

Japan's immigration policies have severely limited the foreign population, which was only about 2 percent in 2012 (figure 3.36). Its policies are due to a strong cultural desire to preserve the ethnic homogeneity of the country. Close to half of the immigrant population is of Korean and Chinese descent. This number includes children of these immigrants, who have not become naturalized citizens. As an example of Japan's quest to preserve its homogeneity, ethnic Japanese from

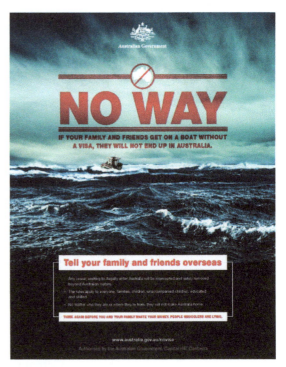

Figure 3.35. Australian counter-people smuggling poster. A 2016 poster warning illegal immigrants to stay away. Image by Australian Department of Immigration and Border Protection.

Brazil were admitted to the country when workers were needed in the 1970s and 1980s. Today, very few refugees are accepted, and immigration is based on high level skills.

Many of the Persian Gulf states have migration policies that focus on temporary migrant workers, as opposed to permanent migrants with the possibility of obtaining citizenship. Qatar has a highly unusual population structure that consists largely of males of working age. The high number of males is attributable to the unique migration policies of Qatar and other countries of the Gulf Cooperation Council (Bahrain, Kuwait, Oman, Qatar, Saudi Arabia, and the United Arab Emirates). When oil was discovered in the region during the 1970s, many of these countries had small populations with limited levels of education. Labor was needed to extract oil and help build their new oil-fueled economies, so they turned to foreign temporary workers.

Figure 3.36. Kyoto, Japan. Japan has maintained a very homogenous population through strict immigration controls. Photo by Savvapanf Photo. Stock photo ID: 547181851. Shutterstock.

These temporary migrants are largely men from Bangladesh, Indonesia, and the Philippines, who work for low wages in construction and other low-skilled service occupations. They are not given the option of naturalization, and thus they do not bring their families to live in the region. In addition, a smaller number of female migrants are employed as domestic servants. The proportion of temporary migrant workers is so large that they constitute nearly the entire workforce of the private sector, while the majority of Gulf State citizens work in well-paid and secure public-sector positions.

While policies for temporary migration have benefited Gulf countries with low-cost labor and have benefited foreign workers with jobs that pay more than in their home countries, a number of problems have arisen. With the private sector dominated by low-wage foreign workers, many youths in these countries struggle to find employment. And with government budgets impacted by unstable oil prices, employment in the public sector is less assured than in the past. In regard to the foreign workers, there are myriad reports of abuse, including unpaid wages, confiscated passports, dangerous working conditions, and poor-quality food and housing. In addition to these risks, female domestic workers face the risk of sexual assault.

Impacts of migration

The nature of spatial interaction is that people and ideas come together from different places. In the case of migration, this interaction can elicit strong reactions, given the intimate nature of people living in the same space. Both the places that send emigrants and the places that receive immigrants can have beneficial and detrimental outcomes. These impacts are best understood by dividing them into two categories: economic and cultural.

Economic

One of the most commonly debated impacts is how immigration affects the economy of receiving places. Most immigration is driven by the push and pull forces of economic opportunity. Immigration means more workers, which can have several positive impacts. First, more workers mean more economic activity. Employers can more easily hire workers when a plentiful supply is available, and thus businesses can more easily expand operations. Business expansion leads to additional job creation and a growing economy. This expansion is further fueled by an increase in consumer demand. Immigrants not only work but also spend their wages on local goods and services. At the same time, business growth and consumption by immigrants contributes to an increase in tax revenues. These revenues can be used to support health and pension plans and upgrade infrastructure. In places with low birth rates and shrinking populations, immigrants can fill the gap in a shrinking workforce and maintain economic activity.

For these reasons, some see immigration as the best way to save Europe's economies. Immigration has prevented population decline, as positive net migration has counterbalanced negative natural increase. However, the counterbalance will not be enough to prevent shrinking European populations—it will only delay it slightly. Immigration in Europe will also slow aging of the population and lower dependency ratios by a small amount. By 2050, with current levels of immigration, there will be forty-eight people over age 65 per 100 instead of fifty-one people per 100 if there was no immigration. So, while immigration will help somewhat with offsetting a declining workforce and aging population, it is unlikely to be the region's salvation. Even higher rates of immigration will be required, or birth rates will have to increase substantially.

But the economic impacts of immigration are not always positive. If an economy cannot absorb the immigrant population and grow at a fast enough pace, tax revenue gains may be offset by an increase in government expenditures. The children of immigrants require more spending on education, which may include extra services for language acquisition and special tutoring. Police and fire services may need to expand to cover a growing immigrant population. Health-care systems can become overburdened. If a place has generous welfare benefits and the immigrant population is lower income, then state expenses can rise. Another negative impact can be employment competition with the native population. If immigrants are willing to work for a lower wage, some native workers may have a harder time finding work or may have to accept lower wage to compete.

Empirical research shows that in the US there is a small, positive effect on the economy due to immigration; however, some negative impacts can affect specific segments of the population. On the positive side, immigrants contribute to economic growth, which raises overall wages for American workers. Immigrants also increase the purchasing power of the native population by lowering the costs of child care, landscaping, cleaning services, and other goods and services from immigrant-heavy sectors. There is a net positive impact on the federal budget when considering taxes paid versus services used by immigrants and their children.

While in the long term, immigrants do not lower native employment rates, there can be a loss of some jobs for natives in the short run, especially among those with lower levels of education. Those with a high school education or less can sometimes face slightly lower wages as a result of immigration as well. Furthermore, some research has shown slightly lower wages for college graduates when immigrants are brought in through high-skilled worker programs. And while there are net benefits to the federal budget, state budgets can be negatively impacted by increases in services to immigrants and their children in the short run. Nevertheless, as immigrants' children grow and join the workforce, these short-term negative financial impacts disappear.

Migration also has a significant impact on places of origin, which can be either positive or negative. This relationship is often referred to as the *migration-development nexus*. In the case of international migration, remittances have the most salient impact. Remittances are money sent by migrants from their destination country to family and friends in their country of origin. Some countries are heavily reliant on remittances, which can make up over 20 percent of the gross domestic product (GDP) in some cases, such as Haiti and Liberia (figures 3.37 and 3.38).

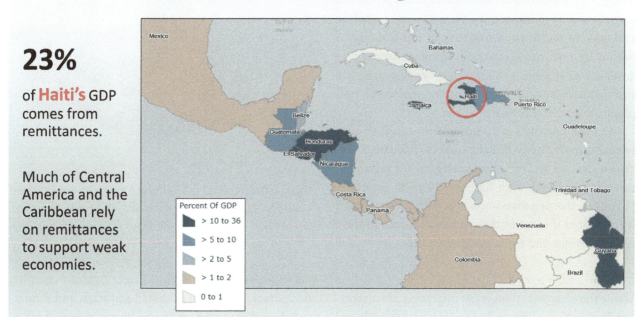

Remittances as a Percentage of GDP, 2014

23%

of **Haiti's** GDP comes from remittances.

Much of Central America and the Caribbean rely on remittances to support weak economies.

Percent Of GDP
- > 10 to 36
- > 5 to 10
- > 2 to 5
- > 1 to 2
- 0 to 1

Figure 3.37. Remittances as a percentage of GDP, 2014. Data source: World Bank.

Figure 3.38. Mobile money transfer in Thailand. Remittances from family members working abroad are an important part of many countries' economies. Photo by John and Penny. Stock photo ID: 607723385. Shutterstock.

In 2014, over 40 percent of Tajikistan's GDP consisted of remittance money. Remittances have been shown to significantly reduce poverty in sending countries, as recipients use the money for daily expenses such as food, clothing, medical care, and education as well as for savings and investment in small businesses.

Immigrants also gain new skills when living abroad, which they can share with friends and family in their country of origin or take with them if they return. Exposure to new technologies, work processes, and management strategies can help improve public and private sector organizations in the country of origin. In best-case scenarios, economic output and government efficiency can increase, while corruption can decrease as returning migrants implement best practices learned abroad.

On the negative side, emigration can lead to a brain drain of highly skilled workers. Education and health care, as well as other sectors of the economy, can suffer if those with higher education emigrate. However, the negative impact of a brain drain can be mitigated if some of these workers return to their place of origin and bring back new skills.

Figure 3.39. Fusion food. The Korean taco with beef, pickled cabbage, and guacamole on a tortilla. Photo by Glenn Price. Stock photo ID: 80246380. Shutterstock.

Cultural

While the economic impacts of migration are important for both sending and receiving places, cultural and social impacts can play a major role in migration debates as well. Through relocation diffusion, immigrants bring culture with them. In some cases, such as religion and language, these cultural characteristics can be very distinct from the native population. At the same time, emigration can have significant social impacts on sending countries, causing family stress and other problems.

On the positive side, immigration allows for people to have contact with different cultures without having to travel. Immigrants bring new ideas, new foods, new music, and more that can enrich the destination place. Fusion of foods, music, and fashion in large, multicultural places makes for interesting and exciting trends (figure 3.39). Also, immigration can benefit the native population by making it easier to learn and practice new languages, which can be of help in a globalized world.

But there are negative cultural and social consequences as well. Western Europe has been facing significant cultural clashes with its large immigrant population from North Africa and the Middle East in recent years. Western Europe has a very strong liberal heritage that favors individual rights, gender equality, democracy, separation of church and state, and secularism. Social attitudes tend to be liberal as well, such as toward homosexuality. In contrast, immigrants from North Africa and the Middle East come from very distinct cultures. While new immigrants can adapt to a new social context, ideas from the home country can remain strong for many. Most come from conservative Muslim countries, where religion and the state have close connections. Laws based on religious texts are common and are seen by many as the natural basis of civil law. Homosexuality in these countries is viewed by majorities as morally wrong, and most people from the region believe a wife should always obey her husband. When immigrants bring these views to Europe, problems can arise. The Muslim veil has become a symbol of the clash of cultures. France, which has a long tradition of strict separation of church and state, banned headscarves and other religious symbols in state schools and more recently banned face-covering veils in all public places. Other local governments in Europe have made similar laws, and several countries have considered it. At the same time, anti-immigrant political parties have gained influence (figure 3.40), pushing for more limits on immigration and a perceived Islamization of the region.

Figure 3.40. Anti-immigrant political party. The Austrian far-right anti-immigrant identitarian movement with banner "Defend Europe" in Vienna. Photo by Johanna Poetsch. Stock photo ID: 435548020. Shutterstock.

Language and immigration can also lead to conflict. For instance, as large numbers of Asians have immigrated to cities just east of Los Angeles, debates over business signage have flared up over the years. Some argue that signs written exclusively in Asian languages risk public safety, since not all police and firefighters can read them. For this reason, there have been proposals to require the inclusion of modern Latin lettering on all business signs. Similar debates over language relate to the extra costs required for bilingual government documents and translation services for immigrants.

High levels of immigration can also lead to a breakdown of social cohesion in a place. When newcomers have a different culture from the native population, people can tend to pull back socially. Levels of trust and social solidarity can decline as people turn to separate and isolated spaces. However, with time, this isolation weakens as immigrants and their children learn the language of their new home.

Sending countries can face negative cultural and social consequences from emigration as well. Family cohesion can be damaged when all are unable to emigrate together. Migrants can face psychological stress and familial bonds can be broken. When parents emigrate but leave their children behind, there can be an increase in risky behavior, with children more involved in drug and alcohol abuse and crime and with lower rates of school attendance. Furthermore, rates of sexually transmitted disease, including HIV, can increase as spouses spend long periods of time in separate locations (figure 3.41).

For all the reasons discussed here, immigration remains a hotly contested political topic in countries around the world. Depending on the data points that one chooses, immigration can help economies or hurt them, bring vitality to a culture or threaten its stability and cohesion. Nevertheless, the push and pull forces that induce people to seek a better life for themselves and their families will continue to drive migration and shape the human geography of our planet.

Go to ArcGIS Online to complete exercise 3.4:
"Does immigration increase unemployment?"

Figure 3.41. HIV/AIDS information posters for migrant workers in India. The bottom poster says in Hindi, "Living Away from Home in the City." Images by National AIDS Control Organisation, Ministry of Health & Family Welfare, Government of India.

References

American Immigration Council. 2016. "How the United States Immigration System Works." https://www.americanimmigrationcouncil.org/research/how-united-states-immigration-system-works-fact-sheet.

Amnesty International. 2013. "North Korea: New Satellite Images Show Continued Investment in the Infrastructure of Repression." https://www.amnesty.org/en/documents/ASA24/010/2013/en.

———. 2014. "North Korea: The Inside Story." https://www.amnesty.org.uk/North-Korea-prison-camp-officials-raped-women-killed-secret.

———. 2015. "Libya: Horrific Abuse Driving Migrants to Risk Lives in Mediterranean Crossings." https://www.amnesty.org/en/latest/news/2015/05/libya-horrific-abuse-driving-migrants-to-risk-lives-in-mediterranean-crossings.

Association of American Geographers. 2011. "Migrant Selectivity—Migration Conceptual Framework: Why Do People Move to Work in Another Place or Country?" AAG Center for Global Geography Education.

Australian National Maritime Museum. 2014. "Australia's Immigration History—Waves of Migration." https://www.anmm.gov.au/discover/online-exhibitions/waves-of-migration.

BBC News. 2017. "The Islamic Veil across Europe." http://www.bbc.com/news/world-europe-13038095.

Bilak, A., G. Cardona-Fox, J. Ginnetti, E. J. Rushing, I. Scherer, M. Swain, N. Walicki, M. Yonetani, and J. Lennard. 2016. *Global Report on Internal Displacement.* Geneva, Switzerland: International Displacement Monitoring Centre.

Broomfield, M. 2016. "Pictures of Life for Turkey's 2.5 Million Syrian Refugees." *The Independent,* April 5, 2016.

Castles, Stephen, Hein de Hass, and Mark J. Miller. 2014. *The Age of Migration,* 5th ed. New York: The Guilford Press.

CBS News. 2016, August 3. "UN: ISIS genocide of Yazidis in Iraq 'ongoing.'" http://www.cbsnews.com/news/united-nations-says-isis-yazidi-genocide-ongoing-in-iraq.

Central Intelligence Agency. 2018. "Refugees and Internally Displaced Persons." https://www.cia.gov/library/publications/the-world-factbook/fields/2194.html.

CNN. 2018, August 30. "Hurricane Katrina Statistics Fast Facts." http://www.cnn.com/2013/08/23/us/hurricane-katrina-statistics-fast-facts.

Cohn, D'Vera. 2015. "How U.S. Immigration Laws and Rules Have Changed Through History." Pew Research Center. http://www.pewresearch.org/fact-tank/2015/09/30/how-u-s-immigration-laws-and-rules-have-changed-through-history.

Costa, D., D. Cooper, and H. Shierholz. 2014. "Facts About Immigration and the US Economy: Answers to Frequently Asked Questions." Economic Policy Institute. http://www.epi.org/publication/immigration-facts.

Doran, K., A. Gelber, and A. Isen. 2016. "The Effects of High-Skilled Immigration Policy on Firms: Evidence from Visa Lotteries." *Journal of Political Economy.* Goldman School of Public Policy, University of California, Berkeley. https://gspp.berkeley.edu.

The Economist. 2016 "Oh, Boy. Are Lopsided Migrant Sex Ratios Giving Europe a Man Problem?" *The Economist,* January 16, 2016. http://www.economist.com/news/europe/21688422-are-lopsided-migrant-sex-ratios-giving-europe-man-problem-oh-boy.

The Editors of Encyclopædia Britannica. 1998. "Sun Belt." *Encyclopædia Britannica.* https://www.britannica.com/place/Sun-Belt.

Eisenbrey, R. 2015. "H-1B Visas Do Not Create Jobs or Improve Conditions for US Workers." *Working Economics Blog.* Economic Policy Institute. http://www.epi.org/blog/h-1 b-visas-do-not-create-jobs-or-improve-conditions-for-u-s-workers.

European University Institute. n.d. "Neo-Classical Economics and the New Economics of Labour Migration." Return Migration and Development Platform European University Institute. http://rsc.eui.eu/RDP/research/schools-of-thought/neo-classical-economics-nelm.

Human Rights Watch. 2016. "World Report 2015: North Korea " https://www.hrw.org/world-report/2015/country-chapters/north-korea.

———. 2016. "World Report 2015: Qatar." https://www.hrw.org/world-report/2015/country-chapters/qatar.

Jalabi, R. 2014. "Who Are the Yazidis and Why Is ISIS Hunting Them?" *The Guardian,* August 7, 2014. https://www.theguardian.com/world/2014/aug/07/who-yazidi-isis-iraq-religion-ethnicity-mountains.

Krogstad, J. M., J. S. Passel, and D. V. Cohn. 2017. "5 Facts about Illegal Immigration in the US." *FACTANK: News in the Numbers.* Pew Research Center. http://www.pewresearch.org/fact-tank/2017/04/27/5-facts-about-illegal-immigration-in-the-u-s.

Labrador, R.C., and D. Renwick. 2016. "Central America's Violent Northern Triangle." Council on Foreign Relations. http://www.cfr.org/transnational-crime/central-americas-violent-northern-triangle/p37286.

Lanzarotta, M. 2008. "Robert Putnam on Immigration and Social Cohesion." *Global*

Economic Symposium. Harvard Kennedy School. https://www.hks.harvard.edu/news-events/publications/insight/democratic/robert-putnam.

Lee, Everett. 1966. "A Theory of Migration." *Demography* 3, no. 1: 47–57.

Lindstrom, D. P., and A. L. Ramirez. 2010. "Pioneers and Followers: Migrant Selectivity and the Development of US Migration Streams in Latin America." *The Annals of the American Academy of Political and Social Science* 630, no. 1: 53–77.

Martin, Philip. 2013. "The Global Challenge of Managing Migration." *Population Bulletin* 68, no. 2: 2–15.

National Archives of Australia. 2013. Immigration Restriction Act 1901 (Commonly Known as the White Australia Policy). National Archives of Australia. http://www.naa.gov.au/collection/a-z/immigration-restriction-act.aspx.

National Drought Mitigation Center. n.d. "Drought in the Dust Bowl Years." https://drought.unl.edu/dustbowl/Home.aspx.

National Geographic Genographic Project. n.d. "Map of Human Migration." https://genographic.nationalgeographic.com/human-journey.

Needham, S., and K. Quintiliani. 2011. "Why Long Beach?" Cambodian Community History & Archive Project. http://www.camchap.org/why-long-beach.

Organisation for Economic Co-operation and Development (OECD). 2018. "Earnings: Real Minimum Wages." OECD Employment and Labour Market Statistics (database). doi: https://doi.org/10.1787/data-00656-en.

———. 2018. "Average Wages (Indicator)." doi: http://doi.org/10.1787/cc3e1387-en.

———. 2018. "Wage Levels (Indicator)." doi: http://doi.org/10.1787/0a1c27bc-en.

Pedersen, P. J., M. Pytlikova, and N. Smith. 2008. "Selection and Network Effects—Migration Flows into OECD Countries 1990–2000." *European Economic Review* 52, no. 7: 1160–86.

Pew Research Center. 2011. "The American-Western European Values Gap." Pew Research Center's Global Attitudes Project. http://www.pewglobal.org/2011/11/17/the-american-western-european-values-gap/.

———. 2015. "Modern Immigration Wave Brings 59 Million to US, Driving Population Growth and Change Through 2065." Pew Research Center's Hispanic Trends Project. http://www.pewhispanic.org/2015/09/28/modern-immigration-wave-brings-59-million-to-u-s-driving-population-growth-and-change-through-2065.

Plumer, B. 2013. "Congress Tried to Fix Immigration Back in 1986. Why Did It Fail?" *Washington Post*, January 30, 2013. https://www.washingtonpost.com/news/wonk/wp/2013/01/30/in-1986-congress-tried-to-solve-immigration-why-didnt-it-work/?utm_term=.7b80deed35a7.

Ratha, D., S. Mohapatra, and E. Scheja. 2011. *Impact of Migration on Economic and Social Development: A Review of Evidence and Emerging Issues*. Washington, DC: World Bank.

Ravenstein, E. G. 1885. "The Laws of Migration." *Journal of the Statistical Society of London* 48, no. 2: 167–235. doi: http://doi.org/10.2307/2979181.

Richardson, V. 2015. "Texas Emerges as Top Destination for Californians Fleeing State." *Washington Times*, August 31, 2015. http://www.washingtontimes.com/news/2015/aug/31/texas-emerges-top-destination-californians-fleeing.

Schreck, C. 2015. "Russian Applications for U.S. Asylum Skyrocket In 2015." Radio Free Europe/Radio Liberty. http://www.rferl.org/a/russia-us-asylum-applications-skyrocket/27345403.html.

Shyong, F. 2013. "Monterey Park Abandons Sign Law Requiring Some 'Latin lettering.'" *Los Angeles Times*, December 7, 2013. http://articles.latimes.com/2013/dec/06/local/la-me-1207-chinese-signs-20131207.

Syrian Refugees. 2016. "Syrian Refugees: A Snapshot of the Crisis—in the Middle East and Europe." http://syrianrefugees.eu.

United Nations. 2016. *International Migration Report*. New York: United Nations.

United Nations High Commissioner for Refugees. "Refugees." UNHCR. http://www.unhcr.org/pages/49c3646c125.html.

———. "The Mediterranean Refugees/Migrants Data Portal." http://data2.unhcr.org/en/situations/mediterranean.

US Department of Homeland Security. 2017. *Yearbook of Immigration Statistics 2014*. Washington, DC: US Department of Homeland Security, Office of Immigration Statistics.

Winckler, O. 2010. "Labor Migration to the GCC States: Patterns, Scale, and Policies." Middle East Institute. http://www.mei.edu/content/labor-migration-gcc-states-patterns-scale-and-policies.

World Bank. 2015. "Migration and Remittances Data." http://www.worldbank.org/en/topic/migrationremittancesdiasporaissues/brief/migration-remittances-data.

Wormald, B. 2013. "The World's Muslims: Religion, Politics and Society." Pew Research Center's Religion & Public Life Project. http://www.pewforum.org/2013/04/30/the-worlds-muslims-religion-politics-society-overview/.

Yeginsu, C. 2016. "In Turkey, a Syrian Child 'Has to Work to Survive.'" *New York Times*, June 4, 2016. https://www.nytimes.com/2016/06/05/world/europe/in-turkey-a-syrian-child-has-to-work-to-survive.html?_r=1.

Zelinsky, W. 1971. "The Hypothesis of the Mobility Transition." *Geographical Review* 61, no. 2: 219.

Chapter 4
Race and ethnicity

One of the most common, and yet most controversial, ways in which people view each other is through the lens of race and ethnicity. For centuries, and even millennia, humans have used this lens to determine who is part of their group and who is considered an outsider. Differences in the way people look, talk, dress, eat, and behave have, sadly, been tied to war, slavery, and all types of human exploitation. But at the same time, these differences contribute to the wonderful diversity of our world. As people have migrated and mixed, new ideas have evolved, from culinary innovation, to religion and philosophy, to architecture and design, and much more.

Because of the importance of race and ethnicity in our perceptions of the world, geographers are interested in understanding their spatial distribution and how these distributions impact social cohesion and conflict, how they contribute to cultural change, and how they influence our landscapes. In this chapter, we explore what race and ethnicity are, where different groups live in the United States, segregation and integration, how race and ethnicity spatially relate to other socioeconomic characteristics, and how they impact the way we create our landscapes.

What is race and ethnicity?
Throughout history people have sought to categorize and organize features of the world to better make sense of it. The world is a very diverse place, and categories help people simplify and make it manageable for understanding. Along with myriad physical and natural features of the earth, such as plant and animal species and soil types, people for millennia have classified humans into different categories. These categories reflect group identities, helping to create bonds between people of the same group, but can also serve to reinforce an "us versus them" attitude by excluding people from different groups. Often, these categories have had a geographical origin, with people classified according to where they come from.

Among the most common ways of classifying humans are through the concepts of race and ethnicity. Ethnicity groups people by a shared cultural heritage. This shared heritage can include belief and value systems, language, food and clothing, cultural traditions, and a shared ancestral homeland. Going back thousands of years, the ancient Greeks and Romans differentiated the people of the Mediterranean region, as well as more distant Germanic and Persian people, among others, by geographically based ethnic classifications. The Bible mentions myriad ethnic groups from the ancient Roman Empire, such as the Hittites, Amorites, Canaanites, Perizzites, Hivites, and Jebusites, often alluding to conflict and disdain between these ancient cultures.

Moving to the present, we commonly use ethnic classifications such as Latino or Hispanic, Chinese, or Arab, among many more, that reflect the region or country from which people come. In each of these cases, people identified (either by themselves or others) as part of an ethnic group are likely to have common sets of cultural characteristics. It must be noted that ethnicity differs from nationality; whereas nationality refers to the nation to which a person currently belongs or used to belong, ethnicity refers to cultural heritage. Thus, there are ethnic Chinese living throughout the world who are not citizens of China and have never set foot in their ancestral homeland.

In contrast to ethnicity, race has traditionally been tied to differences in physical traits of humans, such as skin color, eye shape, bone structure, and hair texture. As with ethnicity, racial categories tend to be grouped by geographical regions. The idea of race is much more recent than that of ethnicity and gained traction as European colonialism in the sixteenth century began an intense process of globalization. While people and places of the earth were brought together culturally, economically, and politically, natural philosophy and taxonomy, or the science of classifying organisms, was growing in influence. As part of the growing trend in taxonomy, early classifications divided humans into four or five racial groups. These groups roughly aligned with the continents, resulting in "races" of Asians, Africans, Europeans, and Native Americans. Not all classifications corresponded perfectly with the continents, however. Some saw a single race running from Europe and North Africa through Persia and into central India and parts of Indonesia, while others used different classifications, for instance, splitting the northern Scandinavia Lapp people apart from the rest of Europe and Asia.

Ultimately, the categories of race and ethnicity are socially constructed by humans. They are not fixed, scientifically determined biological categories but rather are based on how different groups of people view themselves and others. For instance, classifying Europeans as a single race or ethnicity may face resistance from those who say a typically tall, blond Swede is physically and culturally very distinct from a typically shorter and darker Greek. Certainly, Nazi Germany viewed the Aryans as distinct from other Europeans. The same holds true for Asians from Cambodia and Japan and for Africans from South Sudan and Nigeria.

Measuring race and ethnicity

Human geographers are interested in mapping and analyzing the spatial patterns of race and ethnicity. By understanding where different groups of people live and how they interact with other groups in other locations, geographers can better understand how different elements of culture, be it language, religion, food, fashion, music, architecture, or other, interact and blend over space and time to create the earth's cultural landscapes. Furthermore, by quantifying and mapping race and ethnicity, policymakers can better analyze which groups are struggling to be equal partners in a society. If some racial or ethnic groups face greater disadvantages than other groups in a society, then public and private organizations can target programs to help improve their socioeconomic situation.

The question then becomes, which groups should be counted and measured? As discussed, race and ethnicity are socially constructed. There are no fixed scientific categories that can be used for classifying people. Consequently, there has been great variation over space and time regarding which groups are enumerated by national governments.

The US Census

Countries around the world collect data on their populations via censuses. In the United States, the Constitution requires that a census of the population be taken every ten years in order to allocate taxes and legislative representatives among states on the basis of population size. As part of this population count, the US government has also collected demographic data on various characteristics of the population, including race and ethnicity.

Given that the United States was founded partly on the basis of slavery, censuses from 1790 through 1850 classified people exclusively according to whether they were white or black and free or enslaved (figure 4.1). At later points in time, additional general categories where added, which as of 2010 included White, Black, American Indian/Alaska Native, Asian, Hawaiian/Pacific Islander, Other, and Hispanic. Within each of these general categories were more specific categories

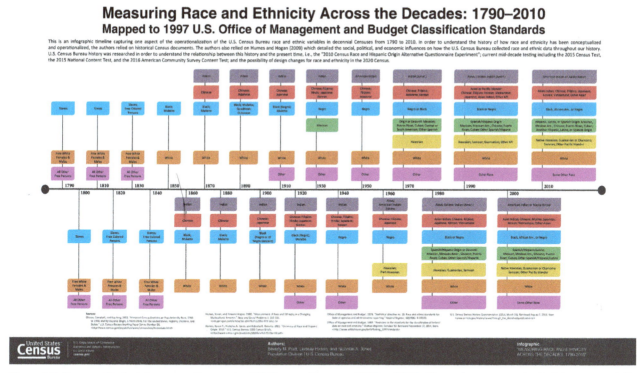

Figure 4.1. Race and ethnicity according to the US Census Bureau, 1790 to 2010. Given that race and ethnicity are socially constructed concepts, the way in which people are classified changes over time and space. Data source: US Census.

based on country of origin. These specific categories change over time, depending on migration flows and other social perceptions of race and ethnicity.

Remember that the concepts of race and ethnicity are socially constructed—humans create categories, which people must then be placed in. But trying to fit people into a limited set of categories can cause problems. Prior to 1960, census takers would determine which category to put people in. Thus, race and ethnicity were imposed externally by US government census takers. Since the 1960 census, however, individuals have been able to select their own race or ethnicity, thus allowing people to record the group they most strongly self-identify with.

Yet when faced with a limited number of choices, some people still struggled to classify themselves. Through the 1990 census, people could choose only one race or ethnicity, so someone with a Chinese

mother and Vietnamese father would have to pick just one on the census form. The same was true for someone with a white mother and black father (such as former President Obama)—you could be black or white, but not both. This limitation changed with the 2000 census, which began to allow respondents to choose one or more races. Because of this, since 2000, it is possible to count the number of multiracial residents of the United States.

Figure 4.2 shows the section of the 2010 census form that asks about race and ethnicity. It includes five race categories: White, Black or African American, American Indian or Alaska Native, Asian (with several national subcategories), and Native Hawaiian or Other Pacific Islander (also with several national subcategories). A sixth category includes Some other race. But note that Hispanic, Latino, or Spanish origin is a separate category.

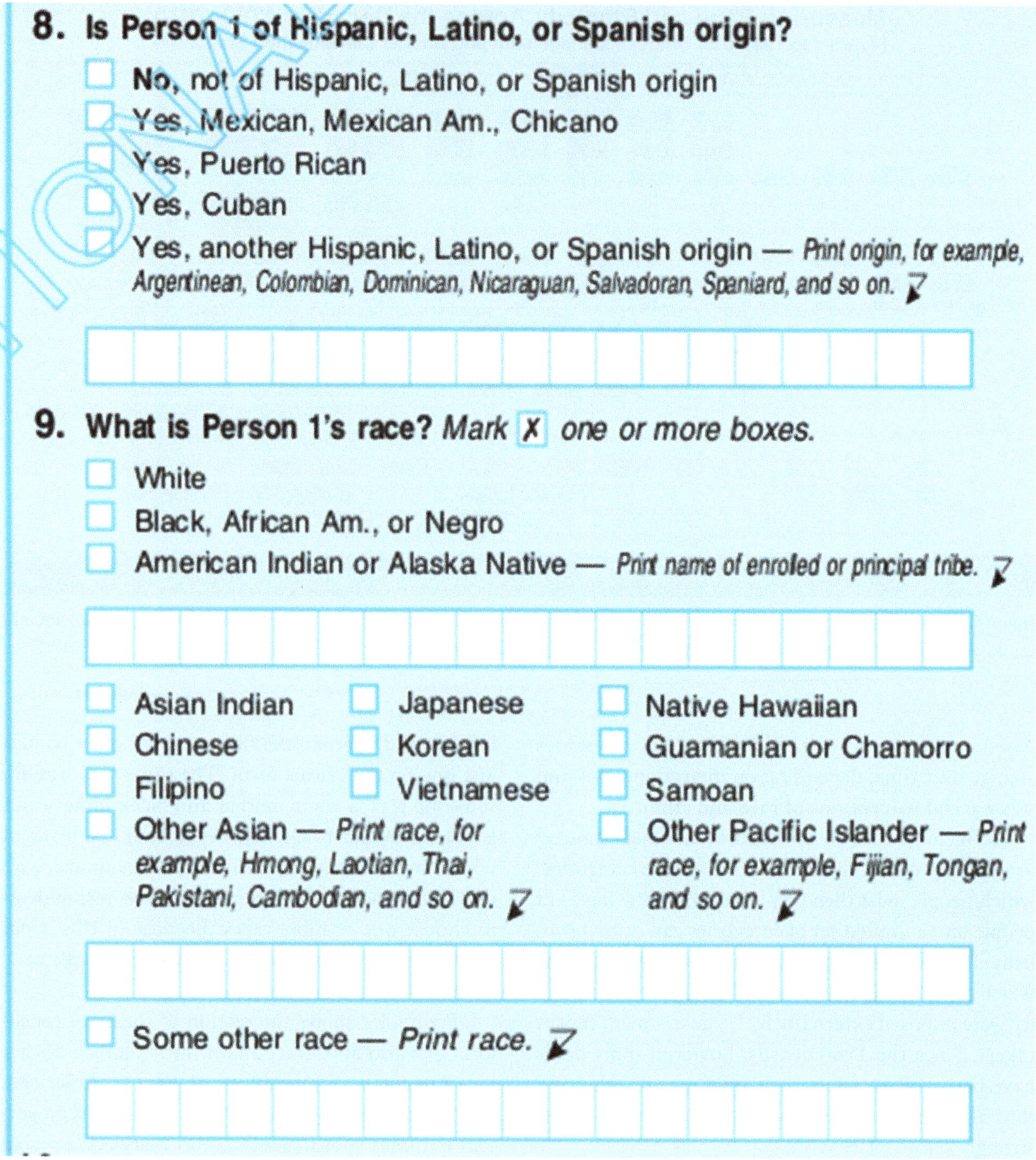

8. Is Person 1 of Hispanic, Latino, or Spanish origin?

☐ **No**, not of Hispanic, Latino, or Spanish origin

☐ Yes, Mexican, Mexican Am., Chicano

☐ Yes, Puerto Rican

☐ Yes, Cuban

☐ Yes, another Hispanic, Latino, or Spanish origin — *Print origin, for example, Argentinean, Colombian, Dominican, Nicaraguan, Salvadoran, Spaniard, and so on.* ⟍

9. What is Person 1's race? *Mark* ☒ *one or more boxes.*

☐ White

☐ Black, African Am., or Negro

☐ American Indian or Alaska Native — *Print name of enrolled or principal tribe.* ⟍

☐ Asian Indian   ☐ Japanese   ☐ Native Hawaiian

☐ Chinese   ☐ Korean   ☐ Guamanian or Chamorro

☐ Filipino   ☐ Vietnamese   ☐ Samoan

☐ Other Asian — *Print race, for example, Hmong, Laotian, Thai, Pakistani, Cambodian, and so on.* ⟍   ☐ Other Pacific Islander — *Print race, for example, Fijian, Tongan, and so on.* ⟍

☐ Some other race — *Print race.* ⟍

Figure 4.2. Options for race and ethnicity on the 2010 US census form. Some people struggle to decide exactly which box(es) they best fit in. Data source: US Census.

This still presents many residents of the United States with a challenge when trying to decide where they fit. Many Latinos consider their Latino heritage to be their primary racial/ethnic identity, yet the US census form requires them to say they are Latino (question 8), but then choose a race in addition (question 9). In 2010, the majority of Latinos (53 percent) listed their race as white, with another 36 percent choosing "Some other race."

Since Hispanic or Latino is counted separately from race, when analyzing and mapping data, geographers often separate them out from their racial category. For example, non-Hispanic white is often used when mapping the white population of the United States. Hispanic or Latino (all races) can then be mapped separately. If all people who classified themselves as white are mapped, a large number of Hispanics will be included in the data.

Other groups struggle to identify themselves in the census as well. People from the Middle East and North Africa, according to US census definitions, fall under the White racial category. Yet many Arabs, Persians, and Turks feel that it is a poor fit. As a group that sometimes faces discrimination and hate crimes, many do not feel that they are white in the same way that those of European descent are.

Because of the difficulty some people face when classifying themselves by US census categories, Some other race was the third most commonly selected category (7 percent) after White alone (72 percent) and Black alone (13 percent) in the 2010 census (figure 4.3). The majority of those who selected Some other race were of Hispanic origin.

Other countries

As illustrated by changes in US census classifications of race and ethnicity, categories are heavily dependent on the social and historical context of each country.

While Americans traditionally saw race in terms of black and white, eighteenth-century Spaniards in the Americas saw many gradations, with different names for different combinations of Spaniard, Amerindian, African, and white (figure 4.4). This detailed classification scheme has since given way in Mexico to no census questions on race or ethnicity, with only a related question that asks if a person speaks an indigenous language.

The historical and social nature of racial and ethnic categories can be seen in other countries as well. In the Brazilian census, people can write in any ethnic group but must select a race that includes white, black, yellow (for Asian), and mixed. It is interesting to note that in the US, people could not select more than one race until the year 2000, whereas the mixed category in Brazil was first used in the 1872 census. This reflects a greater mixing of races throughout Brazil's history,

Figure 4.3. Puerto Rican Day parade in Manhattan. Forced to pick a race in addition to their Hispanic ethnicity, over half of Latinos in the US identified themselves as white in the 2010 census. Another 36 percent said they are "Some other race." Photo by Lev Radin. Shutterstock. Stock photo ID: 197628662.

Figure 4.4. Eighteenth-century painting showing sixteen racial combinations of colonial Latin America. Data source: Museo Nacional del Virreinato, Tepotzotlán, Mexico. Artist unknown.

as compared to an ideology of strict racial segregation in the US.

In Canada, as in Brazil, one can write in any ethnicity as well. Residents then select from a range of racial and ethnic categories, such as Latin American, Arab, Chinese, and West Asian (i.e., Iranian, Afghan), as well as white and black. The United Kingdom also allows for writing in one's ethnicity; in addition, it has categories that relate to its colonial past, including Black Caribbean, Pakistani, and Bangladeshi. It also includes Gypsy or Irish Traveller as a subcategory of white (figure 4.5).

France asks about one's nationality in its census form, but only for those born as citizens of another country. For those born French, no information is collected on race or ethnicity. This follows the French notion of *egalité*, or equality, and its emphasis on French identity over ethnic or racial identity. While this "colorblind" ideology has its appeal, in reality, France is a very multicultural society with stark socioeconomic differences between traditional French and minority racial and ethnic communities. A lack of data on race and ethnicity makes understanding the spatial and social patterns of these differences difficult to study.

Spatial distribution of race and ethnicity

As seen in the previous section, there are myriad ways to classify humans by race and ethnicity. Consequently, it is difficult to talk about spatial distributions at a global scale; people from different countries would create maps based on wildly different categories. Therefore, it is more useful to look at race and ethnicity at larger scales, from countries down to local communities, where a single country's classification can be used.

Mapping distributions at different scales illustrates different processes of spatial interaction. For instance, racial and ethnic patterns mapped by state can be useful for showing international immigration flows and how they are impacting state laws on everything from driver license requirements to bilingual education legislation. It can also reflect interstate migration flows and how economic conditions in each state contribute to these movements. Zooming in to county-level patterns, a finer-grained picture can be seen. Whereas a state may have a large population of a specific race or ethnicity, it may become clear that urban counties have much higher concentrations of that group than

Figure 4.5. A vintage Romani (also known as gypsy) caravan at the Appleby Horse Fair in Appleby, Cumbria, UK. The UK census has subcategories for white that include Gypsy or Irish Traveller. Photo by David Muscroft. Shutterstock. Stock photo ID: 262638497.

rural counties, reflecting the gravitational pull of urban economies. Moving to an even larger scale, processes of integration and segregation can be viewed within a city to understand the degree to which different races and ethnicities interact or do not interact at the neighborhood level.

As of 2015, the US Census Bureau estimated that the majority of the United States was white non-Hispanic, followed by Hispanic (all races), then black or African American (non-Hispanic) and Asian (non-Hispanic) (table 4.1). The smallest two groups consisted of American Indian and Alaska Native (non-Hispanic) and Native Hawaiian and other Pacific Islander (non-Hispanic).

These numbers have changed significantly over the course of US history, although exact comparisons over time are difficult due to modifications in how race and ethnicity have been defined in the US census. The white non-Hispanic population has fallen steadily from nearly 90 percent in the first half of the twentieth century to just under 80 percent of the US population in 1980 and finally to 62 percent as of 2015 (figure 4.6). Change in the black or African American population has been less significant, fluctuating around 10 to 12 percent through most of the twentieth century and constituting 13 percent in 2015.

While the white non-Hispanic population has seen the largest decrease, Hispanic and Asian populations have increased the most. Asians made up only 0.2 percent of the US population through 1950, but by 1980, they constituted 1.5 percent of the population, and by 2015, they constituted 6 percent. Likewise, the Hispanic population has grown substantially. Not consistently counted as a separate group until the late twentieth century, the Hispanic population grew from 6.4 percent of the US population in 1980 to 18 percent in 2015.

The following sections examine spatial distributions and spatial interaction for the four largest racial and ethnic groups in the United States: white (non-Hispanic), Hispanic (all-races), black or African American (non-Hispanic), and Asian (non-Hispanic).

State-level patterns
White (non-Hispanic)

At a state level of analysis, broad regional patterns of each major group can be observed. Whites still dominate the majority of the country, at 62 percent of the total population, especially as one moves away from

Race/Ethnicity	U.S. Population (Percent) 2015
White (Non-Hispanic)	203,787,565 (62%)
Hispanic (all races)	58,501,475 (18%)
Black or African American (Non-Hispanic)	42,816,387 (13%)
Asian (Non-Hispanic)	19,967,708 (6%)
American Indian and Alaska Native (Non-Hispanic)	4,266,951 (1%)
Native Hawaiian and Other Pacific Islander (Non-Hispanic)	1,147,356 (less than 1%)

Table 4.1. Race and ethnic population of United States, 2015. Data source: US Census.

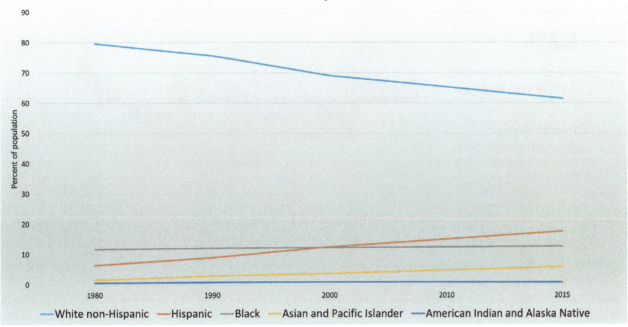

Figure 4.6. US race/ethnicity, 1980 to 2015. The most dramatic changes nationally have been a decline in the white non-Hispanic population and an increase in the Hispanic and Asian populations. Data source: US Census.

the southernmost states. White concentrations in many states are high, with the top quintile reaching 82 to 93 percent (figure 4.7). States with high proportions of whites are the result of early immigration and settlement patterns that predominated from Europe during the seventeenth through twentieth centuries that have not been the focus of more recent immigration flows from other regions.

Between 2000 and 2014, the white non-Hispanic population fell in every state but not in the District of Columbia. This decline can be attributed to low fertility rates among the white population and immigration of nonwhites from other countries. In 2015, non-Hispanic whites had the second-lowest fertility rate after Asians. At 1.71 children per woman, the fertility rate was significantly below replacement level. Furthermore, as shown in chapter 3, immigration from Asia and Latin America accounted for significant population gains in recent years, reducing the proportion of whites in many parts of the country.

Latino or Hispanic (All-Races)

When viewing the Hispanic population by state, it is useful to see the spatial patterns of both counts and rates. In figure 4.8, it is clear that the largest Hispanic clusters are in states along the southern border, such as California, Texas, and Florida, as well as in New York. In fact, California, Texas, and Florida alone account for over 50 percent of all Hispanics in the United States. Slightly different patterns are seen when viewing Hispanics as a percentage of the total population. California and Texas stand out, but so too does New Mexico, all of which have Hispanic populations of at least 31 percent.

Again, these distributions can be explained by fertility rates and immigration. Between births and immigration, Hispanics accounted for over half of the total population increase in the United States between the 2000 and 2010 censuses. States along the southern border have absorbed substantial Latino immigration for many decades, with large numbers of Mexicans

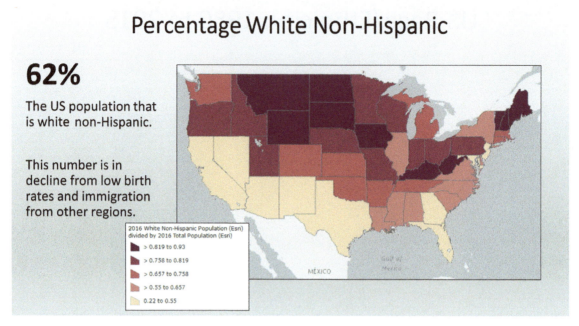

Figure 4.7. Percentage of white (non-Hispanic), 2016. The white population is the largest single racial/ethnic group in the United States, but its distribution varies significantly by state. Explore this map at http://arcg.is/2ldqHzT. Data sources: Esri, US Census Bureau.

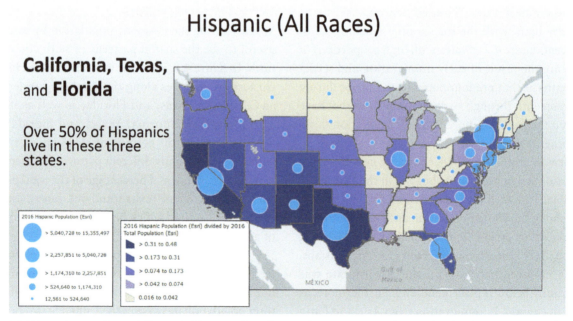

Figure 4.8. Hispanic population (all races), 2016. Much of the southwestern United States has high proportions of Latinos, while California, Texas, Florida, and New York have the largest number of Latinos. Explore this map at http://arcg.is/2ldy0Ye. Data sources: Esri, US Census Bureau.

and Central Americans immigrating to California and Texas, while Cubans and Puerto Ricans have immigrated to Florida. New York also has a large number of Latino immigrants, with many from Puerto Rico and the Dominican Republic.

In 2015, Hispanics in the United States had the highest fertility rate of all major racial or ethnic groups, at 2.53 children per woman, which further expands their demographic footprint. However, it must be noted that there is a large difference by generation. First-generation Hispanics have a fertility rate of 3.35 children, but by the third generation, it falls to 1.98—below replacement level. With a decline in immigration from Latin America in recent years, fewer Hispanics will be first generation, resulting in fertility rates that are expected to closely match those of whites and other major groups in coming decades.

Black or African American (non-Hispanic)

At the state scale, the black or African American population as a percentage of population is clearly concentrated in the southern states (figure 4.9). However, when looking at count data, the population is much less spatially concentrated. For instance, New York and Texas have lower proportions of blacks than do Louisiana and Alabama. However, in raw numbers, both New York and Texas have larger black populations. Likewise, the map shows California, Nevada, and Arizona with similar proportions of blacks, but as a count, California has a much larger black population than its two neighbors. These discrepancies can be attributed to the large total population of states such as California, Texas, and New York. While there are large numbers of African Americans in those states, there are even larger numbers of other races and ethnicities.

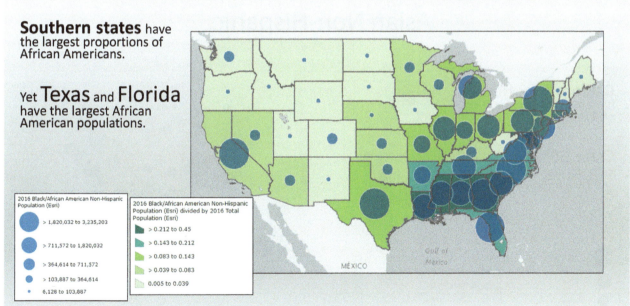

Figure 4.9. Black or African American distribution. While southern states stand out with high percentages of African Americans, large numbers of African Americans can be seen in a wider geographic distribution. Explore this map at http://arcg.is/2ldz26O. Data sources: Esri, US Census Bureau.

Clustering of the black population in the South has fluctuated throughout US history. While slavery ended in 1876, Southern blacks faced limited economic opportunity as sharecroppers. This, combined with legal segregation and Ku Klux Klan violence, meant that poverty was difficult to escape and mobility was limited. In the early twentieth century, around 90 percent of blacks lived in the South, but this began to change with the outbreak of World War I in 1914. The war cut the flow of migrant workers from Europe, resulting in recruitment of Southern blacks for industrial jobs in northern cities such as Chicago, Detroit, and Philadelphia. Active job recruitment in Southern black newspapers led to migration flows from the rural South to the urban North. Known as the Great Migration, this flow led to a massive shift in the distribution of blacks in the United States. By its peak around 1970, less than one half of blacks lived in the South. But as economic conditions improved in the South, and as deindustrialization began to shutter factories in the North, blacks began to return to what some refer to as the black cultural homeland of the southern states. Because of this, by the year 2010, 55 percent of the black population lived in the South.

Asian (non-Hispanic)

The Asian population is largely concentrated in a handful of states (figure 4.10). About 31 percent of Asians live in California alone. Adding just New York, Texas, and New Jersey accounts for over 50 percent of the Asian population. When viewed at a proportion of the population, Hawaii stands out at 36 percent. After that, the next largest concentration is in California, with 14 percent. Growth of the Asian population has been primarily due to immigration. As pointed out in the previous chapter, Asian immigration, highly restricted until the mid-twentieth century, in recent years has been higher than for all other groups, including Hispanics, although it is growing from a much smaller base. Births have a minimal impact on Asian

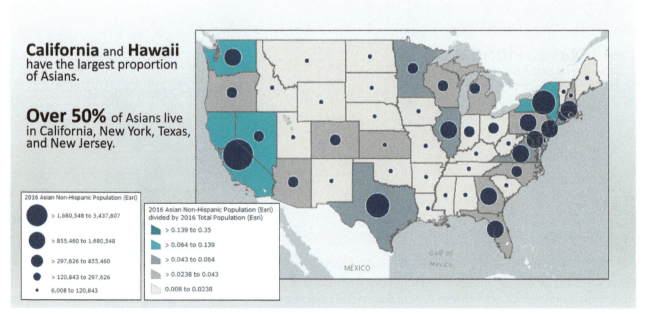

Figure 4.10. Asian population distribution is more concentrated among a limited number of states, generally along the coast and border regions of the US. Explore this map at http://arcg.is/2ldv267. Data sources: Esri, US Census Bureau.

growth. The fertility rate for Asians is the lowest of all major racial and ethnic groups, at 1.66 children, well below replacement level.

County-level patterns

White (non-Hispanic)

At a county scale of analysis, additional spatial patterns can be discerned for each major racial or ethnic group. As of 2014, the average white proportion was lowest in large and medium-sized metropolitan counties and highest in the most rural counties. For instance, figure 4.11 shows that the white population in and around cities such as Chicago, Detroit, Indianapolis, and Milwaukee is significantly lower than in nearby rural counties.

But while rural areas tend to be more white than urban areas, many rural counties are more diverse than they used to be. As economic opportunity has shifted to urban areas, young whites have been migrating from nonmetropolitan counties to urban areas. In some cases, the only people willing to take on jobs in rural economies, such as on dairy farms, slaughterhouses, and labor-intensive agriculture, are nonwhite immigrants, typically from Latin America and Asia.

As the demographics of the United States shift away from being predominately white, the number of counties with nonwhite majorities is growing. By 2014, there were 431 counties where the white population was no longer a majority. Of these, 210 were large metropolitan counties, but another 221 were nonmetropolitan counties with smaller urban and rural populations.

Latino or Hispanic (all races)

County maps also paint a different picture of the spatial distribution of Hispanics compared to state-level analysis (figure 4.12). While California and Texas have large Hispanic populations, those populations are not evenly distributed across the states. In California,

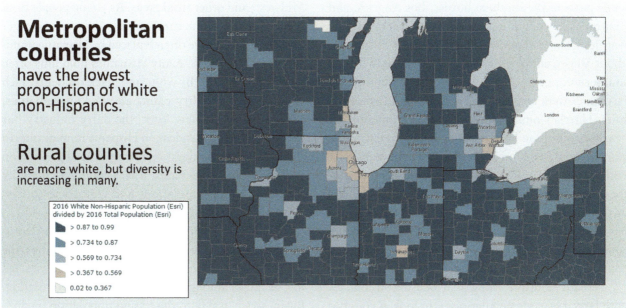

Figure 4.11. White population proportion by county, 2016. Urban counties tend to be more racially diverse, while most rural counties have higher proportions of whites. Explore this map at http://arcg.is/2lYHif0. Data sources: Esri, US Census Bureau.

Hispanic (All Races) by County

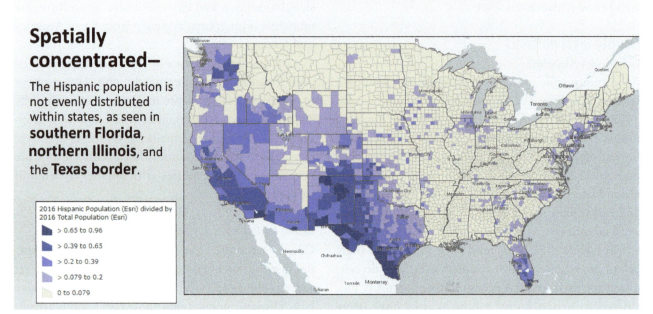

Spatially concentrated—

The Hispanic population is not evenly distributed within states, as seen in **southern Florida**, **northern Illinois**, and the **Texas border**.

2016 Hispanic Population (Esri) divided by 2016 Total Population (Esri)

- \> 0.65 to 0.96
- \> 0.39 to 0.65
- \> 0.2 to 0.39
- \> 0.079 to 0.2
- 0 to 0.079

Figure 4.12. Hispanics, 2016, rate by county. While California, Texas, Florida, and New York have large Hispanic populations, they are not evenly distributed throughout each state. Explore this map at http://arcg.is/2lYKCqm. Data sources: Esri, US Census Bureau.

northern counties have lower percentages of Hispanics, while parts of the southern border, Los Angeles, and Central Valley counties have higher percentages. Los Angeles County alone has over 4.9 million Hispanics, about 50 percent of the county population. In Texas, southern counties have much higher percentages of Hispanics than counties to the north and east of the state. Harris County (which includes Houston) has nearly two million Hispanics, over 40 percent of the county population. In Florida, Miami-Dade stands out with nearly two million Hispanics (over 66 percent of the county population), while the northern panhandle has relatively low percentages. New York State, which has a large Hispanic population, shows the most spatial concentration, with very few high-percentage Hispanic counties outside of the southern portion of the state.

Perhaps more interesting than focusing on counties with large Hispanic populations is to look at counties with fast-growing Hispanic populations (figure 4.13). The fastest-growing Hispanic populations are no longer

in the heavily-Latino Southwest but rather in the South, Midwest, and other rural areas. As young people from predominately white rural counties migrate to larger urban areas, they are often replaced by Hispanic immigrants. In the South, many Hispanics hold jobs often seen as undesirable by white and black residents, such as working in poultry plants and picking tomatoes, while in Upstate New York, the dairy industry has become heavily dependent on Hispanic immigrants. In some counties, immigration merely slowed population decline, but in other cases, it was enough to actually increase county populations and contribute to new vitality in otherwise struggling places.

But while Latino immigrants may be a demographic and economic salvation for some counties, the rapid increase in their numbers has, in some cases, led to a backlash among the traditional white population. Businesses catering to the Latino population, including posting signage in Spanish, in traditionally white small towns has led some to feel that their culture is

Hispanic Population Change (All Races) by County

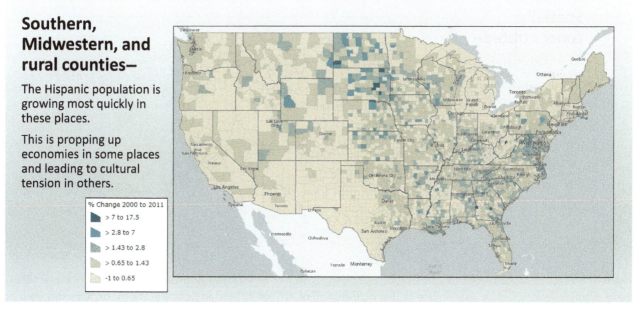

Southern, Midwestern, and rural counties—

The Hispanic population is growing most quickly in these places.

This is propping up economies in some places and leading to cultural tension in others.

% Change 2000 to 2011

> 7 to 17.5
> 2.8 to 7
> 1.43 to 2.8
> 0.65 to 1.43
-1 to 0.65

Figure 4.13. Hispanic population change, 2000 to 2011. The most dramatic growth in the Latino population has been in much of the South and Midwest, not in traditional Latino clusters in the Southwest. Explore this map at http://arcg .is/2jXlVu1. Data sources: Esri, US Census Bureau.

being displaced by new arrivals. At the same time, school districts have had to adapt by adding English as a second language classes—something unheard of prior to rapid ethnic change. Consequently, support for more restrictive immigration policies has increased in places that were long immune to ethnic concerns. This does not imply that all white residents of counties facing change feel anxiety or fear, but it does illustrate how ethnic change is not always a smooth and conflict-free process.

Black or African American (non-Hispanic)

The black or African American population is highly concentrated at the county scale of analysis. Figure 4.14 shows a band of high-proportion black counties running from the southern portion of the Mississippi River through Virginia. This area, often referred to as the Black Belt, was traditionally the center of cotton and tobacco production that historically relied on black labor.

As discussed previously in this chapter, there has been a return migration of blacks to the South. In line with most migration in the US, this return migration tends to consist of urban-to-urban moves, as blacks leave cities in the North for cities in the South. This process can be seen in counties comprising the Atlanta metropolitan area, which have seen a substantial increase in the black population, along with other Southern urban and suburban counties. Outside of the South, black populations are concentrated in larger metropolitan counties, such as those containing New York and Chicago.

Asian (non-Hispanic)

When viewing the Asian population by county, it is clear that they are clustered in a relatively small number of areas (figure 4.15). One of the heaviest concentrations is found in the Silicon Valley region around the San Francisco Bay Area. Coastal Southern California

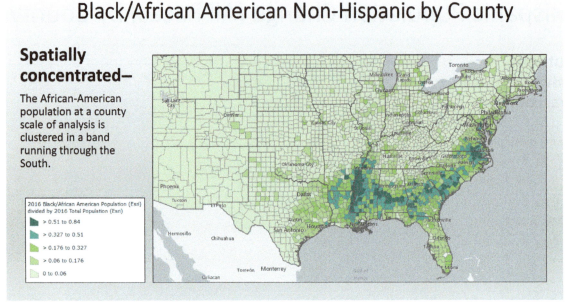

Figure 4.14. Black or African American proportion by county, 2016. The historic footprint of tobacco and cotton plantation slave economies can still be seen in the spatial distribution of African American concentrations. Explore this map at http://arcg.is/2IYCDcR. Data sources: Esri, US Census Bureau.

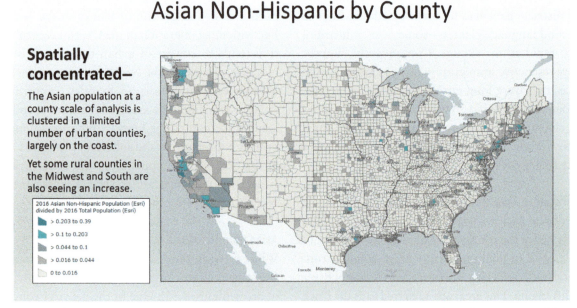

Figure 4.15. Asian proportion of population by county, 2016. Asians are more concentrated in urban regions of the United States. Explore this map at http://arcg.is/2IYJXW2. Data sources: Esri, US Census Bureau.

has high concentrations as well. In Washington, the Seattle area stands out, while in Texas, the urban areas of Dallas, Houston, and Austin contain clusters. Large urban areas in the Northeast, from Boston south through Washington, DC, also have significant Asian proportions, as do a handful of other counties in the Midwest and South.

Clusters of Asians in urban areas are largely due to the pull forces of metropolitan economies. Asians have higher levels of education, such as bachelor's and graduate degrees, than the general US population, as well as higher incomes. Thus, they tend to concentrate in urban areas that rely on more educated workforces.

But not all Asians are highly educated urban dwellers and, as with Latinos, some rural counties are pulling in less-educated Asian immigrants to replace declining white populations. This can be seen in counties scattered through the Midwest and South. As an example, Huron, South Dakota, a small town of 13,000 people in Beadle County, has attracted hundreds of people from the Karen ethnic group of Myanmar in Southeast Asia. As a lower-educated group, many enthusiastically work in the local turkey-processing plant for wages substantially above the minimum wage. Recruitment of Karen refugees began as the plant struggled to attract and retain US-born residents, who do not want to work in meat-processing plants or prefer to pursue employment opportunities in larger urban areas. This pattern can be found in a number of agricultural and food-processing counties.

Census tract-level patterns

Moving to an even finer scale of analysis, that of census tracts (small geographic units of approximately 2,500 to 8,000 people), it becomes possible to measure the degree of racial and ethnic clustering and segregation at the local level.

People tend to be sorted by neighborhood in myriad ways, from communities with young families to those of retirees; from blue collar workers to "creative class" technology and media professionals; from the homeless of skid row to loft-dwelling artists. As you may guess from the subject of this chapter, one of the most salient ways in which people are sorted is by race or ethnicity. For much of US history, blacks and whites lived in separate neighborhoods. Then, as new immigrant groups arrived, ethnic enclaves formed, creating Chinatowns, Little Italys, Little Tokyos, Koreatowns, Latino barrios, and many more (figure 4.16).

Figure 4.16. Chinatown, San Francisco. Ethnic groups often form enclaves that cater to their cultural traditions, be it food, clothing, festivals and celebrations, or more. Photo by Andrey Bayda. Royalty-free stock photo ID: 91115285. Shutterstock.com.

Ethnic enclaves are places occupied by a relatively homogenous ethnic population. They are characterized by businesses catering to the resident ethnic group, such as restaurants, food markets, and clothing stores, with products from the ethnic culture. Signage may use the ethnic group's language, and business names may reflect important people and places from the homeland. Festivals and celebrations tied to the ethnic culture are typically held in these enclaves as well.

Some ethnic enclaves go back a century or more, while others are relatively newly formed. For instance, San Francisco's Chinatown formed well over one hundred years ago, while Little Bangladesh in Los Angeles was officially designated much more recently, in 2010.

Historically, ethnic enclaves formed closer to the inner core of cities, in older and cheaper housing close to busy warehousing and industrial land uses. While many are still located in older central city places, increasingly they can be found in what is known as *ethnoburbs*, suburban communities that were traditionally dominated by whites (figure 4.17). This ethnic transformation of the suburbs has been dramatic in many places. In fact, while the suburbs in 1990 were over 80 percent white, by 2010, they were only 65 percent so, roughly similar to the white population share of the US.

While many minorities live in ethnic enclaves and ethnoburbs, ethnic groups can also be more spatially dispersed throughout a city. Some ethnic enclaves form as a new group immigrates to a city, but then fade as members blend in with the majority culture through spatial assimilation. *Spatial assimilation* is the process by which an ethnic group blends in with the broader US society, in terms of both cultural behaviors and residential location, as people move out of ethnic enclaves to dispersed locations. For instance, Little Italy in cities such as New York and San Francisco now have few Italian residents. Most Italians from these places assimilated into American culture and moved to scattered locations throughout the city and suburbs, leaving only a few Italian restaurants in the original enclave (figure 4.18). The same has happened with most early-twentieth-century ethnic immigrants, be they Italian, Irish, German, Polish, or others.

Another dispersed ethnic settlement pattern has been described as *heterolocal*. With heterolocal settlement, an ethnic group is spatially separate, living in scattered residential areas, but maintains strong cultural connections nevertheless. This can be done via telecommunications, such as cell phones and social media, as well as community organizations, religious establishments, and ethnic-catering businesses.

Figure 4.17. Ethnoburb: Falls Church, Virginia. The Eden Center strip mall serves a large Vietnamese American population in the suburbs of Washington, DC. Photo by Nicole S Glass. Royalty-free stock photo ID: 1040070511. Shutterstock.com.

Figure 4.18. Little Italy in San Diego. Many ethnic groups that immigrated in the early-twentieth century, such as the Italians, have gone through the process of spatial assimilation, whereby they no longer live in ethnic enclaves. What remains in many of their former enclaves are ethnically oriented businesses, such as restaurants. Photo by Gabriele Maltinti. Royalty-free stock photo ID: 601397378. Shutterstock.com.

Measuring ethnic clusters: The location quotient

Identifying racial or ethnic clusters can be done in different ways. Hot spot analysis, where clusters are identified with a degree of statistical significance, was introduced in chapter 1. Another tool commonly used to identify clusters is the location quotient. Economists frequently use this tool to identify places with higher than average employment in specific industries, such as high tech or mining. But the location quotient is also useful for geographers in identifying places with higher than average concentrations of a racial or ethnic group. The advantage of using the location quotient over a simple choropleth map showing ethnic populations is that it calculates a standardized value that allows for comparison of different racial groups and different time periods.

The location quotient consists of a simple formula:

$$\text{Location quotient for a census tract} = \frac{\begin{array}{c}\text{Proportion of ethnic group}\\\text{in the census tract}\end{array}}{\begin{array}{c}\text{Proportion of the ethnic group}\\\text{in the study area}\end{array}}$$

If the proportion in the census tract is the same as the proportion of the ethnic group in the study area (i.e., the city, county, state, or nation), then the location quotient is 1 (figure 4.19). A value less than 1 means the proportion of the ethnic group in that census tract is lower than in the overall study area. If it is over 1, then the proportion is higher. For instance, if the location quotient for Somalis in a census tract in Minneapolis, Minnesota, is 0.5, then the proportion of Somalis in that tract is one-half that of their overall share in the state. Somalis are underrepresented in that tract. If the location quotient is 2 in a tract, then the proportion of Somalis is twice the overall share of the state. Somalis are overrepresented in that tract.

Measuring segregation: The index of dissimilarity

Given the history of racial and ethnic segregation in the US, it is useful to view their spatial patterns not only in terms of clustering but also in terms of segregation. While the location quotient identifies places with concentrations of specific racial or ethnic groups, other tools compare how integrated or segregated two groups are.

The *index of dissimilarity* is a commonly used tool for calculating segregation that measures how evenly two groups are distributed throughout an area. The index ranges from 0 to 100, where 0 means perfect integration and 100 means total segregation. If black-white

Location Quotient: Somalis in Minneapolis

Minneapolis Somalis cluster in some neighborhoods that are **20 to 34 times** more concentrated than found overall in the state of Minnesota.

The cluster near the center includes the Cedar-Riverside neighborhood, often referred to as "**Little Mogadishu**" after the capital city of Somalia.

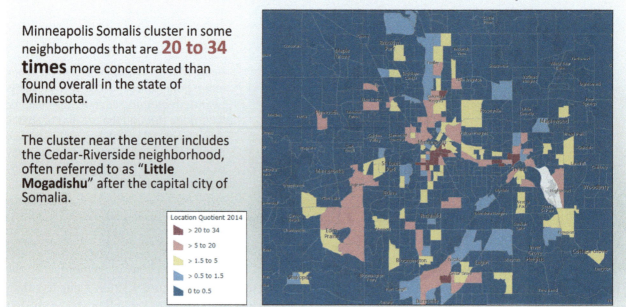

Location Quotient 2014
- > 20 to 34
- > 5 to 20
- > 1.5 to 5
- > 0.5 to 1.5
- 0 to 0.5

Figure 4.19. Somali concentrations that are twenty to thirty-four times that found overall in the state of Minnesota can be seen in red. The cluster near the center includes the Cedar-Riverside neighborhood, often referred to as "Little Mogadishu" after the capital city of Somalia. Data source: US Census.

segregation is being measured at the census tract level, a score of 0 would indicate that blacks and whites are randomly distributed throughout all census tracts and none would have to move in order to achieve perfect integration. A score of 100 would indicate that blacks and whites live in completely different census tracts and 100 percent of one group would have to move in order to achieve integration.

The index of dissimilarity can be used to compare any two racial or ethnic groups, but given the numerical dominance of the white population in the US, whites are typically used as the reference group.

Black-white segregation, as measured by the index of dissimilarity, peaked in US metropolitan areas at 79 during 1960 and 1970. Since then, there has been a steady decline in segregation, reaching 59 by the 2010 census. But while overall segregation has decreased, many cities with large black populations remain highly segregated (figure 4.20). For example, the Detroit, Milwaukee, and New York metropolitan regions, as of 2010, still had dissimilarity scores of over 79, meaning that more than 79 percent of the black population would have to move in order to reach full integration. In contrast, the Las Vegas metropolitan region had a score of less than 36, representing very little segregation.

Hispanic-white segregation is less dramatic than black-white segregation, fluctuating around 50 from 1980 to 2000 and dropping slightly to 48 in 2010. But again, scores vary by metropolitan region. As places with large Hispanic populations, the Los Angeles and New York regions had scores over 63. In these cities, the sheer number of Hispanics means that large segregated communities can form, where the vast majority of residents are of the same ethnic background. Where Hispanic populations are smaller, large homogenous communities are less likely to form. For instance, Seattle and Portland, with fewer Hispanics than cities on the southern border, had scores under 35.

Index of Dissimilarity: Detroit vs. Las Vegas

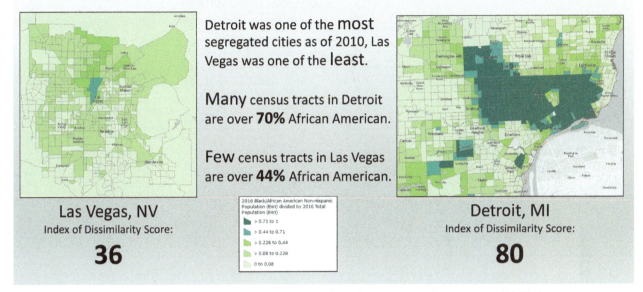

Detroit was one of the **most** segregated cities as of 2010, Las Vegas was one of the **least**.

Many census tracts in Detroit are over **70%** African American.

Few census tracts in Las Vegas are over **44%** African American.

Las Vegas, NV
Index of Dissimilarity Score:
36

2016 Black/African American Non-Hispanic Population (Esri) divided by 2016 Total Population (Esri)

> 0.71 to 1
> 0.44 to 0.71
> 0.228 to 0.44
> 0.08 to 0.228
0 to 0.08

Detroit, MI
Index of Dissimilarity Score:
80

Figure 4.20. Index of dissimilarity: Detroit vs. Las Vegas. Explore these maps at http://arcg.is/2lYGx5x. Data sources: Esri, US Census Bureau.

The lowest levels of segregation are found between Asians and whites. Between 1980 and 2010, scores remained around 41, meaning that only about 41 percent of Asians would have to move in order to achieve integration. At the high end, metropolitan regions such as New York and Houston had 2010 scores of close to 49, while Denver, Phoenix, and Las Vegas had scores at the low end of 30 or less.

Causes of clustering and segregation

To understand the reasons behind racial and ethnic clustering and segregation, it is essential to consider several different factors: racial and ethnic attitudes, financial resources, laws and government policies, and discrimination.

Beginning in the early twentieth century, as blacks began moving from southern states to cities in the North, whites in northern cities began to move to different neighborhoods, away from newly arriving blacks.

Various factors contributed to this white flight, the movement of whites away from racially diversifying neighborhoods. Some white residents moved from pure racial animosity, while others moved due to concerns over changing property values, crime rates, and school quality. As a neighborhood begins to change from white to nonwhite, it can reach a tipping point, at which the pace of white flight increases. In some cases, the white population of a neighborhood will accept a limited number of nonwhite residents, but at a certain point, that number will be perceived as too high, prompting an increase in the rate of whites who move to another place.

To ensure that the neighborhoods to which whites moved would remain racially homogenous, restrictive covenants were used (figure 4.21). These legally enforceable documents restricted who could purchase, rent, or occupy a property based on race, ethnicity, or religion. African Americans, Latinos, Jews, Asians, and other were prevented from living in neighborhoods across the US because of these documents. While racially restricted covenants were ruled illegal by the US Supreme Court in 1948, individuals could still

Restrictive Covenants: Seattle, 1908

Segregation was enforced in thousands of housing developments across the country through restrictive covenants.

Only in 1948 did the US Supreme Court overrule them.

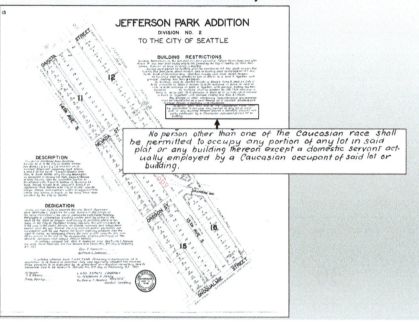

Figure 4.21. Restrictive covenants: Housing development in Seattle. Image source: King County Public Records.

refuse to sell or rent on the basis of discriminatory criteria until the 1968 Fair Housing Act.

While racially restrictive covenants and white flight clearly led to segregation, the actions of real estate agents also contributed via *racial steering* and *blockbusting*. Racial steering is when real estate agents guide clients to neighborhoods of the client's own race or ethnicity, regardless of their income or neighborhood preference. For example, a black homebuyer may be shown homes in black neighborhoods but not homes of equal value in white neighborhoods. Up through the 1940s, racial steering was part of the real estate agent's code of ethics, which stated that an agent "should never be instrumental in introducing into a neighborhood a character of property or occupancy, members of any race or nationality, or any individual whose presence would clearly be detrimental to property values in that neighborhood."

Through blockbusting, real estate agents would play on whites' racial fears by encouraging them to sell in neighborhoods at risk of racial change. In a sense,

this strategy played on the idea of a tipping point. By selling a home on a city block to a nonwhite family, for instance, an agent could then more easily convince white residents that the neighborhood was changing and that it was a good time to sell. This naturally benefitted the agents, who make money from the buying and selling of houses.

Federal Housing Administration policies from 1934 through 1968 also contributed to segregation. During this time, urban neighborhoods were ranked from A through D on the basis of mortgage credit risk (figure 4.22). Neighborhoods with A ratings were considered the lowest risk. These places were defined as "homogenous" and not yet fully built up. In essence, these were the new suburban areas whites where fleeing to. Neighborhoods in the B category were "still desirable." In contrast, neighborhoods ranked C were aging, with expiring racial restrictions and "infiltration of a lower grade population." These places had higher credit risks, and banks were encouraged to limit lending there. Finally, neighborhoods with a D rating were the

Redlining: Richmond, VA, 1923

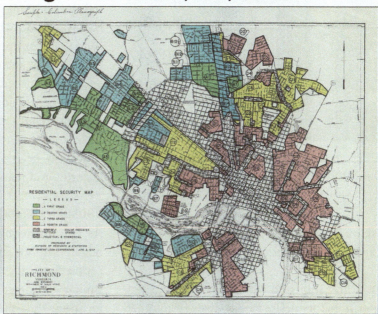

Mortgage risk maps such as this were used to determine where loans should be made and where they should not.

A and B were newer and whiter neighborhoods.

C and D were older, poorer, and more minority neighborhoods.

Figure 4.22. Mortgage risk map for Richmond, Virginia. Beginning in the 1930s, maps such as these were made for many US cities. Older neighborhoods with diversifying populations (yellow and red) received lower rankings than newer, whiter neighborhoods (green and blue). Data source: The U.S. National Archives and Records Administration.

worst for lending. This *redlining*, or the drawing of a red line on a map (sometimes literally and sometimes figuratively) around certain neighborhoods where mortgages were difficult to get, gave whites further reason to sell quickly and move. As loans became more difficult to get in redlined neighborhoods, home prices could only fall.

Changes in housing law now prohibit racially restrictive covenants, racial steering, blockbusting, redlining, and other discriminatory practices, yet the historic footprints of segregated communities still shape the racial and ethnic distributions found in US cities.

Despite changes in housing law, discrimination can still contribute to segregation. Racial minorities attempting to rent sometimes run into landlords who lie about the availability of units or do not inform them of special offers and incentives for moving in. And when

purchasing homes, minorities are rejected for mortgages at higher rates than whites. Nevertheless, housing discrimination is significantly less than in the past.

While discrimination is less prevalent now than in the past, segregation and clustering can still occur for other reasons. Income is a powerful factor in sorting people by residential location, and in the US average incomes vary by race and ethnicity. Table 4.2 shows the 2014 median household income of the four major racial and ethnic groups. While incomes can vary at the individual level—there are low-income Asians and high-income blacks—on average, groups will be sorted along socioeconomic lines. Based on average incomes, Asians will be disproportionately in more expensive neighborhoods, while blacks will be in less expensive neighborhoods. Whites, based on average income, will be in places closer to Asians, while Hispanics will be in places closer to blacks. Even if housing discrimination is

Median Household Income, 2014	
Asian	73,244
White	58,847
Hispanic	42,396
Black	35,600

Table 4.2. Ethnic and racial groups in the US are segregated partially due to differences in median household income. Data source: US Census.

eliminated, we will still find segregation and clustering until income differences between groups is eliminated as well.

Lastly, segregation and clustering can occur not from a discriminatory desire to stay away from other groups but from a desire to be with one's own group. This can be especially true for new immigrants, who look for places where residents share a common language, where local markets stock foods from the homeland, and where connections for jobs and housing can be more easily obtained. As shown above, Asians have high incomes on average. This should allow them to live wherever they want, yet large Asian communities can be found in cities throughout the United States. Research shows that groups other than immigrants also have preferences for neighborhoods of their own race. Whites tend to prefer neighborhoods with more white residents, while blacks and Hispanics also prefer neighborhoods with a substantial proportion of their own group. With that said, it is difficult to disentangle how much residential choice is due to personal preference and how much is due to a sensation of discrimination, or "otherness" when living as a minority in a community.

Go to ArcGIS Online to complete exercise 4.1: "Measuring spatial distribution with population ratios," and exercise 4.2: "Measuring spatial distribution with the location quotient."

Spatial relationships

Geographers are not interested in mapping the spatial distributions of racial and ethnic neighborhoods merely to know where they are. Rather, mapping often helps illustrate important spatial relationships between ethnic groups and quality of life issues. Targeted public and private programs aimed at improving quality of life can be developed by understanding how race or ethnicity relates spatially to a wide range of variables, such as consumer needs, income and housing, levels of education, health, crime, exposure to environmental hazards, and many more.

Consumer markets

As you intuitively know from observing different neighborhoods in different places, similar types of people cluster together. These similarities translate into similar consumer behaviors that vary by geographic location; thus there is a strong spatial relationship between where clusters of similar types of people live and the types of products and services they consume. By understanding this spatial relationship, companies, government agencies, and nonprofit organizations can target their products and services by geographic location, modifying their mix of goods and services and marketing strategies so that they appeal to the people who live in each place.

There are numerous examples of how race and ethnicity relate to consumer markets. When an ethnic group includes a large proportion of recent immigrants, they may prefer smaller local businesses that cater specifically to their tastes in food and clothing over national chains. These behaviors offer opportunities for small entrepreneurs and challenges to mainstream stores, which may have to change their merchandising and marketing strategies to attract customers. Governmental and nongovernmental organizations can also use information on consumer behavior to guide health programs. Consumption of junk food, cigarettes, alcohol, and other unhealthy products can vary by race and ethnicity, and by understanding these differences, health program strategies can be targeted to different groups.

Consumer clusters are identified through market segmentation analysis. This analysis uses a wide range of demographic and lifestyle data to identify neighborhoods with similar characteristics, with the goal of identifying unique consumer market segments. The ArcGIS platform includes Tapestry Segmentation data, which divides the US population into fourteen broad Life Mode segments that are further divided into sixty-seven more detailed segments. Of these, two Life Modes and eleven detailed segments focus on ethnically diverse neighborhood types (figure 4.23).

When mapped, Tapestry segments illustrate the complex geographic patterns of demography and lifestyle that spatially correlate with unique consumption patterns. Figure 4.24 shows the Tapestry Segmentation for Denver, Colorado. A business looking to open in the ethnically diverse census tracts outlined in black, referred to as the Ethnic Enclaves and Next Wave Life Modes, will have to target its marketing campaign and

products to an audience that is significantly different from that in other areas of the city, such as the Affluent Estates or Senior Styles. For instance, residents of the Las Casas segment of the Next Wave Life Mode consume more baby products and children's apparel and rent more than own homes. They also have lower incomes than the US median. Based on this type of information, a second-hand children's clothing store may make sense, while a do-it-yourself home-repair construction supply store may not.

Go to ArcGIS Online to complete exercise 4.3:
"Spatial relationships: Tapestry segmentation,
race, and ethnicity."

Income and housing

Aggregate differences in median income by race and ethnicity (as seen in table 4.2) play out spatially in relation to housing. A simple visual analysis of figures 4.20

Market Segmentation Analysis

LifeMode 7 Ethnic Enclaves	
• Established diversity—young, Hispanic homeowners with families	7A Up and Coming Families
• Multilingual and multigenerational households feature children that represent second-, third- or fourth-generation Hispanic families	7B Urban Villages
• Neighborhoods feature single-family, owner-occupied homes built at city's edge, primarily built after 1980	7C American Dreamers
• Hard-working and optimistic, most residents aged 25 years or older have a high school diploma or some college education	7D Barrios Urbanos
• Shopping and leisure also focus on their children—baby and children's products from shoes to toys and games and trips to theme parks, water parks or the zoo	7E Valley Growers
• Residents favor Hispanic programs on radio or television; children enjoy playing video games on personal computers, handheld or console devices	7F Southwestern Families
• Many households have dogs for domestic pets	

LifeMode 13 Next Wave	
• Urban denizens, young, diverse, hard-working families	13A International Marketplace
• Extremely diverse with a Hispanic majority, the highest among LifeMode groups	
• A large share are foreign born and speak only their native language	13B Las Casas
• Young, or multigenerational, families with children are typical	
• Most are renters in older multi-unit structures, built in the 1960s or earlier	13C NeWest Residents
• Hard-working with long commutes to jobs, often utilizing public transit to commute to work	
• Spending reflects the youth of these consumers, focus on children (top market for children's apparel) and personal appearance	13D Fresh Ambitions
• Also a top market for movie goers (second only to college students) and fast food	
• Partial to soccer and basketball	13E High Rise Renters

Two tapestry segments focus on ethnically diverse neighborhoods.

The mix of goods and services offered in each tapestry segment will differ based on each segment's demographic and lifestyle characteristics.

Figure 4.23. Market segmentation analysis. Data source: Esri.

Tapestry Segmentation: Denver, CO

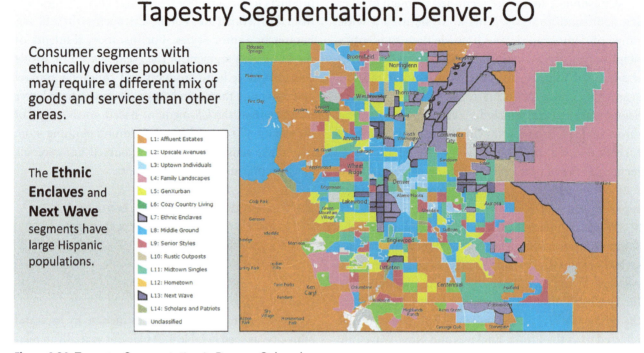

Consumer segments with ethnically diverse populations may require a different mix of goods and services than other areas.

The **Ethnic Enclaves** and **Next Wave** segments have large Hispanic populations.

L1: Affluent Estates
L2: Upscale Avenues
L3: Uptown Individuals
L4: Family Landscapes
L5: GenXurban
L6: Cozy Country Living
L7: Ethnic Enclaves
L8: Middle Ground
L9: Senior Styles
L10: Rustic Outposts
L11: Midtown Singles
L12: Hometown
L13: Next Wave
L14: Scholars and Patriots
Unclassified

Figure 4.24. Tapestry Segmentation in Denver, Colorado. Data source: Esri.

and 4.25 shows a strong spatial relationship between race, income, and the value of homes. Blacks in Detroit are concentrated in neighborhoods with low incomes and, as a result, low home values. Many of these neighborhoods contain overcrowding, substandard housing, and deferred maintenance of homes. By understanding the relationship between spatial concentrations of race, income, and housing, policymakers can begin to explore solutions to Detroit's persistent segregation and urban decay.

For instance, while discrimination, especially in the past, may be partially responsible for these patterns of race, income, and housing, there is evidence that places themselves play a strong role in perpetuating them. Children who spend their whole lives in low-income neighborhoods earn less as adults than those who grow up in higher-income neighborhoods. Better schools, less exposure to crime, and more stable two-parent households provide advantages in higher-income places that translate into higher incomes. For this reason, a government program was established

to allow some residents of poor neighborhoods to use rental-assistance vouchers to move to neighborhoods with low rates of poverty. The income, education level, and family structure of the recipients did not change; the only change was the income level of the neighborhood they moved to. The result was that, as adults, the children who moved to the new neighborhoods earned significantly more than the children who remained in low-income, segregated places. The difference in income was greatest for children who moved at a younger age, meaning the more time spent in a higher-income neighborhood, the greater the increase in earnings as an adult. The difference was much larger for boys than for girls, showing that boys are more negatively affected by places of high-concentration poverty and segregation.

That places matter in the life outcomes of people leads to potential policy solutions. It may make sense to stop providing low-income housing in segregated low-income communities. Instead, housing vouchers and subsidized units could be placed in neighborhoods

Income and Housing Value: Detroit, MI

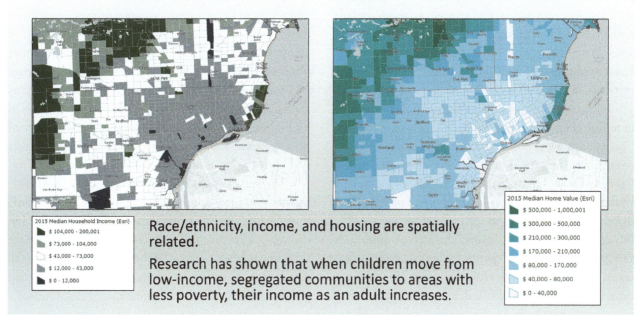

2015 Median Household Income (Esri)
- $ 104,000 - 200,001
- $ 73,000 - 104,000
- $ 43,000 - 73,000
- $ 12,000 - 43,000
- $ 0 - 12,000

Race/ethnicity, income, and housing are spatially related.

Research has shown that when children move from low-income, segregated communities to areas with less poverty, their income as an adult increases.

2015 Median Home Value (Esri)
- $ 500,000 - 1,000,001
- $ 300,000 - 500,000
- $ 210,000 - 300,000
- $ 170,000 - 210,000
- $ 80,000 - 170,000
- $ 40,000 - 80,000
- $ 0 - 40,000

Figure 4.25. Median household income (left) and median home value (right) in Detroit, 2015. Explore these maps at http://arcg.is/2lYXSLv. Data sources: Esri, US Census Bureau.

with low levels of poverty—although this can be very difficult from a political standpoint. But, of course, it is unrealistic to move everyone out of poor segregated neighborhoods. For this reason, a holistic approach may be needed to end multigenerational cycles of poverty in these places. Attempts to provide jobs alone, or to just improve schools, may be insufficient when children are enmeshed in neighborhoods with myriad social and economic problems.

Understanding the spatial relationship between racial and ethnic clusters, poverty, and housing can lead to other group-specific solutions as well. While Asians, on average, have high incomes, there is variation by nationality. The Chinese and Indians tend to have higher levels of education and income. Therefore, spatial clusters of these groups tend to be in higher-value neighborhoods with fewer social problems. However, many Southeast Asians, such as the Vietnamese, Cambodians, and Laotians, have lower incomes and often cluster in lower-income places with high crime

and social ills. Housing vouchers to help poor Southeast Asians move to new neighborhoods may not work when English language skills are limited. Instead, a first step may require more intense English language instruction in local schools. Furthermore, war-related trauma suffered by Southeast Asian immigrants may have an impact on their US born children, which would require specialized resources in local neighborhoods. In either case, the solutions aimed at helping improve the lives of Southeast Asians would likely be different than for blacks in Detroit.

Go to ArcGIS Online to complete exercise 4.4: "Spatial relationships: Income, race, and ethnicity."

Crime

As another example, the spatial relationship between racial and ethnic clusters and crime can help illuminate contemporary concerns over police-minority friction.

Race and ethnicity are not causes of crime, but they have a strong spatial relationship with factors that do cause crime, such as lack of opportunity for people with low incomes and lack of education, single-parent households unable to supervise youth, and other variables. For instance, figure 4.26 shows that overall crime has a strong spatial relationship with the low-income minority communities of Detroit.

High crime rates lead to a heavier police presence in minority communities. In many cities, a disproportionate number of police officers are of a different race or ethnicity than the communities in which they serve. Coming from different neighborhoods and different backgrounds can cause friction and real or perceived discrimination against minority communities.

At least one reason behind the lack of minorities serving as police in their own communities is the higher arrest and conviction rates for minorities, especially for drug use or possession. While drug use among minority communities is not higher than for white communities, arrest and conviction rates are. This can be for a number of reasons. For instance, a lack of private space in homes and yards in poor, minority communities can lead to more illegal drug use in public spaces where risk of arrest is higher. Residents of more affluent communities often have more private space in which to use drugs and therefore may be less likely to get arrested. This can be compounded by higher levels of police patrols in high-crime, low-income minority communities. Racial bias by police and courts can also lead to more arrests and convictions of minorities for drug use.

The result is that more young adults from low-income minority communities have criminal records and do not qualify to become police officers. Even though whites in more affluent areas may also have used drugs, they are less likely to be caught and punished. Thus, many cities struggle to build trust among police and minority communities.

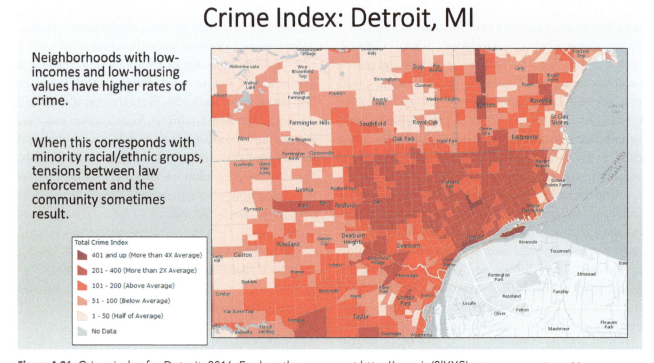

Figure 4.26. Crime index for Detroit, 2016. Explore these maps at http://arcg.is/2lYXSLv. Data sources: Esri, AGS.

Environmental quality

Racial and ethnic clusters also have spatial relationships with health problems and environmental hazards. All three of these come together in the concept of environmental justice. Often, when a polluting factory, toxic landfill, smelly wastewater treatment plant, or waste incinerator is to be built, the site is near low-income, and thus frequently minority, neighborhoods. Rarely are sites near higher-income, and thus disproportionately white, neighborhoods. Because of the disproportionate impact of these types of facilities on low-income minority communities, the US government, in 1994, established environmental justice policy for the "fair treatment and meaningful involvement of all people regardless of race, color, national origin, or income with respect to the development, implementation and enforcement of environmental laws, regulations and policies."

Remaining with the example of Detroit, there are three local examples of environmental justice. One is a waste incinerator located in a predominately black and poor neighborhood. Reports of foul odors and high rates of asthma in the surrounding area have been attributed to the incinerator. The second is a riverfront park, also in a lower-income minority community, closed to the public because of contaminated soils. The third involves a city policy to cut off water to homes of residents who do not pay their water bill. Some consider clean water to be a human right that cannot be denied by government. In each of these cases, environmental threats overlay with the location of low-income, minority communities.

Go to ArcGIS Online to complete exercise 4.5: "Environmental justice."

Race, ethnicity, and cultural landscapes

Cultural landscapes are the material imprint that people make on places. Different cultures create places with distinct sights, sounds, and smells. Traveling place to place, one can experience sensory diversity in types of architecture, businesses and products, languages on signs, soundscapes of traffic, smells of foods, and much more. These unique cultural characteristics reflected in the landscape reveal who settled a place and who lives there now.

While geographers can often identify racial and ethnic clusters with census data, some communities can be found only by exploring the cultural landscape. For instance, Little Italys in many cities now have few Italian residents, so locating them with census data on Italian heritage would be impossible. Yet Little Italys can still be found by identifying cultural landscapes where there are a disproportionate number of Italian restaurants or stores with goods from Italy. Little India and Little Tokyo in the Los Angeles region reflect this problem (figure 4.27). Both populations are heterolocal, with no large cluster in the region, yet Little India and Little Tokyo are thriving districts filled with restaurants, clothing stores, and mini-markets that cater to the ethnic population. These important centers of Indian and Japanese culture in Southern California would not be identified by relying on census data. Only fieldwork examining the cultural landscape can lead to an understanding of this ethnic place.

The concept of sense of place, or the emotional attachment of people with places, was introduced in chapter 1. As people form racial and ethnic clusters, they transform physical and human landscapes to reflect their own cultural tastes and preferences. Through this process, regions and communities can form a unique sense of place tied to their racial or ethnic identity.

As an example, at a broad regional scale, we have seen that Hispanics have historically concentrated in the southwestern United States. Given that much of this area was originally settled by the Spanish, many landscape elements reflect their influence. Cities founded by the Spanish all evolved around a central plaza faced by a church and government building. Cities such as Los Angeles, California; Santa Fe, New Mexico; and San Antonio, Texas, all grew from central plazas such as these, which are now tourist points of interest. Spanish land-use patterns can also be seen in the landscape.

Figure 4.27. Little Tokyo, Los Angeles, California. This heterolocal community comes from around the Los Angeles region for Japanese-themed shops and food. The Japanese cultural landscape is evident in signs, products, and people. Photo by Kit Leong. Royalty-free stock photo ID: 704666428. Shutterstock.com.

Large ranchos from well over a century ago influence city boundaries and street orientation, some of which follow rancho property lines. Architectural styles can also reflect Spanish influence, with terracotta tile roofs, arches, and stucco walls of homes and public buildings. And, of course, Spanish place-names, also known as *toponyms*, are reflected in signage for cities (Los Angeles), streets (Santa Monica Boulevard), rivers (Rio Grande), mountains (Sierra Nevada), and myriad other places in the Southwest.

As Spanish Alta California became part of the United States in the nineteenth century, Anglo settlement quickly superseded the Latino population. Nevertheless, themes of the original Hispanic cultural landscape remained, especially in the Spanish colonial architectural style and promotion of the region with images of Spanish missions surrounded by lush agricultural land.

Outside of architecture and some promotional imagery, the Southwest gradually came to resemble an Anglo cultural landscape. English language toponyms for cities, streets, and other places came to dominate; businesses catered to the consumer tastes of the Anglo population in clothing, food, and other goods; and commercial architecture followed trends out of New York and Chicago.

The cultural landscape of downtown Los Angeles offers a clear example of how race and ethnicity modify places to reflect group identities. As Los Angeles grew rapidly in the early twentieth century, Broadway became the primary street for theaters and shopping. The street bustled with pedestrian traffic during the day and night as a central point of activity for the city. Streetcars linked Broadway with expanding areas of residential growth, allowing for easy access to the city center. But by the end of the 1940s, the city was changing. Suburban growth accelerated in the 1950s and beyond. The predominately Anglo population moved farther from downtown, and the streetcars were removed throughout the region and replaced by freeways. Urban renewal downtown cleared many residential neighborhoods and replaced them with office buildings occupied by workers who commuted daily from the suburbs. Downtown businesses and theaters relocated to new suburban areas as well, vacating streets such as Broadway.

But as the Anglo population vacated Broadway, falling commercial and residential rents became attractive to nearby Latino residents of East Los Angeles, who were prevented from living in many other parts of the city due to restrictive covenants, and attractive as well to new immigrants. As the Latino population steadily grew, Broadway was transformed to a distinctly Latino cultural landscape. Signage came to be nearly exclusively in Spanish; numerous *quinceañera* and bridal shops opened; and street vendors selling tacos, cut fruit with *chile*, and other Mexican foods set up on sidewalks. Music coming from storefronts reflected the latest Ranchero stars. Non-Hispanic business owners, many of whom were immigrants from other parts of the world, learned Spanish in order to sell to their Latino clientele (figure 4.28).

In recent years, Broadway, and downtown Los Angeles in general, have been going through yet another sociodemographic change, with resulting alterations to the cultural landscape. As American central cities have become safer in the past decade, and as some people prefer pedestrian-oriented neighborhoods over freeway-centric development, downtowns have been making a comeback, attracting new residents with higher incomes. New apartments and condominiums are being occupied by young professionals, and businesses are returning to cater to their tastes. The cultural landscape is now reflecting a younger, hipper demographic. Latino businesses on Broadway are being replaced by upscale cafés and bars, higher-end clothing stores, and boutique hotels (figure 4.29). The nearby Grand Central Market, which once offered low-cost meats and produce, now contains dozens of restaurants, including an artisanal wine bar, organic juices, and places with trendy names such as Eggslut.

Los Angeles remains a majority Latino city, so the cultural landscape will reflect that culture even if a higher-income population transforms downtown. For instance, elements of the Latino cultural landscape are being integrated into urban planning regulations for the area, including the legalization of street vending and decorative wall murals. And, of course, food from Latin America remains a central part of the Angelino's diet. Despite changes to the Grand Central Market, there are still several establishments that serve Mexican food, a Salvadoran *Pupusería*, and a Latino dry goods stand offering *chile*s and *mole*s.

Go to ArcGIS Online to complete exercise 4.6: "The cultural landscape."

Racial and ethnic conflict

Unfortunately, a discussion of race and ethnicity cannot omit the concepts of racism and xenophobia. The two concepts are similar in that they both represent antipathy or discrimination against another group. In the

Figure 4.28. Downtown Los Angeles shifted from having an Anglo cultural landscape with many theaters to a largely Hispanic cultural landscape, with jewelry stores and other businesses exhibiting Spanish language signage. Photo by Hayk Shalunts. Royalty-free stock photo ID: 377672158. Shutterstock.com.

Figure 4.29. More recently, trendy clothing stores and higher-end apartments in downtown Los Angeles are pushing out much of the Hispanic cultural landscape.
Photo by Hayk Shalunts. Royalty-free stock photo ID: 377682217. Shutterstock.com.

case of racism, this discrimination is aimed at people of another race, while xenophobia is aimed at people of a different ethnic or cultural group.

Slavery was clearly based on the racist idea that whites were superior to blacks and thus had the right to force them into bondage. This ideology continued even after slavery was abolished, resulting in the patterns of segregation discussed earlier in this chapter.

When taken to an extreme, racism and xenophobia can lead to genocide or ethnic cleansing. Genocide, as defined by the United Nations, is the intent to destroy a group of people based on nationality, ethnicity, race, or religion. The United Nations includes ethnic cleansing as a type of genocidal act, whereby a group is forced to leave a specific territory in order to make it ethnically homogenous.

Within the United States, the Native American population faced genocide at the hands of European settlers. With the diffusion westward of European settlers in the United States in the eighteenth and nineteenth centuries, they came across land already occupied by various Native American tribes. From around the 1770s through 1815, hundreds of Indian towns in the eastern and southern portions of the United States were burned and the residents killed

or forced to flee. A clear process of genocide through ethnic cleansing officially began in 1830 with the Indian Removal Act, which authorized the removal of all Indians east of the Mississippi River for relocation to Indian Territory in Oklahoma and Kansas. As a result, roughly 30 percent of the population of affected tribes died from disease, starvation, and exposure to the elements. The US military used force to ensure compliance and killed many who resisted.

As settlers moved farther west, the genocide continued. In some cases, armed groups of settlers attacked and killed Indians, while in other cases the US Army did the killing as part of the Indian Wars. Hunting of buffalo and other Indian game further reduced Indian numbers through starvation and ensuing disease. From the time of initial European contact with the Native Americans of North America to the twentieth century, ethnic cleansing had removed Indians from nearly all of their territory, and it is estimated that their population declined by 70 to 90 percent.

Later in the twentieth century, there were numerous cases of genocide and ethnic cleansing. The term *genocide* was coined in 1944 to describe Nazi crimes against Jews in Europe and was adopted by the United Nations in 1948. Later, in the 1990s, ethnic cleansing came into wide usage to describe the ethnic violence that tore apart the former country of Yugoslavia in Eastern Europe. During this conflict, ethnic Serb forces killed roughly 100,000 Bosnian and Croatian residents in an attempt to clear the Yugoslav republic of Bosnia-Herzegovina of these groups (figure 4.30). Also in the 1990s, the country of Rwanda in sub-Saharan Africa was torn apart by ethnic conflict. In this case, the majority Hutus killed approximately 800,000 people from the minority Tutsi ethnic group during a three-month period.

Figure 4.30. Genocide in Bosnia. United Nations forensic experts unearth victims from a mass grave in the city of Srebrenica. Photo by Northfoto. Shutterstock. Stock photo ID: 90838403.

A somewhat lesser known case of ethnic cleansing, but one that is important in understanding ongoing conflict in Iraq, was the Arabization policy in northern Iraq. During the 1970s and 1980s, the Iraqi government forcibly removed ethnic Kurds and Turkmen from oil-rich lands in northern Iraq in order to consolidate control of the region. Hundreds of thousands were killed or forced by the Iraqi military to flee, their villages were bulldozed, and property title was turned over to new Arab immigrants from other parts of Iraq. Images of chemical weapons used in this campaign against Kurdish villagers in 1988 were employed by the US government to justify military intervention on human rights grounds in

1990. Ongoing disagreements over the borders of the Iraqi Kurdistan autonomous region, blurred and confused by the Arabization policy, still taint relations between Kurds and the central government of Iraq. Many Kurds wish to secede from Iraq and form an independent Kurdistan, leading to the ongoing question of whether Iraq will continue to exist in its current form.

References

Adamy, J., and P. Overberg. 2016. "Places Most Unsettled by Rapid Demographic Change Are Drawn to Donald Trump." *Wall Street Journal*, Novermber 1, 2016. http://www.wsj.com/articles/places-most-unsettled-by-rapid-demographic-change-go-for-donald-trump-1478010940.

Alba, R. D., J. R. Logan, B. J. Stults, G. Marzan, and W. Zhang. 1999. "Immigrant Groups in the Suburbs: A Reexamination of Suburbanization and Spatial Assimilation." *American Sociological Review* 64, no. 3: 446. doi: https://doi.org/10.1177/0002716215594611

Allen, J., and Eugene Turner. 2002. *Changing Faces, Changing Places. Mapping Southern California*. Northridge, CA: The Center for Geographical Studies.

Blake, J. 2010. "Arab- and Persian-American Campaign: 'Check It Right' on Census." *CNN*, May 14, 2010. http://www.cnn.com/2010/US/04/01/census.check.it.right.campaign.

Chetty, R., and N. Hendren. 2017. "The Impacts of Neighborhoods on Intergenerational Mobility I: Childhood Exposure Effects." *EconPapers*. http://econpapers.repec.org/RePEc:nbr:nberwo:23001.

Constable, P. 2012. "Alabama Law Drives Out Illegal Immigrants but Also Has Unexpected Consequences." *Washington Post*, June 17, 2012. https://www.washingtonpost.com/local/alabama-law-drives-out-illegal-immigrants-but-also-has-unexpected-consequences/2012/06/17/

gJQA3Rm0jV_story.html?utm_term=
.4794790f6f58.

Frey, W. H. 2004. "The New Great Migration: Black Americans' Return to the South, 1965–2000." Brookings, May 1, 2004. https://www.brookings .edu/research/the-new-great-migration-black-americans-return-to-the-south-1965-2000.

Garner, S. 2014. "How the Evolution of L.A.'s Broadway Traces the Life of the City." *Curbed*, November 14, 2014. https:// www.curbed.com/2014/11/14/10023252/ los-angeles-broadway-history.

Gebeloff, J. E. R. 2016. "Affluent and Black, and Still Trapped by Segregation." *New York Times,* August 21, 2016. https://www.nytimes .com/2016/08/21/us/milwaukee-segregation-wealthy-black-families.html?_r=0.

Hardwick, S. W., and J. E. Meacham. 2005. "Heterolocalism, Networks of Ethnicity, and Refugee Communities in the Pacific Northwest: The Portland Story." *The Professional Geographer* 57, no. 4: 539–557. doi: https:// doi.org/10.1111/j.1467-9272.2005.00498.x.

Havekes, E., Bader, M., and Krysan, M. 2016. "Realizing Racial and Ethnic Neighborhood Preferences? Exploring the Mismatches Between What People Want, Where They Search, and Where They Live." *Population Research and Policy Review*, 35: 101–126. doi: http://doi .org/10.1007/s11113-015-9369-6.

Hawthorne, C. 2014. "'Latino Urbanism' Influences a Los Angeles in Flux." *Los Angeles Times*, December 6, 2014. http://www.latimes.com/ entertainment/arts/la-et-cm-latino-immigration-architecture-20141206-story.html.

Human Rights Watch. 2004. "III. Background: Forced Displacement and Arabization of Northern Iraq." In *Claims in Conflict: Reversing Ethnic Cleansing in Northern Iraq.* https:// www.hrw.org/reports/2004/iraq0804/4. htm#_Toc78803800.

James, Michael. 2017, "Race." *The Stanford Encyclopedia of Philosophy* (Spring 2017 edition), Edward N. Zalta (ed.). https://plato. stanford.edu/archives/spr2017/entries/race.

Logan, J., and B. Stults. 2011. "The Persistence of Segregation in the Metropolis: New Findings from the 2010 Census." Census brief prepared for Project US2010.

Madrigal, A. C. 2014. "The Racist Housing Policy That Made Your Neighborhood." *The Atlantic,* May 22, 2014. https://www.theatlantic.com/ business/archive/2014/05/the-racist-housing-policy-that-made-your-neighborhood/371439.

Mejia, B. 2016. "Fiesta Broadway Lives on as the Street Slowly Loses its Latino Heart." *Los Angeles Times*, April 24, 2016. http:// www.latimes.com/local/california/la-me-adv-broadway-latinos-20160424-story.html.

Navarro, M. 2012. "For Many Latinos, Racial Identity Is More Culture Than Color." *New York Times,* January 14, 2012. http://www.nytimes .com/2012/01/14/us/for-many-latinos-race-is-more-culture-than-color.html.

Neil C. Bruce. 2013. *Real Estate Steering and the Fair Housing Act of 1968*, 12 Tulsa L. J. 758. http://digitalcommons.law.utulsa.edu/tlr/vol12/ iss4/8.

Ojmarrh Mitchell, and Michael S. Caudy. 2015. "Examining Racial Disparities in Drug Arrests." *Justice Quarterly* 32, no. 2. doi: http://dx.doi.org /10.1080/07418825.2012.761721.

The Pew Charitable Trusts. 2014. "Changing Patterns in U.S. Immigration and Population. Immigrants Slow Population Decline in Many Counties." Pew. http://www.pewtrusts.org/ en/research-and-analysis/issue-briefs/2014/12/ changing-patterns-in-us-immigration-and-population.

Piccorossi, M. 2015. "What Census Calls Us: A Historical Timeline." Pew Research Center's Social & Demographic Trends Project. http:// www.pewsocialtrends.org/interactives/ multiracial-timeline.

Public Broadcasting Service (PBS). 2016. "South Dakota Town Embraces New Immigrants Vital to Meat Industry." PBS, July 2, 2016. http://www.pbs.org/newshour/bb/south-dakota-town-embraces-new-immigrants-vital-to-meat-industry.

Shertzer, A., and R. P. Walsh. 2015. "Racial Sorting and the Emergence of Segregation in American Cities." National Bureau of Economic Research. http://www.nber.org/papers/w22077.

Sisson, C. K. 2014. "Why African-Americans are Moving Back to the South." *Christian Science Monitor,* March 16, 2014. https://www.csmonitor.com/USA/Society/2014/0316/Why-African-Americans-are-moving-back-to-the-South.

Stille, A. 2014. "Can the French Talk About Race?" *The New Yorker,* July 1, 2014. https://www.newyorker.com/news/news-desk/can-the-french-talk-about-race.

Suh, M. 2015. "Total Fertility Rate for Population Estimates and Projections, by Race-Hispanic Origin and Generation: 1965–1970, 2015–2020 and 2060–2065." Pew Research Center's Hispanic Trends Project. http://www.pewhispanic.org/2015/09/28/modern-immigration-wave-brings-59-million-to-u-s-driving-population-growth-and-change-through-2065/9-26-2015-1-30-23-pm-2.

US Census Bureau. 2002. "Demographic Trends in the 20th Century." *Census 2000 Special Reports.* https://www.census.gov/prod/2002pubs/censr-4.pdf.

———. "Racial and Ethnic Residential Segregation in the United States: 1980–2000." *Census 2000 Special Reports.* https://www.census.gov/hhes/www/housing/housing_patterns/pdftoc.html.

———. 2011. "Overview of Race and Hispanic Origin: 2010." *2010 Census Briefs.* https://www.census.gov/prod/cen2010/briefs/c2010br-02.pdf.

———. "The Hispanic Population: 2010." *2010 Census Briefs.* https://www.census.gov/prod/cen2010/briefs/c2010br-04.pdf.

———. "The Black Population: 2010." *2010 Census Briefs.* https://www.census.gov/prod/cen2010/briefs/c2010br-06.pdf.

US Census Bureau Public Information Office. 2016. "2010 Census Shows America's Diversity—2010 Census." US Census Bureau Newsroom Archive. https://www.census.gov/newsroom/releases/archives/2010_census/cb11-cn125.html.

US Department of Agriculture Economic Research Service. 2013. *Atlas of Rural and Small-Town America.* Washington, DC: USDA ERS. https://www.ers.usda.gov/data-products/atlas-of-rural-and-small-town-america.

Chapter 5
Urban geography

Sometime around 2008, humans became an urban species. For millennia, humans lived in small groups of hunters and gatherers, but with the emergence of agriculture, people began to settle permanently in small villages. Over time, those villages grew in size and population, becoming what we now call cities. By roughly the year 2008, this process led to a majority of humanity living in urban areas. Each year, even larger proportions of the world's population is urban, with 54 percent living in cities and towns as of 2015 (figure 5.1).

Given that we are now an urban species, geographers are interested in understanding the forms and functions of our cities. This is important in that most of us spend most of our time in urban areas. They are where we live our lives, facing issues such as congested traffic, housing affordability, access to open space, availability of goods and services, employment opportunities, and much more.

The biggest force driving the growth of cities is that they are economic powerhouses. People living and working in close proximity drives innovation and efficiency through the intermingling of new ideas and the sharing of urban resources. The wealth created in cities pulls in migrants from rural areas, who see greater opportunities for education and employment in bustling urban areas.

When cities function well, they can be amazing places to live. Economic opportunity can provide financial security for families, while cultural amenities such as parks, theaters, restaurants, museums, and myriad goods and services can make them exciting and satisfying places to be. But when cities do not function well, they can appear as dystopian worlds. Vast swaths of the urban landscape can consist of cramped slum housing with open sewage and garbage. Crime and corrupt city officials can plague neighborhoods and inhibit economic opportunity. The rich must live in isolated and fortified housing, with private security guards and economically segregated shopping malls. Obviously, geographers and others study cities in order to make them work well for all residents, so that economic opportunity and cultural amenities can be enjoyed by everyone without fear.

Figure 5.1. Dubai. United Arab Emirates. Most people now live in urban areas. Population growth, economic activity, and technology feed urban growth, even in the hot, arid deserts of the Arabian Peninsula. Photo by Anna Om. Shutterstock. Stock photo ID: 240176458.

Measuring "urban"

When measuring and mapping urban areas, it is important to first understand what is meant by the word *urban*. As it turns out, there is no fixed definition. Rather, each country, typically via national statistics agencies, establishes its own criteria. These criteria vary and can include population size, population or building density, the proportion of people who work in agriculture, the presence of infrastructure, or any combination of these factors.

For instance, Lithuania, in the Baltics region of Europe, defines urban areas as having closely built permanent dwellings, a population of 3,000 or more, and two-thirds of the workforce in industry and business. Albania, in the Balkans region of Europe, simply defines it as a town or center of more than 400 residents. Peru identifies urban areas as populated centers with 100 or more dwellings, while Egypt simply lists major cities and district capitals. In the United States, urban areas are defined at having 2,500 residents, generally with a density of 1,000 people per square mile or more.

So, while we are an urban species, that does not imply that the majority of people live in large urban areas. Rather, it reflects the fact that human settlements no longer consist of isolated rural homes spread around agricultural regions.

Urbanization refers to the process of urban growth and measures the increase (or decrease) in urban populations. It can be measured in terms of counts and rates, so that for example, between 2014 and 2015, the urban population of the United States grew by 2,591,887 people, resulting in an urban growth rate of about 1 percent. This rate of growth pales in comparison to many developing countries, such as in sub-Saharan Africa, where urban growth rates often exceed 5 percent, or five times that of US cities.

Origin and growth of the city

For most of human history, people lived in small groups of hunters and gatherers. Then, about 10,000 years ago, humans began to settle in fixed places and farm the same land rather than move from place to place in search of wild plants and animals. This was the first step in the development of urban settlements. As agricultural techniques improved, people began to produce surplus food—that above what was needed for themselves and their families. With agricultural surpluses, some members of the community could specialize in activities other than farming. People could take up positions as traders who exchanged surpluses for different foods and items from other farming groups. Others could function as accountants to monitor the types and quantities of traded goods. Still others could take on full-time positions as political leaders of the group, while others could become spiritual leaders. Occupational specialization also included metal and glass workers, soldiers, carpenters, and myriad other activities. These nonfarming functions worked best when people lived close to each other rather than in scattered farmhouses. Thus, human settlements spatially consolidated in denser patterns and evolved into cities.

The earliest cities developed in the Mesopotamia region of modern-day Iraq sometime after 4500 BCE. Cities formed in Egypt around 3000 BCE, and several hundred years later there were cities in the Indus valley of modern-day Pakistan (figures 5.2 and 5.3). Cities first formed in China around 1700 BCE. Later, in Mesoamerica, the Zapotec and Mixtec cultures began building cities around 800 to 700 BCE.

But while cities were evolving in many parts of the world, the proportion of people living in them remained small until very recently. Estimates vary, but up until the 1800s, only about 5 percent or less of the world population lived in urban areas. Until recently, there were few large cities. In 1360 BCE, only Thebes, Egypt, had a population of 100,000 people. By 100 CE, Rome had over 500,000 residents, but no city reached one million until Beijing, China, around 1800.

During the 1800s, cities began to grow rapidly in connection with the industrial revolution. During this time, mechanization in agriculture reduced the need for manual labor in the field, while mass-production industry came to rely on large numbers of urban-industrial

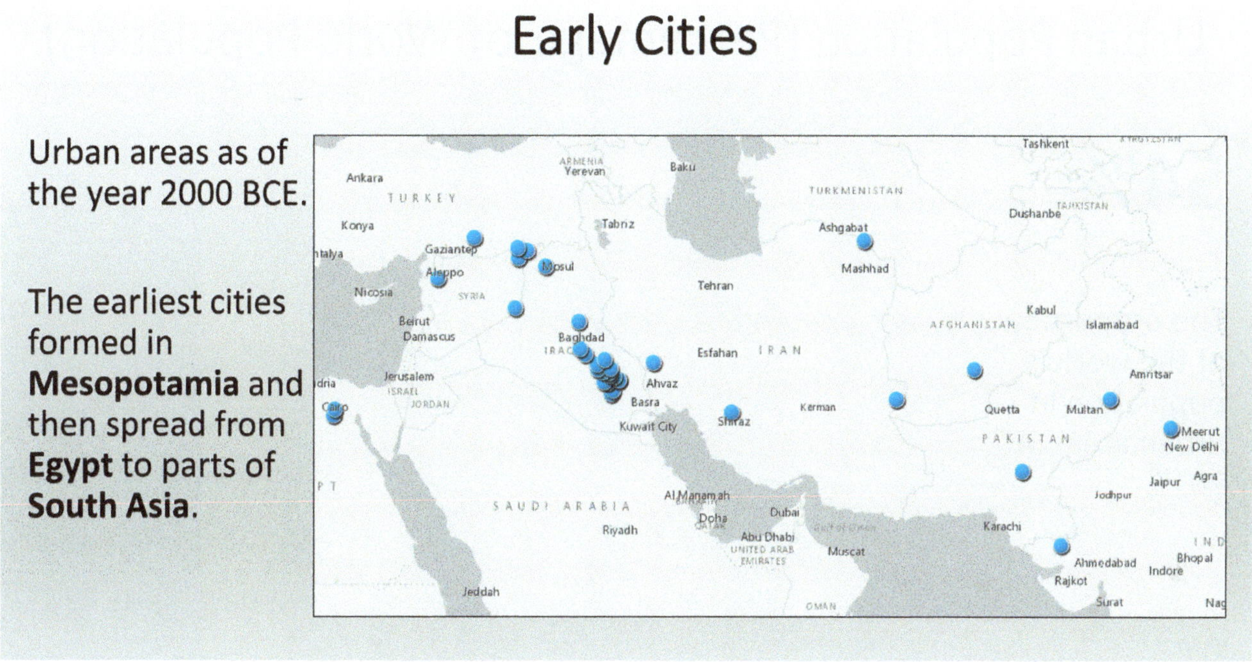

Early Cities

Urban areas as of the year 2000 BCE.

The earliest cities formed in **Mesopotamia** and then spread from **Egypt** to parts of **South Asia**.

Figure 5.2. Early cities. Data source: Reba, M. et al. 2016.

Figure 5.3. The Ziggurat from the ancient city of Ur in modern-day Iraq. This 2,000-year-old structure was part of a temple and administrative complex in one of the world's oldest cities. Photo by Homo Cosmicos. Stock photo ID: 494758876. Shutterstock.

workers. While roughly 5 percent of the world population lived in cities in 1800, by 1900, the number had risen to over 13 percent. Likewise, while one city, Beijing, had a population of one million in 1800, by 1900 there were sixteen cities of that size while four cities had a population of two million, and one, London, of five million.

More reliable data on the urban population is available from the second half of the twentieth century. Since then, the proportion of people living in urban areas has grown steadily, from about 34 percent in 1960 to 54 percent in 2015 (figure 5.4).

Spatial distribution of cities at a global scale

While over half of the world's population now lives in urban areas, the spatial distribution of urban populations is uneven. Given that urban areas are engines of economic growth, high-income countries are also the most urban (table 5.1). Countries with lower levels of per capita income tend to be less urban, as larger proportions of people work in low-wage and low-productivity agricultural jobs.

Urban levels also vary by region. Despite that early cities developed in Mesopotamia and Egypt (Middle East and North Africa), the Indus Valley (South Asia),

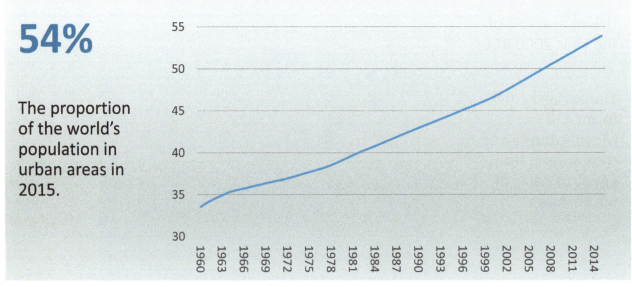

Urban Population (Percentage of World Population)

54%

The proportion
of the world's
population in
urban areas in
2015.

Figure 5.4. Urban population (percentage of world population). Data source: World Bank.

Income and Levels of Urbanization

Higher income countries:

More urban but slower urban growth.

Lower income countries:

Less urban but higher urban growth.

INCOME GROUP 2015	% URBAN	% URBAN GROWTH
WORLD	54	2.05
HIGH INCOME	81	0.84
UPPER MIDDLE INCOME	64	2.13
MIDDLE INCOME	50	2.34
LOWER MIDDLE INCOME	39	2.65
LOW INCOME	31	4.16

Table 5.1. There is a strong correlation between a country's income and its proportion of urban residents. Higher-income countries have larger proportions in urban areas, while lower-income countries have smaller proportions. Data source: World Bank.

and Northern China (East Asia and Pacific), these regions are not the most urban today (table 5.2 and figure 5.5). North America and countries of the European Union are highly urbanized, largely due to rapid industrial growth during the twentieth century. Latin American urbanization occurred somewhat later, so that while the region was less than one-half urban in 1960, by 1990, it had surpassed the European Union as economic opportunity pulled migrants to cities from rural areas.

Urban growth rates vary substantially around the world as well, generally with higher growth seen in places that are less urban. For instance, the largely urbanized European Union saw an urban growth rate of less than 1 percent in 2015 (table 5.2). This stands in stark contrast to sub-Saharan Africa, where cities grew by over 4 percent, more than seven times the rate of European Union countries.

This variation in urban growth rates is due to two major factors: natural increase and migration. Cities grow when the urban crude birth rate is larger than the crude death rate and when there is positive net urban migration.

As discussed in previous chapters, the natural increase rate changes as countries move through the demographic transition. Lower-income countries with fewer urban populations are more likely to be in stage 2 of the demographic transition. Birth rates there are higher, even for urban residents, who still follow the reproductive norms of the countryside. This is the situation in many quickly urbanizing parts of Africa, the Middle East, and Asia.

As economies develop and people shift from agrarian work in rural areas to industrial and service positions in cities, birth rates eventually fall. As you'll recall, a decline in birth rate results from lower infant mortality rates, improved educational and career opportunities for women, and the higher economic cost of raising children in urban areas. With fewer babies being born in urban areas with these characteristics, such as in North America and Europe, growth rates are lower.

Region and Levels of Urbanization

While cities first formed in the Middle East and Asia, these regions are no longer the most urban.

North America, Latin America, and Europe surpassed them as urban industrial opportunities pulled more people in from the countryside.

WORLD REGION 2015	% URBAN	% URBAN GROWTH
NORTH AMERICA	82	1.01
LATIN AMERICA & CARIBBEAN	80	1.41
EUROPEAN UNION	75	0.56
MIDDLE EAST & NORTH AFRICA	64	2.35
EAST ASIA & PACIFIC	57	2.31
SUB-SAHARAN AFRICA	38	4.13
SOUTH ASIA	33	2.66

Table 5.2. Region and levels of urbanization. Data source: World Bank.

Percentage Urban by Country

Sub-Saharan Africa and Asia:

Many countries in these regions are among the least urban in the world.

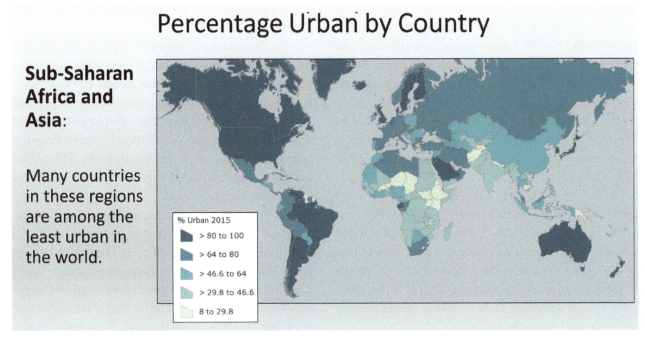

Figure 5.5. Percentage urban by country. Explore this map at http://arcg.is/2lPTiyZ. Data source: World Bank.

Countries that are highly urbanized also have less rural to urban migration. Mass migration from farms to cities in North America and Europe happened early in the twentieth century, while the same process occurred in the later twentieth century in Latin America. But in places that still have large rural populations, migration to cities is substantial. Many countries with low urban percentages are just now following in the footsteps of higher-income countries, with many people moving away from rural agricultural work to urban industrial and service jobs. High rural birth rates further contribute to pressures to migrate to the city.

Dividing urban areas into categories by population size, several trends can be seen. First, *megacities*, those with ten million residents or more, have been increasing steadily in number (figure 5.6). Whereas there were only two megacities in 1950 (New York and Tokyo), by 2015, there were twenty-nine. Furthermore, while megacities were found in more developed countries initially, by the year 2000, the majority were located in developing countries, such as Mexico, Brazil, India, and China.

Second, *large cities*, those with five to ten million inhabitants, have also grown rapidly (figure 5.6). In 1950, there were only five cities of this size (London, Osaka, Paris, Moscow, and Buenos Aires). By 2015, this number had increased to forty-four. As with megacities, the majority of large cities are now located in developing parts of the world. These tend to be the places with large national populations that are coalescing into urban areas as migrants leave the countryside or the places with rapidly growing populations in both rural and urban places.

The gravitational pull of megacities and large cities exerts significant influence on national economies as well as the global economy, drawing in disproportionate amounts of investment and people. Rapid growth in these urban areas presents myriad opportunities and challenges. Economic opportunities tend to be greater in these cities, but so do problems with affordable housing, transportation, income inequality, and too often, crime.

Large Cities and Megacities

Megacities and large cities have increased substantially in number.

Number of Large and Mega-Cities

Most are now found in developing countries.

Figure 5.6. The number of large cities and megacities has increased substantially over time. Explore this map at http://arcg.is/2lPUsKI. Data source: United Nations.

Lagos, Nigeria, Africa's largest megacity, illustrates these issues well (figure 5.7). It dominates Nigeria's economy, comprising over 15 percent of GDP; is a regional banking center and home to the country's business and political elite; and draws migrants in search of a better life from around the region. Yet, at the same time, it has struggled to accommodate growth. Many residents live in self-built housing with limited connections to infrastructure. For instance, in 2008, only 5.4 percent of households had access to

Figure 5.7. Megacity: Lagos, Nigeria. Megacities offer myriad opportunities but also can have vast inequality. The rich live in luxury housing (left), while the poor struggle to survive with self-built homes and a lack of services. Luxury housing photo by Bill Kret. Stock photo ID: 229106317. Shutterstock. Self-built housing photo by Jordi C. Stock photo ID: 392656117. Shutterstock.

piped water, while nearly 44 percent of homes were not connected to sewerage systems. And while the city offers economic opportunity, income inequality is immense. Multimillionaires live close to the urban poor who struggle to find enough daily food.

The third broad trend involves *small* and *medium cities*—those with less than one million residents—which are growing at the fastest rates. While less well known than their larger urban siblings, these are the places where most of the world's urban population lives. As these cities draw in people and investment, they too face increasing challenges in housing, transportation, and the provision of infrastructure and public services.

Go to ArcGIS Online to complete exercise 5.1: "City growth rates: Where are opportunities for businesses and development organizations?"

Spatial distribution of cities at a regional scale

Moving to a larger scale of analysis, geographers have developed different models and theories to explain patterns of cities by size and by spatial location within countries and regions.

Urban hierarchy

Cities can be described as part of an urban hierarchy, typically based on population size. Clearly, cities within a country differ by population. For instance, in the United Kingdom, London is much larger than Birmingham, which is larger than Leeds. Moving further down the urban hierarchy are cities such as Oxford and Cambridge. Finally, at the bottom of the hierarchy are small towns and villages.

Interestingly, the distribution of cities by size tends to follow a regular pattern. The *rank-size rule* says that there is an inverse relationship between city population and its rank in the urban hierarchy. This relationship can be calculated using *Zipf's law*, which posits that the second-largest city in a country will

be half the size of the largest city. The third-largest city will be one-third the size of the largest city, and so on. An inverse linear relationship is clear when plotted logarithmically. This relationship can be seen using United Nations data for urban agglomerations of over 300,000 people. In 2015, the largest urban agglomeration in the United States was New York, with a population of 18,593,000. According to Zipf's law, the second largest urban area, Los Angeles, should have a population that is half the size of New York, which would be 9,296,500. In reality, the population of the Los Angeles area was 12,310,000, somewhat higher than expected. Nevertheless, when 135 US urban areas are plotted using Zipf's law, there is a very close fit to the expected straight line predicted by the model (figure 5.8).

Zipf's law provides a good description of urban areas in many parts of the world, but not all fit the model. Some countries are dominated by a *primate city*. Primate cities are those that are much larger than all other cities in a country. Typically, as per Zipf's law, they are at least twice as large as the next largest city. They have larger populations and are typically the national political, economic, and cultural centers. They contain all important urban functions, such as government, education, arts and entertainment, manufacturing, and business and consumer services.

France provides a good example of the primate city phenomenon (figure 5.9). The Paris urban agglomeration had a 2015 population of 10,843,000. If French cities followed Zipf's law, the next-largest city, Lyon, would have 5,421,500 residents. But, in reality, Lyon had only 1,609,000 inhabitants, which is well below the expected value. Paris dominates the French state as its primate city. It is the political capital of France and the location of most central government functions. While manufacturing is more spread throughout the country, most headquarters are located in Paris. Likewise, it is the center of the financial industry, insurance, technology, and multinational corporations. Top universities that train the country's business and political elite are found in the city, as are the most well-known cultural

Rank-Size (Zipf's Law) for US Urban Agglomerations

Urban agglomerations in the US have a close fit to Zipf's Law.

However, the largest urban agglomerations, such as New York City, are somewhat smaller than predicted by the model.

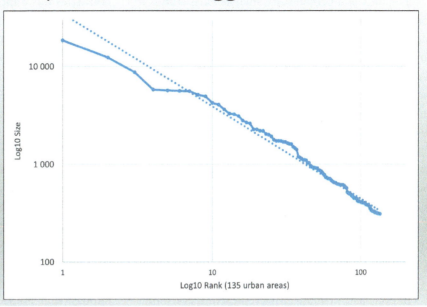

Figure 5.8. Rank-size (Zipf's law) for US urban agglomerations. Data source: United Nations.

Primate City in the Urban Hierarchy: Paris

France's largest city, Paris, is larger than predicted by Zipf's Law.

This is a clear sign of being a primate city.

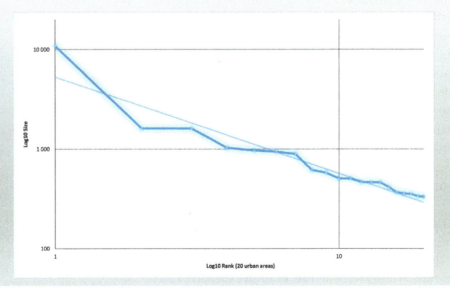

Figure 5.9. Primate city in the urban hierarchy: France. Data source: United Nations.

centers, such as the Louvre museum and other art galleries, theaters, and concert halls.

The idea of a primate city such as Paris stands in contrast to cities in the United States that better match Zipf's law. In the case of the United States, there is no primate city that dominates the country. Rather, Washington, DC, is the political center, New York is the financial center, Los Angeles is the entertainment center, San Jose's Silicon Valley is the technology center, and so on.

Different explanations have been given as to why cities follow a rank-size rule or why primate cities form. Some argue that the rank-size rule reflects countries with well-developed urban systems. These are places where more political decisions can be made at a state or local level and where wealth and economic opportunity is more spatially distributed throughout a country.

In contrast, primate cities represent a consolidation of activity in a single urban place. Often, they form when political decision making is more centralized or authoritarian. A disproportionately powerful political elite may ensure that investment remains concentrated in a single city, while smaller regional cities are neglected or used merely to supply labor and raw materials. From an economic standpoint, urban primacy can also offer benefits. When transportation costs are low, people and investments can more easily move to a single city. By concentrating labor and investment capital in a single large city, efficiencies can be gained via *agglomeration* effects (to be discussed in more detail shortly). On the other hand, when transportation costs are high, people and investment capital are more likely to remain in local regions, leading to a more dispersed pattern of urban areas throughout a country.

Falling transportation costs helps explain why primate cities are more common in developing countries than in more developed regions such as Europe. Urbanization occurred earlier in Europe, when transportation costs were higher. Prior to expansive road networks, automobiles, and trains, regions within countries remained more spatially isolated. Thus, urban areas formed to serve local markets and fewer primate cities formed. Urbanization in less developed countries has happened more recently, when roads, cars, trains, and airplanes allow people to more easily move from place to place. As transportation networks spatially integrate countries, it is easier for workers and firms to agglomerate in a single primate city.

Primate cities formed in Latin American countries in the twentieth century through heavy centralized state roles in economic development. Government promoted industrialization policies focused on developing manufacturing in capital cities (figure 5.10). Limited budgets meant that investments in everything from roads and sewers to universities and hospitals were built in the primate city. Economic and social development in smaller cities and towns was therefore limited. Transportation networks were developed enough that migrant workers from other cities and towns could move to the primate city, concentrating population. As Latin American governments have moved away from state control in the economy, and shifted toward private investment and the export of raw materials and other goods, cities outside of the dominant urban core have been growing at a faster pace in recent years.

Figure 5.10. Primate city: Buenos Aires, Argentina. Many cities in Latin America formed as primate cities, being much more than twice as large as any other urban place. Buenos Aires dominates all aspects of Argentina, being its economic, political, and cultural capital. Photo by Ed-Ni Photo. Stock photo ID: 579932641. Shutterstock.

Central place theory

While the rank-size rule can help describe the size of cities, it does not explain the spatial distribution of urban areas within a country. One model that does attempt to describe these patterns is the *central place theory* developed by German geographer Walter Christaller in 1933. This theory posits that cities, towns, and villages will organize themselves in a regular pattern throughout a region or country. This organization is tied to economic forces, whereby businesses will cluster into urban areas as they search for the best location to maximize their market area and distance themselves from other competitors.

The theory begins with the idea of a central place, a settlement that provides goods or services to the surrounding market area (also known as the *hinterland*). The market area is determined by two variables: *threshold* and *range*. The threshold is the minimum market size required for a business to be profitable, while the range is the distance people are willing to travel to purchase a good or service (figure 5.11).

Businesses can be broken up into *high-order* and *low-order* providers of goods and services, which will have distinct thresholds and ranges. Higher-order businesses provide specialized goods and services that people use on a less frequent basis. Lower-order businesses provide goods and services that are used more frequently.

High-order providers will coalesce into larger urban areas, while lower-order providers will locate in both larger cities and smaller cities and towns. This sorting of high-order providers in large cities and lower-order providers in both large and small cities and towns is due to differences in their *thresholds* and *ranges* (figure 5.12). High-order goods and services have a larger threshold and a larger range. For example, a large hospital that specializes in cardiac surgery has a large threshold; it requires a large number of patients in order to be financially viable. Given that most people never have cardiac surgery, and those who do typically have it only once, the hospital must locate in a place with a large number of people surrounding it.

Market Area

Market Area is based on...

- **Threshold**: minimum number of customers needed to support a business

- **Range**: distance people are willing to travel to a business.

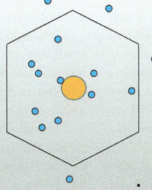

Example:

- Threshold = 8 households minimum.

- Range = 5 miles.

- Ten households fall within the five mile range.

  - They are within the market area.

- Since the threshold is eight households, the business has a viable market area.

- Four households are located beyond the five mile range.

  - They are outside the market area.

Figure 5.11. Market area: Threshold and range. Image by author.

High-Order and Low-Order Places

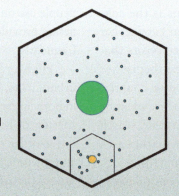

High-Order Places:
- Larger threshold requirements.
  - More people required to support the business.

- Greater range.
  - People are willing to travel farther to use the business.

- A large cardiac hospital.

Low-Order Places:
- Lower threshold requirements.
  - Fewer people required to support the business.

- Smaller range.
  - People are willing to travel shorter distance to use the business.

- A family doctor's office.

Figure 5.12. High-order versus low-order places. Image by author.

It will locate in a large city and serve not only that city's population but also residents of many smaller surrounding settlements. At the same time, a cardiac surgery hospital has a large range. People are willing to travel a greater distance to obtain heart surgery, given that it is a rare occurrence and cannot be done by more local providers.

On the other hand, a family doctor is a lower-order service provider. People visit a family doctor on a more regular basis, and therefore a family doctor has a smaller threshold. There are enough customers in a smaller city or town to make a profit. The range for a family doctor is smaller as well. People will not be willing to travel a great distance if they have only a minor injury or illness.

Through this process, a hierarchy of high- to low-order settlements develops. Higher-order businesses will form in a more limited number of large, high-order cities, while lower-order businesses will form in smaller, low-order cities and towns. High-order cities will have specialized medical care; larger malls and shopping districts; luxury retailers that sell brands

such as Louis Vuitton, Gucci, Rolex, and Maserati; and higher-end entertainment such as major league sports teams, playhouses, and music halls.

Medium-order cities may have a general hospital; more midrange retail establishments; nonluxury car dealerships; and common entertainment such as movie theaters, community playhouses, and possibly a minor-league sport team.

Finally, low-order cities will have a family doctor, small local markets and a limited number of general merchandise retail outlets, possibly a movie theater, and common services such as auto mechanics and hairdressers.

To simplify his model, Christaller proposed that given a flat landscape, with equal ease of transportation in all directions and evenly distributed consumer purchasing power, different types of businesses cluster into high-, medium-, and low-order settlements that form a regular spatial pattern (figure 5.13).

There will be a small number of high-order settlements, which have a large market area that includes many surrounding medium- and low-order settlements.

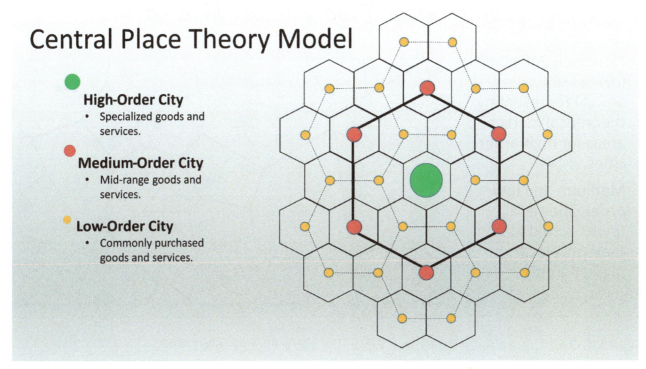

Figure 5.13. Central place theory model. Image by author.

Within this market area, there are enough people within range to support the high thresholds of specialized goods and services providers. Each high-order settlement will be spaced far enough from another high-order settlement that their market areas do not overlap.

Medium-order settlements will surround the larger, high-order settlement. The market areas of these settlements encompass many low-order settlements that surround them. Again, they are spaced so that their market areas do not overlap. Medium-order providers cluster in these settlements, drawing customers from nearby lower order towns to semi-specialized goods and services.

Finally, there is an even larger number of small, low-order settlements, evenly spaced so that their market areas do not overlap either. People consume low-order goods and services in their local communities but travel to higher-order cities to purchase more specialized items.

In reality, few urban areas follow the pattern predicted by the central place theory. It was developed on the basis of German towns, and some studies have shown it to work moderately well on the flat plains of the US Midwest (figure 5.14), but in most cases, physical landscapes, transportation routes, and socioeconomic characteristics vary too much for the model to fit well.

Site and situation

The distribution of cities actually can be idiosyncratic, with the location of urban areas tied to the unique characteristics of places and their relative location. This can be done with the concepts of *site* and *situation*.

Site refers to the local characteristics of a place, while situation refers to its relative location. Given that many cities formed as a result of agricultural development, favorable site characteristics for urban areas often include the presence of agricultural land for farming, with ample precipitation or fresh water for irrigation. For example, many towns in the American Midwest formed as market centers for surrounding agricultural hinterlands. Other urban areas developed near natural

Central Place Theory in the Midwest

"Thiessen Polygons" (created with ArcGIS Pro software) representing theoretical market areas for **high-order** cities.

Medium- and **low-order** cities and towns would have their own local market areas.

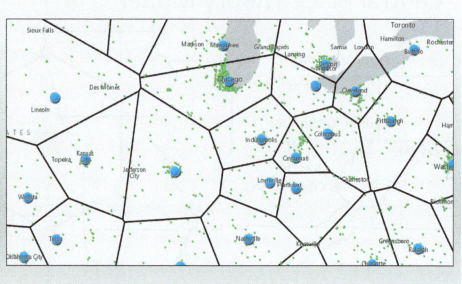

Figure 5.14. Central place theory in the Midwest. Explore this map at https://arcg.is/1X4vbP. Data sources: Esri, HERE, Garmin, NGA, USGS, NPS.

resources, such as mineral deposits. For instance, Denver, Colorado, formed in the mid-1800s following the discovery of gold along the South Platte River.

Cities have also been founded at defensible sites, such as hilltops or sloping land that offers views of approaching invaders, as well as small islands and peninsulas. Paris was founded on the Île de la Cité, a small island in the Seine River that offered protection from outside invaders (figure 5.15).

Other site characteristics include the presence of natural harbors or navigable rivers. Singapore, for example, was identified by British colonial leaders as an ideal site in Southeast Asia for a trading hub due to its deep natural harbor.

Figure 5.15. Site characteristic: Defensible space. Paris was founded on the Île de la Cité, a small island in the Seine River that offered protection from outside invaders. Photo by Kiev.Victor. Stock photo ID: 339038354. Shutterstock.

More recently, urban areas have grown on the basis of physical amenities or desirable features. Cities throughout Southern California grew rapidly in the twentieth century due in large part to a pleasant year-round climate. Cities such as Boulder, Colorado, and Portland, Oregon, while initially founded for other reasons, have grown substantially as people are attracted to their abundance of outdoor recreation opportunities (figure 5.16).

The situation of cities has quite possibly played an even greater role in the location of urban areas. *Situation* refers to the relative location of a place, or its proximity to and spatial connections with other places. For instance, while small mining towns formed in the California Sierra Nevada mountain range during the Gold Rush of the late 1800s, other cities such as Sacramento and San Francisco grew even more. Although they were not located at the site of gold discoveries, they were located along transportation and trade routes that linked the Sierra Nevada range with global migration and trade routes. People and supplies would move through San Francisco from Pacific trade routes and along the Sacramento River to Sacramento, just below the foothill and mountain location of Gold Country. More money was made in these urban areas from trade and supplying miners than was earned from gold mining itself.

While the site characteristics of a natural harbor can influence the location of a city, even more important are the situational linkages to world markets. Singapore has a good natural harbor, but its most important feature for colonial Britain was its situation in the Strait of Malacca, a key trade route linking East Asia with Europe. As another example, New Orleans thrived as a port city at the end of the Mississippi River, which linked the interior of the United States with the Gulf of Mexico and the world. Cities such as New York and Buffalo expanded greatly with the construction of the Erie Canal, which linked the Midwest with Atlantic trade routes via the Great Lakes. Cities far from the ocean, such as Chicago, could further thrive as their situation—linkages with the outside world—changed in response to construction of this canal (figure 5.17).

The significance of site and situation on urban success can be seen when these factors change for the worse. As natural resource booms dry up or as trade routes shift, the fate of urban areas can change dramatically. Ghost towns can be found throughout the world, such as those abandoned after the California Gold Rush or the end of the nitrate era in northern Chile. Great cities of Central Asia fell off the map,

Figure 5.16. Site characteristic: amenities. Boulder, Colorado, offers a wide range of outdoor and cultural activities, which draw people to live there. View of the University of Colorado Boulder campus and the Rocky Mountains. Photo by Jeff Zehnder. Stock photo ID: 556896937. Shutterstock.

Figure 5.17. Situation characteristic: transportation connections. Canals and other types of transportation connections can drive urban development. The Erie Canal helped link New York to interior cities along the Great Lakes. View of Lock 23 of the Erie Canal. Photo by georgejpatt. Stock photo ID: 526635568. Shutterstock.

figuratively, once sea trade replaced overland routes of the Silk Road.

Thus, the spatial distribution of urban areas has been highly dependent on site and situation characteristics. Arguably, situation has been the most important of the two for most of urban history, as great cities have tended to lie along important trade routes, be it those linking Gold Rush towns to the Pacific, those linking interior agricultural markets to global markets via the Mississippi River or Erie Canal, or those along land routes of the Silk Road in Central Asia that tied together China with the Middle East and Europe.

The megalopolis

As urban areas have grown rapidly in the past hundred years or so, a new type of spatial pattern has evolved at the national and regional scale, the *megalopolis*. The megalopolis is a massive urban region that consists of multiple cities. These cities are linked via transportation and communication infrastructure, creating an interlinked urban-economic system; in essence, these are a type of functional region, as described in chapter 1. While traveling through a megalopolis, it is possible to spend hours driving through urban landscapes. In the United States, the urban area stretching from Boston to Washington, DC, was one of the first places labeled

as a megalopolis, but since then, other urban areas have grown and consolidated along the same pattern.

Megalopolises have become powerful economic engines for their countries and contribute a disproportionate amount to national economic output. Due to their size, they attract people and capital investment that intermingle to create innovation in science, technology, and business. Typically, there is functional specialization in different parts of the megalopolis, each of which contributes to the economic strength of the whole. The Pearl River Delta megalopolis in the Guangdong Province of China, has been called the largest megalopolis in the world (figure 5.18). The region has a population of around sixty million, and its economy accounts for over 9 percent of China's total GDP. Its growth has been based on specialization, whereby Hong Kong functions as the location for multinational corporate headquarters and a conduit for foreign capital, while the nine mainland Chinese cities of the region focus on manufacturing. Each of the nine cities has further specialized, so that one concentrates on autos and auto parts, while others produce electronics and computers components, and still others make textiles and furniture.

Other megalopolises have been identified in the United States (figure 5.19). Each of these areas is tied

Figure 5.18. Megalopolis: Guangdong, China. The Guangdong province is a vast urban-industrial area and one of the most powerful economic engines of China. Canton Tower cable car photo by Lao Ma. Stock photo ID: 581066062. Shutterstock. City skyline photo by mrwood. Stock photo ID: 418251925. Shutterstock.

Megalopolises in the United States

Massive urban regions that are powerful economic engines for their regions.

These **functional regions**, referred to by their creators as "megaregions," were identified by commuting patterns.

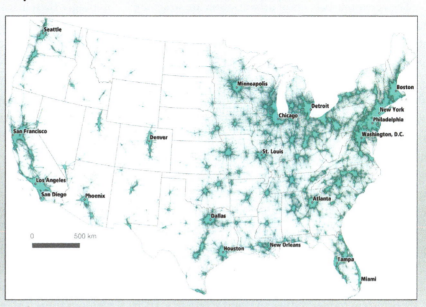

Figure 5.19. Megalopolises in the United States. These functional regions, referred to by their creators as "megaregions," were identified by commuting patterns. Image by Dash and Rae, 2016.

to a key city and a signature industry. For example, Cascadia is tied to Seattle and the aerospace industry, while the Gulf Coast is linked to Houston and the energy industry.

Go to ArcGIS Online to complete exercise 5.2: "Zipf's Law and primate cities," and exercise 5.3: "Site and situation: Why is my city here?"

Economic power of the city

Cities grow because of the economic advantages they offer to people. As discussed in chapter 3, people are often pulled to urban areas because of opportunity. While life for a rural-to-urban migrant frequently can be hard in the city, the migration would not have occurred unless life in the city promised to be better than in the rural village.

Within cities, productivity and incomes are higher, and innovation is greater than in rural areas. Fully 80 percent of global GDP is produced in cities, while only 54 percent of the world's population live in urban areas. Innovation, as represented by the number of patent applications granted, is disproportionately produced in urban areas. Within countries of the Organization for Economic Cooperation and Development, a group of mostly rich countries, 70 percent of patents came from metropolitan areas. At the same time, in regions around the world, higher urban incomes and employment result in lower levels of poverty. In Mexico, 50 percent of urban residents, but fully 62 percent of rural residents are poor. Nigeria's urban population has a poverty rate of about 15 percent, while 50 percent of its rural population is poor. Likewise, Vietnam has an urban poverty rate of under 4 percent and a rural rate of over 18 percent.

Urbanization's benefits also diffuse outward, bene-fiting those who remain in rural areas. Urban demand for rural goods can increase incomes for farmers. Rural incomes can also rise as surplus workers migrate to cities, reducing the oversupply of people searching for work in the countryside. Remittances sent from urban workers back to family members in rural areas further benefit the rural hinterland.

The economic benefits of urbanization are so strong that while many governments and development organizations used to try to reduce rural-to-urban migration, they now actively try to make it more efficient.

Basic and nonbasic industries

An important reason behind the economic strength of cities relates to the spatial interaction of people and businesses in urban areas. Employment, and thus wealth, can first be understood in terms of *basic* and *nonbasic industries*. Basic industries are those that drive the economy by exporting goods or services to a wider region, bringing money into the urban area. Basic industries are often related to the site and situation of a city. If a city's site is tied to mining or agriculture, for instance, then selling products from these industries likely brings in money. Other site characteristics, such as pleasant amenities and large research universities can drive basic industries, as in the case of tourism and technology in the San Francisco Bay Area. Likewise, if a city's situation places it along important trade routes, then logistics service industries at a port, rail yard, or highway-adjacent distribution center will bring in money. In other cases, proximity to raw materials can fuel manufacturing as a basic industry.

Basic industries are key components of urban growth, but they alone are not sufficient. Nonbasic industries offer goods and services to local consumers, both individuals and firms. These include auto mechanics, hairdressers, restaurant staff, retail workers, small consulting and business-to-business service providers, local parts suppliers, and many more. As cities grow beyond simple one-employer company towns, nonbasic employment quickly outnumbers basic employment.

The creation of nonbasic employment is the result of the multiplier effect. Basic industries bring money in from outside of the region. When basic-sector workers spend their wages around town, their spending drives demand for nonbasic goods and services, diffusing economic benefits throughout the city. For example, a port worker spends money on rent, which creates demand for a property manager and maintenance crew to care for the apartment building. The property manager then hires an accountant, as well as an occasional lawyer for legal disputes. The accountant and lawyer, along with the port worker, property manager, and maintenance workers, shop at the grocery store, creating demand for workers at the supermarket, and so on. If the multiplier effect for a basic industry is five, for example, then five additional nonbasic jobs are created for every new basic job. It is because of the multiplier effect that cities often compete with each other to attract a large, new employer. A new factory or corporate headquarters will create basic employment, but as a result of the multiplier effect, it will create many more nonbasic jobs in the city.

Thus, urban employment rapidly expands from an often small number of basic industry jobs. People from other areas are drawn to these jobs, leading to demographic and economic expansion.

Urban agglomeration

The economic power of cities is further enhanced by the *agglomeration* effect. Agglomeration involves the clustering of people and firms in an urban area. When goods, people, and ideas are in close proximity, economic efficiencies are gained and innovation thrives. These lead to greater wealth creation, higher incomes, and more jobs in cities.

There are several ways in which agglomeration improves economic efficiency. First, the costs of urban infrastructure can be shared among many people and firms. Streets, highways, ports, airports, water, sewers, electricity, and telecommunications are expensive to build and maintain. But when people and firms cluster in urban areas, the costs of this infrastructure can be shared by many. If a company locates in a rural setting, it may have to pay for upgraded infrastructure to support its needs, such as increased electrical capacity or high-speed internet connections. It will also face increased transportation costs to move people and goods to airports and logistics hubs.

Second, when firms cluster together in urban areas, they have access to a wide range of specialized workers and suppliers. A large pool of specialized workers can reduce training costs, allowing firms to put employees to productive work right away. A movie studio in Los Angeles can draw from a wide range of experienced workers, such as actors, light and sound specialists, film editors, and entertainment-focused financial experts. At the same time, firms can out-source nonessential functions to local service providers. Again, a Los Angeles movie studio can contract with local firms that specialize in the latest digital special effects rather than having to develop its own in-house special effects division. The same can be done with accounting or human resources functions. Specialty subcontractors that focus on the entertainment industry in Los Angeles thus allow the movie studio to focus on its core competency of moviemaking.

Third, in addition to lowering business costs and thus increasing productivity, urban agglomeration benefits workers. Workers in an urban agglomeration can more easily switch jobs without facing the added costs of moving to a new place. Workers can pursue opportunities that make best use of their skills and offer the most job satisfaction.

People (and firms) also benefit from the transmission of ideas in urban settings. Despite diffusion of communications technology around the globe, ideas and innovation still thrive when people interact face to face. Serendipitous interactions at coffee houses, bars, office lunch rooms, university dorm rooms, and on the street, can lead to new ideas that are unlikely to occur when people are not physically in the same place. As the author Matt Ridley colorfully stated in a book on how ideas develop, innovation occurs "when ideas have sex." The more people interact, sharing their thoughts, knowledge, problems, and dreams, the more likely a new innovation will develop. High-density urban settings make the probability of this interaction happening much higher than if people are spread out in isolated rural communities.

It must be noted, however, that agglomeration dis-economies can also be found in urban areas. As cities grow, traffic congestion rises, offsetting the benefits of shared urban transportation infrastructure. Pollution can lead to illness and lost workdays, while crime can reduce the satisfaction of urban residents and lead to extra expenditures on security for households and firms. Also, a high demand for urban housing can lead to high home and rent prices, offsetting urban gains in wages for workers.

World cities

Urban economic power is not evenly distributed among the cities of the world. Rather, a handful of cities have disproportionate influence over the global economy. These *world cities*, sometimes also called *global cities*, are globally connected, and thus their influence is not bound by national borders. They are home to large multinational corporations that determine where to deploy investment for production, distribution, and consumption around the world. Production of goods in factories, as well as services in office buildings, can be located on one side of the world for consumers on the other, but the central command functions that coordinate this system cluster in world cities.

Different world cities specialize in specific functions, drawing on the benefits of agglomeration. Through the concentration of skilled workers and specialized subcontractors, some cities focus on financial services, while others focus on advertising and media, energy, consumer goods, or technology.

Often, the economic fates of world cities are tied more to each other than to the country in which they are located. In recent years, concerns over slow economic growth and income inequality has risen in the United States and other countries. But the spatial distribution of growth and wealth has been uneven. Many world cities have been thriving, with fast-rising incomes for people with the right skills, while wages for those without globally demanded skills and those in smaller cities and towns have stagnated or declined. World cities capitalize on the benefits of globalization, functioning as command and control centers with distributed networks of offices that find the best locations for low-cost production of goods and services and for selling in scattered places of economic growth. The success of world cities can be seen in the high demand for housing by well-paid workers that drives up home prices. In New York, London, Paris, and other global cities, median home values are well over half a million dollars. If one focuses just on the core areas of global cities such as these, median home values are over one million dollars. These values are often three times or much more than the median home value for the countries in which the world city lies.

One well-cited ranking of world cities looks at the global connectivity of cities in terms of advanced producer services such as finance, insurance, accounting, advertising, law, and management consultancy. These encompass key sectors of economic activity that drive the global economy and that cluster in important urban areas (figure 5.20).

The world cities are ranked by *alpha*, *beta*, and *gamma* levels. Alpha cities are the most highly integrated into the global economy. Beta cities are important at a more regional scale, integrating their regions or countries into the global economy. Gamma cities are those that link smaller regions or countries into the world economy or those that focus on industries other than advanced producer services.

Spatial organization of the city

Cities are of most interest to people at a local scale; that is, the scale in which we live our lives, as we travel from home to work and school, go shopping, and spend leisure time around town. Because of this, geographers and others have attempted to describe and explain how different parts of the city are organized, which helps us understand why housing costs are high in some parts of town and low in others, why traffic and congestion wastes so much of our time, and why some neighborhoods are safe, clean, and thriving, while others are dangerous, dirty, and deteriorating.

Urban models

Urban morphology describes the spatial organization of the city whereby different parts of the city contain different land uses and serve different functions. Some

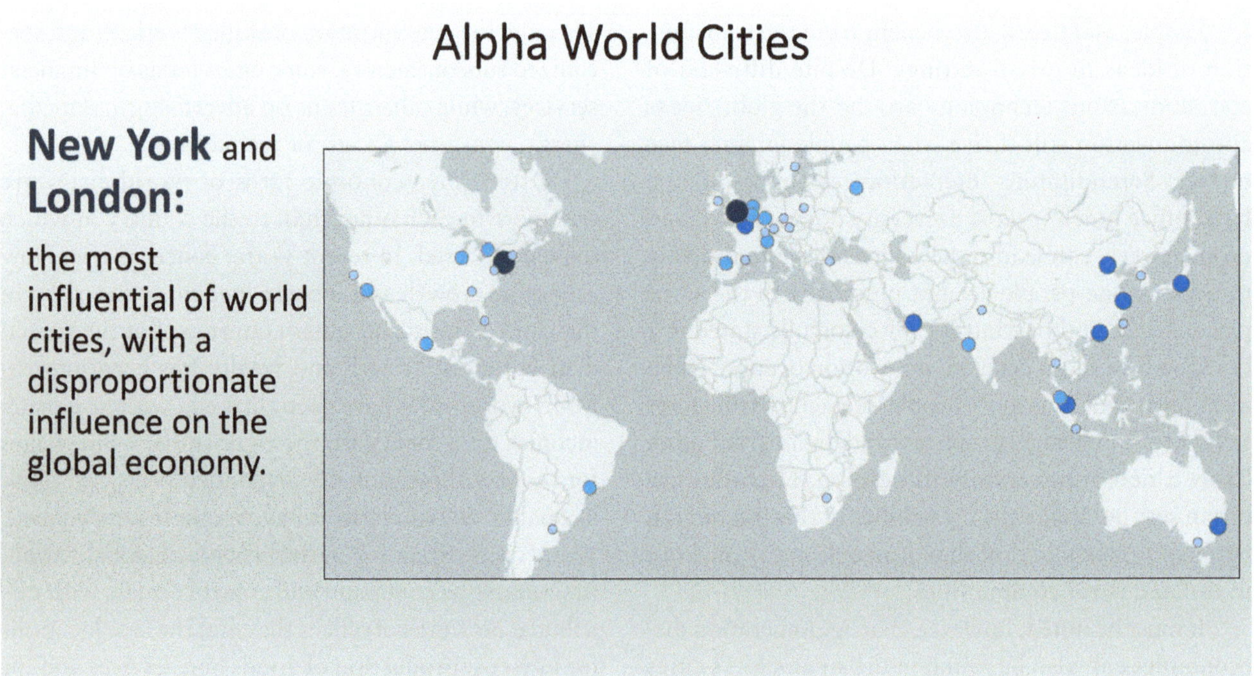

Alpha World Cities

New York and **London:**

the most influential of world cities, with a disproportionate influence on the global economy.

Figure 5.20. Alpha world cities. Data source: Globalization and World Cities (GaWC) Research Network. The data was produced by G. Csomós and constitutes Data Set 26 of the GaWC Research Network (http://www.lboro.ac.uk/gawc/) publication of inter-city data.

parts have business and commercial uses, while others have residences, industry, or open space. Various urban models have been developed that attempt to describe the morphology of cities. As models, these are simplifications of real-world cities that aim to uncover the general forms and processes of urban areas.

Concentric ring model

One of the first and most well-known urban models is the *concentric ring model*, developed by Canadian urban sociologist Ernest Burgess in 1925. This model describes the city as a series of concentric rings emanating from a *central business district* (CBD), more commonly known as downtown. The CDB contains retail, offices, banks, hotels, and museums and is a central point for transportation networks. Just beyond the CBD is an area of wholesale businesses, many of which serve downtown businesses (figure 5.21).

The next ring is known as the *zone in transition*. This area is characterized by older housing and slums,

with poverty, deteriorating buildings, and a mixture of light manufacturing facilities. Typically, this zone is inhabited by recent immigrants.

Moving farther from the CBD is the *zone of independent workingmen's homes*. Residents of this zone often are second-generation immigrants who have improved their economic situation enough to move out of the zone in transition.

The *zone of better residences* comprises the next ring. Here, single-family homes and high-class apartment buildings are located. In many cases, this is where restrictive covenants aimed to keep racial and ethnic minorities out.

The last and outer ring is known as the *commuter's zone*. High-class homes are located in this area, and residents commute to the CBD for work on a daily basis.

The concentric ring model contains a dynamic component, whereby groups of people move outward to progressively higher-income rings over time. This process of invasion and succession functions as new

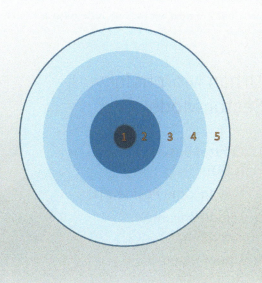

Urban Morphology: Concentric Ring Model

1. Central Business District (CBD)

2. Zone in Transition/Manufacturing

3. Zone of Independent Workingmen's Homes/Low Income Residential

4. Zone of Better Residences/Middle Income Residential

5. Commuter's Zone/Upper Income Residential

Process of change: Invasion and succession.

Figure 5.21. Urban morphology: Concentric ring model. Image by author.

groups of low-income immigrants "invade" the zone in transition. With time, these groups improve their economic situation and move to a ring farther out, and the zone in transition then becomes occupied by a new group of low-income immigrants.

Sector model

In 1939, American land economist Homer Hoyt developed the *sector model*. This model, while influenced by the concentric ring model of Burgess, described the city in terms of sectors. Whereas the concentric ring model assumes a flat surface, this model emphasizes how sectors are influenced by topography and transportation. Topography, such as hills and mountains, rivers, lakes, and oceans, constrain and guide the form of the city. The location of transportation routes further influences urban morphology.

In the sector model, different social classes live in *wedges*, or sectors that emanate outward from the CBD (figure 5.22). These originate early on, with the rich on one side of the CBD and the poor on another. From there, each social group expands outward in a wedge shape.

High-income residential areas grow outward along transportation axes radiating from the CBD to the periphery. These sectors include high-amenity characteristics such as better views or cleaner landscapes and often stretch toward high-ground open country or homes of community leaders. Other affluent communities can form as isolated nuclei along commuter rail lines. Adjacent to the high-class sectors are middle-class communities.

Low-income neighborhoods also radiate outward from the CBD toward the periphery but tend to be located in entirely separate parts of the city on land that is less desirable.

Multiple nuclei model

A third influential model, the *multiple nuclei model*, was developed in 1945 by American geographers Chauncy Harris and Edward Ullman. Unlike the concentric zone and sector models, which include just one central nuclei, the CBD, this describes the city with multiple nodes.

Multiple nuclei form for several reasons. First, different activities have different site and situation requirements. Retail districts require good accessibility from residential districts, while ports need to be at waterfront sites and manufacturing nodes need ample land and transportation connections. Second, distinct nuclei form due to agglomeration effects. Retail clusters benefit from busy consumer traffic, while financial

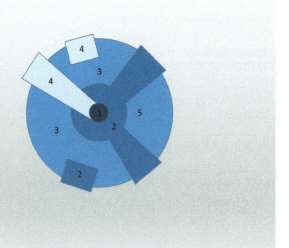

Urban Morphology: Sector Model

1. Central Business District (CBD)
2. Low Income Residential
3. Middle Income Residential
4. Upper Income Residential

Figure 5.22. Urban morphology: Sector model. Image by author.

service clusters can take advantage of person-to-person professional networking and deal making. Third, some types of nodes repel each other, resulting in dispersed patterns. High-end housing will be far from industry, while wholesale districts need ample loading spaces clear of heavy pedestrian and car traffic. Finally, land values push different uses to different nodes. Land with amenities such as waterfront views will have a high value with expensive housing, while land close to industrial land uses may have lower values and low-end housing.

In this model, the point of highest land values lies in the *retail section* of the central business district. It is the point of best accessibility, allowing people from throughout the city to shop at this location. Due to the high demand afforded by this accessibility, businesses are willing to pay top dollar for space in this zone (figure 5.23).

The next distinct district is the *wholesale and light manufacturing zone*. It is often near the CBD, but more important, it lies along transportation networks, such as rail lines and roads, where goods can easily be moved in and out of warehouses.

The *heavy industry district* locates further from the CBD. Due to noxious odors, noise, and hazardous materials, this zone forms away from the main center of the city. It too will locate near transportation networks, be it rail, road, or water.

Residential districts, as with the other models, are spatially sorted by social class, this time into distinct nuclei. Higher-income neighborhoods form in areas that are well-drained and not subject to flooding as well as on high land that is separated from industry and rail lines. In contrast, low-income neighborhoods locate closer to factories and warehousing, both for access to jobs and because of lower land values.

Other minor nuclei can form around myriad land uses. For example, a university can serve as the nuclei for housing, retail, and entertainment. At the same time, a park and golf course may form the nuclei for an upper-class housing development. Outlying business and light manufacturing nodes may form further from the CBD as well, as the city expands further from its original center.

Lastly, suburb and satellite communities form on and beyond the edge of the city. While suburbs have residents that commute into the city for work, satellites generally are too far away for regular commuting, although they tend to be closely tied to the city economically.

Urban Morphology: Multiple Nuclei Model

1. **Central Business District (CBD)**
   a. Minor node: University
   b. Minor node: Business/Commercial
2. **Light Manufacturing/Warehousing**
3. **Low Income Residential**
4. **Middle Income Residential**
5. **Upper Income Residential**
6. **Heavy Manufacturing**

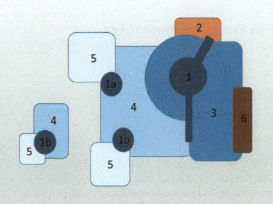

Figure 5.23. Urban morphology: Multiple nuclei model. Image by author.

No city fits perfectly with any of the three models discussed thus far. However, it has been noted that certain land uses can sometimes fit well with specific models. Socioeconomic status, such as income and level of education, often align with the sector model. In these cases, wedges of affluent residences lie in one part of the city, while wedges of lower-income residences lie in other parts of the city. Other variables, such as family structure or life cycle, often form in concentric rings. Young adults often live in higher-density housing closer to employment downtown, while families with children live in suburban rings further out. Finally, racial and ethnic patterns can correspond with the multiple nuclei model. Pockets, or clusters, of different racial and ethnic groups can be scattered in various parts of the city, forming places such as Little Korea, Little Arabia, and so on.

Urban transportation model

Later urban models further built on the multiple nuclei model, describing urban form in response to growing use of the automobile and expansion of suburban communities. In 1970, geographer John Adams described US cities in terms of *urban transport eras*. In this model, population density, urban form, and housing and commercial types changed in response to the prevailing mode of transportation. Four eras were identified (figure 5.24):

- The walking/horsecar era: Up to the 1880s
- The electric streetcar era: 1880s to 1920
- The recreational auto era: 1920s to 1941
- The freeway era: 1945 onward

In the walking and horsecar era, cities were very dense. Due to limited mobility, all urban functions had to be performed within a relatively small area, creating high population densities. People lived, shopped, and worked in and around the CBD. Lower-income urban residents lived in multistory apartment buildings (there were no elevators for taller buildings), while the middle class occupied multistory row houses. Only the rich could afford single-family homes.

By the late nineteenth century, streetcars were being built in cities, leading to significant changes

Urban Morphology: Urban Transportation Model
Generalized Application to San Diego, CA

1. **The Walking/Horsecar Era:** **Downtown is compact** and densely populated.

2. **The Electric Streetcar Era:** San Diego **expands along trolley lines** by 1910.

3. **The Recreational Auto Era:** San Diego expands along roadways through 1940.

4. **The Freeway Era:** San Diego expands further outward along freeways by 1960.

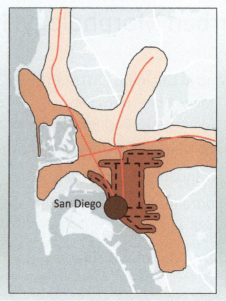

Figure 5.24. Urban morphology: Urban transportation model. Image by author.

in urban form. Population densities began to decline as streetcar lines opened up new land for housing development. Single-family homes on narrow lots were built within walking distance of streetcar lines. For the first time, a larger segment of the population could leave the dense, congested core of the city for more suburban locations. Cities took on a star form, as development radiated outward along the streetcar lines. Nevertheless, commercial activity still focused on the CBD.

In the 1920s, more families began purchasing automobiles, and cities entered the third transport era. The automobile opened up large tracts of land that were previously inaccessible. With a substantial increase in the supply of land, people were able to purchase larger lots and build single-family homes. This further decreased population densities as people moved to auto-oriented suburban neighborhoods. Around this time, some commercial activity began to expand outside of the central business district. People no longer had to travel downtown for all purchases but rather could hop in their car and drive to a new auto-oriented shopping center near home. In this era, urban development filled in the spaces between the star from the streetcar era.

Finally, after World War II in the second half of the 1940s, US cities entered the freeway era. This era represents the low-density, single-family, auto-oriented development pattern that has been the dominant form of urban growth ever since. Vast tracts of land consist of single-family homes on large lots on the ever-expanding periphery of the city. Auto-oriented shopping centers dominate the commercial landscape, and most people no longer have to travel to the CBD for shopping or employment.

Galactic metropolis and edge cities

With the dominance of the freeway and automobile, the idea of multiple nuclei was expanded and described by American geographer Peirce Lewis as a *galactic metropolis*, whereby shopping centers, residential developments, and industrial parks appear to "float in space" like stars and planets with empty space in between when viewed from above. These detached urban nuclei do not depend on the CBD of another city but rather function on their own, providing housing, entertainment, employment, and shopping for residents. Similar urban forms have been described as *edge cities*, suburban places with offices, retail, and housing, located along freeways and designed around the automobile.

Go to ArcGIS Online to complete exercise 5.4:
"Urban models: Which best represents my city?"

Suburban growth and urban decline

As cities have evolved from being focused on a single central business district to being multinodal and suburban, the fates of suburban and inner-city neighborhoods have shifted. Through the walking and horsecar and the streetcar eras, transportation, and thus economic activity, was focused on the CBD. But with the automobile era, and more so with the freeway era, this changed dramatically. New homes in auto-oriented suburban locations were purchased by the middle and upper classes—those who could afford cars and new property. The federal government helped subsidize this process by expanding access to home mortgages through the GI Bill for World War II veterans and federally backed mortgage guarantees. This began to pull wealth away from the inner city. Soon, commerce followed the money and relocated to the suburbs as well. Small mom-and-pop grocers in the inner city could not compete with the efficiency of large, new suburban supermarkets, while department stores left downtown for new peripheral shopping malls. At the same time, industry moved away from the city, as suburban land accessible for trucks via freeways was cheaper on the outskirts of the city than in multistory urban buildings. Commercial and industrial jobs, along with the more affluent segment of the urban population, moved away from the older inner city and into new suburban developments.

This left the inner city in a state of decline. Downtown business districts saw an increase in boarded-up storefronts. Employment opportunities for those unable to move to the suburbs declined. This often impacted racial and ethnic minorities disproportionately. They tended to have lower income and were sometimes prevented from moving to the suburbs by racially restricted covenants. As the economy of older cities declined, tax revenues fell. Older cities were left with a lower-income population as well as fewer resources to finance schools, provide police and fire protection, maintain streets, and maintain parks and public spaces. In contrast, new suburban cities saw an increase in tax revenues, which served a more affluent population and financed new, lower-maintenance infrastructure.

Detroit is one of the most visible examples of this decline. When the J. L. Hudson Company department store left downtown for the suburbs in the early 1950s, many other businesses followed. Twenty years later, Detroit's commercial center was decimated. The city's population declined as higher-income whites moved to new suburban cities and poor blacks remained. This pattern of inner-city decline and rising segregation was repeated in cities across the United States.

Suburban decline and urban growth

For many decades after World War II, geographers and others identified a fairly clearly defined dichotomy between the more affluent (and whiter) suburbs and the poorer (and minority) inner city. But in recent years, this dichotomy has become less clear, as pockets of poverty form in suburban neighborhoods and pockets of wealth grow in inner-city areas.

One study found that poverty between 2000 and 2011 grew by 64 percent in US suburban places. While the poverty rate is higher in inner-city neighborhoods, poverty growth in suburban areas was twice the rate of that in inner-city areas (figure 5.25). This meant that for the first time in US history, the suburbs had a larger number of poor residents than big cities.

In a sense, increases in suburban poverty would be expected. Given that employment had shifted to suburban areas, when economic crises such as a decline in US manufacturing strike, the impact is most strongly felt in suburban communities near employment centers. In addition, some suburban housing, especially on the far edges of cities (sometimes referred to as *exurbs*), offers an affordable means of living the American dream of a single-family home. However, people attracted to this affordable housing, such as immigrants and lower-skilled workers, have a more tenuous financial situation, making them more likely to fall into poverty.

This pattern was clearly visible during the 2008 mortgage crisis in fast-growing exurbs of cities in Arizona, California, Florida, and Nevada. In these places, many new homes were purchased with higher-interest subprime loans. When the economy turned to recession, many owners of these homes lost their jobs and were unable to make payments on their mortgages, resulting in high rates of foreclosure.

Figure 5.25. Suburban poverty and decline. For many decades, the suburbs saw nearly uninterrupted growth as the middle class left inner cities for auto-oriented neighborhoods. But in recent years, some suburban, and especially exurban, communities have experienced economic problems and rising poverty. Photo by rSnapshot Photos. Stock photo ID: 88256701. Shutterstock.

As suburban poverty is a relatively new phenomenon, it presents a unique set of challenges for the suburban poor. Most social services agencies, such as food banks, government social service offices, and charity groups, are in inner-city neighborhoods where poverty has traditionally been concentrated. The suburban poor can find themselves without the social safety net that inner-city poor may have. Suburban schools may lack specialized services to assist poor students. Furthermore, public transportation is limited in suburban areas, so if a poor family cannot afford a car, they may struggle to find employment outside of their immediate neighborhood.

While suburban poverty has increased in many cities, some inner-city neighborhoods have been seeing an increase in home prices and higher income residents. The process whereby higher income and better-educated people move into lower-income inner-city neighborhoods is known as *gentrification*. Typically, it involves changes to a neighborhood's demographics, real estate markets, land use, and character. Each of these changes can have a negative effect on existing residents, who may be displaced by newcomers with more money. In terms of demographics, gentrification often results in a decline in the minority population and an increase in the white population. With changes in the real estate markets, home values and rents increase. While this can benefit lower-income minority homeowners, the majority who rent can quickly get priced out of their neighborhood. In terms of land use, old warehouses and factory buildings get converted to art galleries, coffee shops and restaurants, and live-work lofts. Stores can switch from those that serve a low-income population, such as dollar stores and liquor stores, to those that serve higher-income residents, such as trendy chain stores and wine shops (figure 5.26). All of these changes mark a dramatic shift in the culture, or sense of place, of the community, as when a working-class neighborhood transforms into a "hipster" community.

The reasons behind gentrification vary. In many cities, crime has fallen substantially since its peak in the 1980s and early 1990s. New York City had 2,200 murders in 1990 and only 352 in 2015. This pattern has been followed by cities throughout the Western world. In addition, young people, especially the college educated, are marrying and having children at older ages. Young, childless people enjoy the activity of the city and do not have to worry about often lower-quality public schools. These factors, combined with growth of jobs that require brains over brawn, mean that more people want to live in denser neighborhoods, where, again, innovation and creativity flourish "when ideas

Figure 5.26. Gentrification in Manchester, England. A tapas and wine bar in this working-class city is a clear sign of gentrification. Photo by Alastair Wallace. Stock photo ID: 477205234. Shutterstock.

have sex." Often, the original gentrifiers are artists and musicians looking for cheap places to pursue their creative dreams. But as they transform neighborhoods, they create the conditions for higher-paid professionals to move in.

In recent years, gentrification has been a hot topic in San Francisco, among other cities. Being a beautiful city with myriad cultural amenities, interesting neighborhoods, and a year-round mild climate, many people wish to live there. As the Silicon Valley technology boom, centered south of San Francisco, has expanded and created numerous well-paying jobs, the demand for housing has increased even more. In 2016, median rent for a one-bedroom apartment in San Francisco was about $3,500, while a two-bedroom was over $4,500. Clearly, this price range is beyond the reach of most workers. As nontechnology workers face rising rents that they are unable to pay, demands for different housing solutions have been called for. However, there is little agreement on how to best address rising rents and gentrification. Some want more government-subsidized housing, while others call for fewer restrictions on developing new housing units in the city.

Similar patterns of gentrification have been found in cities around the world, from New York and Los Angeles to London, Paris, and Mexico City. Even Tallinn, Estonia, in former communist Eastern Europe has seen gentrification.

It must be noted, however, that some researchers dispute the negative impact that gentrification has on minority residents. Often, minority residents move from gentrifying neighborhoods at a lower rate than from nongentrifying neighborhoods, and many benefit from falling crime rates and an increase in retail services. Furthermore, places that gentrify do not always have a large number of residents. Rather, they are often old industrial and warehouse districts that are underutilized. As these places get converted to residential and retail land uses, economic and cultural benefits help the city more than they hurt.

Go to ArcGIS Online to complete exercise 5.5:
"Suburban poverty and gentrification."

Market influences on urban morphology

We have seen that urban form can take the shape of concentric rings, sectors, and multiple nuclei. We have also seen that the form and size of a city can be influenced by changing modes of transportation. But one of the forces underlying all of these is economic. In most cities in most of the world, the market concepts of supply and demand strongly influence how cities grow and the shapes that they take.

Bid-rent theory

In urban economic geography, *bid-rent theory* is used to explain how different land uses form in relation to a central place. In essence, different urban functions will bid different amounts of money to rent (or buy) land, considering the tradeoffs between accessibility and land costs. Bid-rent curves slope downward from the central place, so land is more expensive near the center and cheaper as one moves toward the periphery of the urban area. This is because the central place is highly accessible from all directions, resulting in a higher demand and higher prices.

Higher land prices near the center also create a negatively sloped density gradient. Skyscrapers and other multistory buildings will be found closer to the central place, since high land costs justify the added expense of building vertically. On the other hand, as land costs fall toward the periphery, development takes a more horizontal, low-density character.

The bid-rent theory is most clearly understood when tied to the Burgess central place theory (figure 5.27). In this case, the CBD, lying at the center of an urban area, is the point of best accessibility, since people from all directions can travel to it. Businesses that place a high value on accessibility will locate in the CBD. These will tend to be office and commercial functions that need to be easily accessible for employees and customers. Next, warehouse and light manufacturing functions that need central access to downtown business customers will make the highest bids for land. From there come bids for residential land uses. Inner rings, due to

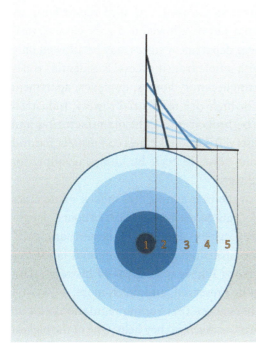

Bid-Rent Theory and the Concentric Ring Model

1. **Central Business District (CBD)**
   - Highest bids from commercial and retail.

2. **Zone in Transition**
   - Highest bids from warehouses and light manufacturing.

3. **Zone of Independent Workingmen's Homes**
   - Highest bids from higher density apartment developers.

4. **Zone of Better Residences**
   - Highest bids from suburban housing developers.

5. **Commuter's Zone**
   - Highest bids from large-lot housing developers.

Figure 5.27. Bid-rent theory and the concentric ring model. Image by author.

higher land costs, will consist of apartment buildings, while outer rings, where land costs are lower, will have lower-density single-family homes.

The bid-rent theory offers an explanation as to why land uses change in cities. When the demand for one type of land use increases, its bid-rent curve will shift upward. This leads to an expansion of its zone and more intense use of land. For instance, when demand for commercial space downtown increases, adjacent zones of warehouses and lower-income apartments can be redeveloped for higher-density office or retail establishments. These same places can be converted to higher-end housing, such as luxury apartments and lofts, if demand for downtown space by more affluent households increases. Likewise, peripheral farmland can be converted to new housing subdivisions if demand for single-family homes increases.

The bid-rent concept can also be used on non-monocentric models, with land values highest at easily accessible central nodes and sectors and lower as one moves away from these more desirable locations. In all cases, land will be allocated on the basis of competition between groups with different abilities and willingness to pay. Those with more money will occupy the most desirable locations, and those with less money will occupy less-desirable locations.

Government influences on urban morphology

In an unregulated market economy, bid-rent theory would be sufficient to explain much of the spatial form of cities. However, all urban development is also influenced, in varying degrees, by government regulations and policies.

Zoning

One of the most commonly used tools that governments use to influence the shape of cities is *zoning*. Zoning involves regulations that govern the use, intensity, and form of urban development. For instance, a zoning code can delineate where commercial, residential, and industrial land uses are allowed. It can establish allowable densities, as in the number of housing units per parcel. It can also control the size and shape of buildings, setting maximum heights and distances from property lines.

Established in the early twentieth century in US cities, zoning was intended to prevent incompatible land uses and development patterns from damaging quality of life in urban areas. In New York City, residents were concerned that too many tall buildings were limiting sunlight and the flow of air through the city, resulting in zoning that controlled the height of buildings and setbacks of upper-level floors. It also intended to keep the garment industry from encroaching into the commercial area of Fifth Avenue.

Today, zoning ordinances can be complex and detailed. Broad categories of commercial, residential, and industrial areas are still used, but each of these is further broken into more detailed categories

(figure 5.28). Commercial zones can differ by the type of business allowed. An auto repair shop or nightclub may not be allowed in a zone close to residences. Large department stores may be allowed only in zones with ample transit access. Residential zones can allow single-family homes, duplexes, apartment buildings, or high-rise residential towers. Industrial zones can be broken into light manufacturing and heavy industry. Additional zones can protect natural areas, scenic views, historic buildings, or allow for mixed land uses.

Many argue that zoning has indeed made cities more livable. Few people have to worry about a slaughterhouse opening next door to their home today, parkland can be protected from commercial land uses willing to pay a higher rent, areas of historic architecture can be preserved, and much more.

But others argue that zoning regulations end up harming cities as much as they help. One of the biggest

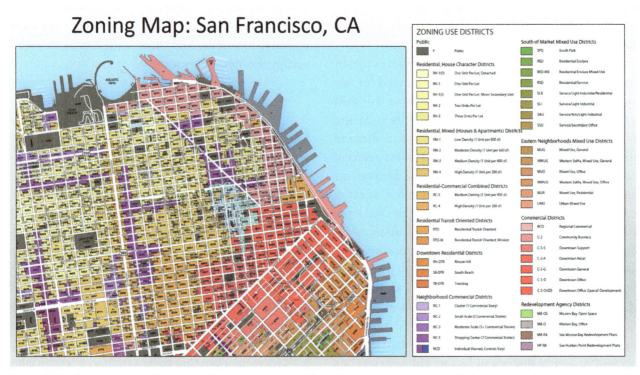

Figure 5.28. Zoning map. This map shows a section of San Francisco and some of the zoning districts that guide development in the city. Image source: City and County of San Francisco. Map by Michael Webster.

criticisms is that by restricting housing development, housing costs are unnaturally high in many cities, and segregation is intensified. As described by bid-rent theory, if there is a demand for more housing in a certain area, then the amount of land devoted to housing will increase, as will housing density. But if zoning laws restrict the horizontal and vertical growth of housing, prices must increase. Demand will outstrip supply.

California is a prime example of how restrictions on housing development can increase real estate prices. From 1980 through 2010, housing development in high-demand coastal California was significantly below the rate of development for other US metropolitan areas. Consequently, housing prices in coastal California cities are among the highest in the nation. Existing residents of single-family homes resist zoning changes that allow for higher-density multifamily housing. Also, city governments zone more land for commercial uses, knowing that sales and hotel taxes go directly to city accounts, whereas residential property taxes go to the state government.

The San Francisco Bay Area best illustrates the impact of limited housing development on prices. In 2015, 64,000 jobs were added to the regions, while only 5,000 new homes were built. Because of intense demand and limited supply, developers in San Francisco now build a typical unit aimed at a family of four with an annual income of $270,000. This gentrification is creating what some are calling a segregated city of affluent technology workers, where the working class and lower-income minorities can no longer afford to live.

Urban renewal

The form of urban areas in the United States has also been influenced by government-sponsored urban renewal policies. Going back to the concentric ring model from 1925, residential areas around the CBD were described as lower income and the loci of poverty and social problems. As time progressed and American cities expanded, becoming even more suburban, concern grew over how to best reverse inner-city decline.

Beginning in the 1930s, but gaining strength through the late 1940s and 1950s, the federal government assisted cities in clearing slums and blighted areas. Government-backed loans and grants were used by cities to purchase large tracts of blighted inner-city communities through the use of *eminent domain*, whereby the government legally expropriates private property for public use, paying compensation at market rates.

The goal was to revitalize downtown and inner-city communities so that some of the people and investment lost to the suburbs would return. But in most cases, this goal was not achieved. Large tracts of housing for lower-income residents, as well as many small mom-and-pop businesses, were destroyed and replaced with office buildings, hotels, convention centers, retail establishments, and sports stadiums. This type of development failed to enliven central city communities. The 24-hour life of neighborhoods with a mixture of people and businesses was replaced with large-scale, single-use facilities.

Freeways were built to link the suburbs to downtown zones so that people could more easily work and shop there. But this further destroyed the urban fabric as freeways cut neighborhoods in two or, as in Seattle and San Francisco, destroyed urban amenities by blocking views and access to bays. People would simply drive from the suburbs to work downtown, or possibly see a ballgame or symphony concert, and then drive back to their suburban homes.

In terms of the inner-city poor, new housing failed to be built to replace that which was demolished. Millions of mostly black and Latino residents were forced to move to new housing. When new housing was provided, it tended to concentrate poverty in large housing projects (figure 5.29). This concentrated poverty resulted in lower property values in surrounding areas. By the twenty-first century, high crime and lack of maintenance resulted in the demolition of many of these projects.

In the end, urban renewal policies failed to revitalize inner-city neighborhoods but rather gave people more reasons to live in suburban parts of the city.

Figure 5.29. Public housing project. Too often, city neighborhoods were destroyed as part of urban renewal policies, and new public housing served to concentrate poverty even more. Photo by trekandshoot. Stock photo ID: 82422766. Shutterstock.

Urban morphology in the developing world

The focus of this chapter has been on cities in North America, but in many parts of the world, urban form and processes function in different ways. As mentioned earlier, urbanization has been occurring at a faster rate in lower-income countries. This has presented serious challenges as governments struggle to accommodate a rapid influx of new urban residents. All too often, the demand for housing and urban infrastructure—water, sewer, electricity, and roads—has exceeded the ability of governments to supply and coordinate its development.

Consequently, cities in developing countries often have high rates of informal housing construction. Informal housing is that which is built without planning or permits and without the involvement of professional building contractors. Instead, housing is self-built by families, often on land that is not formally owned by them. Informal housing development results in large squatter settlements, or shantytowns, that are common throughout much of the developing world.

The Latin American city model

The *Latin American city model* serves as a good illustration of how cities in developing countries differ from those in the United States and other developed regions (figure 5.30). Again, going back to the 1925 concentric ring model, it was pointed out that as people's economic situation changes, they move to new neighborhoods. Through the process of invasion and succession, older immigrant groups would move out of the zone in transition to higher income rings farther out. Similarly, more recent descriptions of cities have described how people leave older communities for new homes in the suburbs. But in the Latin American city model, the form of the city and the way in which neighborhoods change are very distinct.

The Latin American city model begins with a CBD, part of which consists of modern office buildings and hotels but with another section that has more traditional street-oriented markets for the wider population. Extending outward from the CBD is a commercial spine that follows a major transportation route. This is bordered by the *elite residential sector*, similar to that found in the Hoyt sector model.

Beyond the CBD and elite sector, however, the model quickly diverges from models based on US cities. Surrounding the CBD is the *zone of maturity*, which consists of middle-class homes and apartments. One reason that more middle-class residents stay closer to downtown is for the public transportation and infrastructure services. Traditionally, residents of developing countries own fewer cars than in the developed world, instead relying on public transit. Because of more limited transportation options, the middle class is willing to pay more for housing closer to the CBD. Also, full infrastructure in developing countries is not available farther from the city center. Water, sewer, electricity, and paved roads are fully developed in this section (and along the elite spine), unlike in zones farther out where incomes begin to fall.

The zone of *in situ accretion* contains a more working-class population. Housing appears only partially complete, so there may be homes that are built partially of cement block, partially with wood or other

Latin American City Model

1. Commercial. CBD, commercial spine, and peripheral shopping mall.

2. Downtown market.

3. Industrial zone.

4. Zone of maturity.

5. Zone of in situ accretion.

6. Zone of peripheral squatter settlements.

7. Elite residential sector.

8. Gentrification.

9. Middle class residential tract.

Figure 5.30. Latin American city model. Copyright © 1996 by the American Geographical Society.

materials, and with exposed rebar of unfinished room additions. In this zone, infrastructure is only partially complete. Main roads may be paved, but side streets are still dirt; electricity may be available, but piped water and sewer may not (figure 5.31).

On the edge of the city lies the *zone of peripheral squatter settlements.* Typically, residents do not have formal land tenure, or ownership. Here, housing is built of a wide range of materials, such as plywood, tarps, scrap metal, and just about anything else. This is the lowest-income part of the city, where transportation to jobs downtown via public transit is slow and expensive. Infrastructure is mostly absent. Roads are dirt, few have electricity unless they can make illegal connections to main power lines, water is available only by well or water truck, and sewage runs directly into gullies.

When comparing the Latin American city model with models of North American cities, a couple of significant differences stand out. First is that the poor live on the periphery, while the middle class lives closer to the CBD. The second major difference is the process

of neighborhood change. As people's lives improve in the Latin American city, they are more likely to stay in the same place but upgrade their homes. The reason the zone of in situ accretions appears to be unfinished

Figure 5.31. Zone of in situ accretion. Favela Santa Marta in Rio de Janeiro, Brazil. Here, self-built homes are in various stages of completion. Some infrastructure has been added, as seen by the electrical power lines in the background.
Photo by lazyllama. Stock photo ID: 432647785. Shutterstock.

is that it is—people are continuously adding space to their homes as their families grow and incomes rise. With time, more homes become more complete, and the community pressures the government to add better infrastructure. Over time, this zone transforms into a zone of maturity. At the same time, squatter settlements upgrade as well. A house made of scrap metal and plywood may add a new room made of solid cement block. From there, more solid construction replaces the original materials. The city may begin to pave a few major streets, and electricity may be added to the neighborhood. Over time, the squatter settlement transforms into a zone of in situ accretions. Meanwhile, a new zone of peripheral squatter settlements appears farther out on the edge of the city. This process is ongoing, so the city grows organically outward, in direct contrast to the formally built housing of cities in the developed world.

More recently, Latin American cities are seeing more professionally built development in suburban locations. As economies improve, allowing for a wider distribution of urban infrastructure, and as more families purchase private automobiles, suburban middle-class residential housing tracts are forming, along with suburban industrial parks. In some ways, this growth is moving in the direction of North American urban development, but the traditional organic growth of many Latin American cities still gives them a distinct character.

Solutions to informal housing

There are different strategies for dealing with informal housing developments. Sometimes, when these developments are on land that is needed for other uses, squatters can be forcibly removed. This happened in Brazil for construction of some 2016 World Cup and Olympic sports facilities. But generally, governments do not want to face the conflict that can arise from forced evictions. Instead, regularizing land tenure is often done, whereby squatters are given legal title to their land, sometimes for a manageable fee and sometimes at no cost. Once legal ownership is ensured, the process of in situ accretion can speed up, as people

feel more secure in spending to improve their homes. Governments can also provide site and service solutions, whereby a legal plot of land and basic infrastructure is provided, allowing for families to then build their own homes over time.

Livable cities

As cities grow, there is always a tension between the economic benefits of agglomeration and urban quality of life. In both developed and developing countries, most people can relate to being stuck in traffic jams, paying too much for rent, breathing polluted air, seeing open space lost to development, and having waterways polluted from urban runoff.

Smart growth

By controlling the form of urban growth, governments attempt to mitigate this tension so that cities remain engines of economic opportunity but are also pleasant places to live. One way of promoting livability is through *smart growth* (figure 5.32). The aim of smart growth is to build communities so that people can live and work in pleasant places near employment, schools, and shopping.

There are several key strategies for creating smart-growth developments. First, by mixing land uses, people are more likely to travel shorter distances for their daily activities. In traditional urban development, zoning was set to allow one type of land use. In that case, if the zoning allowed for only single-family homes, then there could not be any commercial, business, or other type of land use. This restriction forces people to travel greater distances to shop or work. When land uses are mixed, stores and offices can be blended with residential areas, much like in older neighborhoods, where corner stores and shops are more common alongside housing, and people can walk, bike, or take shorter trips by car or transit.

Second, mixed land uses allow for more diversity in housing types. When a community has single-family

Principles of Smart Growth

- Mix land uses.

- Take advantage of compact building design.

- Create a range of housing opportunities and choices.

- Create walkable neighborhoods.

- Foster distinctive, attractive communities with a strong sense of place.

- Preserve open space, farmland, natural beauty, and critical environmental areas.

- Strengthen and direct development toward existing communities.

- Provide a variety of transportation choices.

- Make development decisions predictable, fair, and cost effective.

- Encourage community and stakeholder collaboration in development decisions.

Figure 5.32. Principles of smart growth. Image by author.

homes, apartments, and townhouses of different sizes, a more diverse population can live there. Young singles can live in apartments. When they marry and have children, they can move to a larger single-family home with a yard. Then as they grow older, they can move to a smaller townhouse. All of these moves can be within the same neighborhood, allowing people to maintain social ties in the same place as they move through stages of their life.

Another component of smart growth is higher density. Apartment buildings and townhomes allow for more units per square mile, as do single-family homes if lot sizes are smaller than in many typical suburban neighborhoods. With more housing units per square mile, more land can be used as open space, such as parks and nature preserves. Density also promotes walkability. When housing and land uses are more closely built (and when combined with mixed land uses), distances are shorter and biking and walking become viable transportation alternatives.

Another significant feature of smart growth is to foster a diversity of transportation options. Denser neighborhoods and mixed land uses can facilitate walking and biking as well as public transit use. This is enhanced by transit-oriented development whereby denser housing is located adjacent to commuter rail stations and public transit centers. Street design, with wider sidewalks, good street lighting, landscaping, segregated bike lanes, and well-located transit stops all build on the benefits of density and mixed land use.

Lastly, smart growth can include developing a strong sense of place or a unique character so that people want to spend their time in the community. This can be done by creating public gathering places such as town squares (figure 5.33), or plazas or creating a nightlife district with restaurants and theaters. Public art can enliven the landscape and give communities a unique character. Interesting paths, such as well-designed streets, that link distinct nodes of a neighborhood can further enhance a positive sense of place.

Figure 5.33. CityPlace in West Palm Beach, Florida, has many characteristics of smart growth. It includes housing and commercial space, where people can walk from their homes to stores and restaurants. It also includes ample plaza space for informal gathering and socializing. Photo by Ritu Manoj Jethani. Stock photo ID: 360099605. Shutterstock.

The connected city

For many people in developed countries, it is hard to imagine driving though the city without live traffic updates on a cell phone. We are also accustomed to finding a nearby coffee place or restaurant, customer reviews included, at the touch of a button. As information and communication technology has diffused through urban areas around the world, it is increasingly being used to make cities more livable. Private companies offer mapping services to help us travel through the city and find the goods and services we are looking for. Increasingly, city governments are using technology to improve services as well.

The first step in creating a connected city is to build the requisite information and communication infrastructure. Private companies in many places have established extensive cell coverage, allowing those with the financial resources to benefit from access to the internet. But many with fewer resources are left out. For this reason, some cities are building public Wi-Fi sites so that anyone with a smart phone can access the internet without having to pay for data (figure 5.34). In Rwanda and Brazil, some city buses now offer these services, and in New York City, free public Wi-Fi links are being placed throughout its five boroughs.

Cities not only are linking people to information and communication infrastructure but also are linking a wide range of urban features, allowing for complex data analysis and real-time response. Traffic sensors can feed information to traffic lights so that vehicle flows are maintained, sensors in water and sewer lines can warn engineers when replacement parts are needed or leaks form, parking garages can now inform drivers how many spaces are available and in which locations, air pollution monitors are placed on street lights, and cameras can monitor the movement of people for security purposes.

Cities are offering free applications that allow residents to report graffiti or illegal dumping, flooding and downed power lines during storms, mosquito breeding grounds, unsafe street conditions, and much more. Through embedded sensors and citizen crowdsourcing, city governments are collecting vast amounts of data that can be used to improve efficiency and make cities safer, cleaner, and more pleasant places to live.

In addition to collecting data for internal use, many cities are now moving toward open data, whereby governments publicly share data that was once difficult for residents to acquire. This is allowing individuals and companies to create applications and analyze data in ways that few cities can with their limited resources.

Local governments are continuously adding new applications and datasets that are open to the public. With this information, residents can be more informed about government projects, and innovative new applications can be developed by the public with city data.

Figure 5.34. Free Wi-Fi in Tokyo, Japan. Digital connections in cities can help residents without regular cell phone data plans benefit from digital services such as bus arrival times, weather, social media communication, and much more. Photo by Ned Snowman. Stock photo ID: 698470531. Shutterstock.

Go to ArcGIS Online to complete exercise 5.6: "Smart vs. suburban growth," and exercise 5.7: "Open data."

References

Adams, J. S. 1970. "Residential Structure of Midwestern Cities." *Annals of the Association of American Geographers* 60, no 1: 37–62. doi: https://doi.org/10.1111/j.1467-8306.1970.tb00703.x.

Alamo, C., and B. Uhler. 2015. *California's High Housing Costs: Causes and Consequences.* Legislative Analyst's Office, State of California. http://www.lao.ca.gov/reports/2015/finance/housing-costs/housing-costs.aspx.

Bairoch, P., and C. Braider. 1988. *Cities and Economic Development from the Dawn of History to the Present.* Chicago: University of Chicago.

Buntin, J. 2015. "The Myth of Gentrification." *Slate Magazine*, January 14, 2015. http://www.slate.com/articles/news_and_politics/politics/2015/01/the_gentrification_myth_it_s_rare_and_not_as_bad_for_the_poor_as_people.html.

Cooper, M. 2014. "China's Pearl River Delta: Tying 11 Cities into a Megaregion." *Urban Land Magazine*, September 22, 2014. https://urbanland.uli.org/industry-sectors/infrastructure-transit/chinas-pearl-river-delta.

Dash, Nelson G., and A. Rae. 2016 "An Economic Geography of the United States: From Commutes to Megaregions." *PLoS ONE* 11, no. 11: e0166083. doi: https://doi.org/10.1371/journal.pone.0166083.

The Economist. 2013. "What Is Driving Urban Gentrification?" *The Economist,* September 17, 2013. https://www.economist.com/blogs/economist-explains/2013/09/economist-explains-5.

Federal Reserve Bank of San Francisco. 2016. "Can the SF Bay Area Solve Its Affordable Housing Crisis?" (blog). http://www.frbsf.org/our-district/about/sf-fed-blog/can-san-francisco-bay-area-solve-affordable-housing-crisis.

Fischer, C. 2014. "The Public Housing Experiment." *Boston Review,* January 14, 2014. https://bostonreview.net/blog/fischer-public-housing-experiment.

Ford, L. R. 1996. "A New and Improved Model of Latin American City Structure." *Geographical Review* 86, no. 3: 437.

Globalization and World Cities (GaWC) Research Network. 2012. "The World According to GaWC 2012." http://www.lboro.ac.uk/gawc/world2012t.html.

Gottmann, Jean. 1957. "Megalopolis or the Urbanization of the Northeastern Seaboard." *Economic Geography* 33, no.3: 189–200. doi: https://doi.org/10.2307/142307.

Harris, Chauncy D., and Edward L. Ullman. 1945. "The Nature of Cities." *The Annals of the American Academy of Political and Social Science: Building the Future City* 242: 7–17. doi: https://doi.org/10.1177/000271624524200103.

HKTDC Research. 2016. "PRD Economic Profile." http://china-trade-research.hktdc.com/business-news/article/Fast-Facts/PRD-Economic-Profile/ff/en/1/1X000000/1X06BW84.htm.

Hoyt, H. 1939. *The Structure and Growth of Residential Neighborhoods in American Cities.* Washington, DC: Federal Housing Administration.

Kazeem, Y. 2016. "Lagos Is Africa's 7th Largest Economy and Is About to Get Bigger with Its First Oil Finds." *Quartz Africa,* May 5, 2016. https://qz.com/676819/lagos-is-africas-7th-largest-economy-and-is-about-to-get-bigger-with-its-first-oil-finds.

Kneebone, E., and A. Berube. 2014. *Confronting Suburban Poverty in America.* Washington, DC: Brookings Institution Press.

Lewis, Peirce. 1983. "The Galactic Metropolis." In Liberman, A. (ed.), *Beyond the Urban Fringe: Land Use Issues in Nonmetropolitan America.* Minneapolis: University of Minnesota Press.

Moreno, J. C. C. E., and J. Clos. 2016. *Urbanization and Development: Emerging Futures.* New York: UN Habitat.

National Association of Regional Council (NARC). *Livability Literature Review: A Synthesis of Current Practice.* Washington, DC: NARC.

Organisation for Economic Cooperation and Development (OECD). 2016. "Patent Activity in Metropolitan Areas." In *OECD Regions at a Glance 2016.* Paris: OECD Publishing. doi: http://dx.doi.org/10.1787/reg_glance-2016-31-en.

Portes, Alejandro, and Bryan Roberts. 2005. "The Free-Market City: Latin American Urbanization in the Years of the Neoliberal Experiment." *Studies in Comparative International Development* 40, no.1: 43–82. doi: http://dx.doi.org/10.1007/BF02686288.

Puga, D. 1996. "Urbanisation Patterns: European vs. Less Developed Countries." Center for Economic Performance. Discussion Paper No. 305.

Reba, M., Reitsma, F, and Seto, K. 2016. "Spatializing 6,000 Years of Global Urbanization from 3700 BC to AD 2000." *Scientific Data* 3: 160034. doi: http://dx.doi.org/10.1038/sdata.2016.34.

Ridley, M. 2011. *The Rational Optimist: How Prosperity Evolves.* New York: Harper Perennial.

Sassen, Saskia. 2005. "The Global City: Introducing a Concept." *Brown Journal of World Affairs* 11, no. 2: 27–43.

Schildt, C., N. Cytron, E. Kneebone, and C. Reid. 2013. *The Subprime Crisis in Suburbia: Exploring the Links Between Foreclosures and Suburban Poverty.* Community Development Working Paper 2013-02. San Francisco: Federal Reserve Bank of San Francisco.

United Nations. 1980. *Patterns of Urban and Rural Population Growth.* Population Studies, No. 68. New York: United Nations.

———. 2014. "World's Population Increasingly Urban with More Than Half Living in Urban

Areas." UN DESA Department of Economic and Social Affairs. http://www.un.org/en/development/desa/news/population/world-urbanization-prospects-2014.html.

UN-Habitat. 2014. *State of African Cities 2014, Re-imagining Sustainable Urban Transitions.* State of Cities—Regional Reports. https://unhabitat.org/books/state-of-african-cities-2014-re-imagining-sustainable-urban-transitions.

———. 2016. *World Cities Report.* United Nations. http://wcr.unhabitat.org/.

US Environmental Protection Agency. 2017. "About Smart Growth." https://www.epa.gov/smartgrowth/about-smart-growth#smartgrowth.

Von Boventer, E. 1969. "Walter Christaller's Central Places and Peripheral Areas: The Central Place Theory in Retrospect." *Journal of Regional Science* 9, no. 1. doi: https://doi.org/10.1111/j.1467-9787.1969.tb01447.x.

Weiss, M. 1990. "The Origins and Legacy of Urban Renewal." In J. Paul Mitchell (ed.) *Federal Housing Policy and Programs: Past and Present.* New Brunswick, NJ: Center for Urban Policy Research.

World Bank. 2015. *East Asia's Changing Urban Landscape: Measuring a Decade of Spatial Growth.* Washington, DC: World Bank.

———. "World Bank Report Provides New Data to Help Ensure Urban Growth Benefits the Poor." Press release. http://www.worldbank.org/en/news/press-release/2015/01/26/world-bank-report-provides-new-data-to-help-ensure-urban-growth-benefits-the-poor.

Chapter 6
Food and agriculture

The world population in 2017 was over 7.3 billion and will reach 11.2 billion by the year 2100. Much of this growth will be in the developing countries of Africa, the Middle East, and parts of Asia, where many poor people struggle to obtain adequate food and nutrition. At the same time, urban areas are growing rapidly. While in 1950 there were two cities with over ten million people, by 2015 there were twenty-nine, most of which were in developing countries with substantial poverty. As human populations grow and as urban populations swell, food production must increase substantially. If it does not, the planet faces risk of a Malthusian crisis with rampant food shortages, urban food riots, and social upheaval.

Great challenges lie in increasing the quantity, quality, and sustainability of food production (figure 6.1). Quantity can be increased with technology that increases farm yields, including much-maligned GMOs (genetically modified organisms). Improvements in agricultural infrastructure, such as irrigation, road and shipping networks, and proper storage facilities, are also necessary so that the food that is produced can reach consumers without being lost in transit. Quality matters as well. Increasingly, affluent urban residents demand food that is not only plentiful at an affordable price but also offers a wide variety of choices. Food that provides adequate nutrition is also essential so that people can lead long and healthy lives. Given that agriculture inherently requires the use of natural resources, it can have impacts on natural habitats,

climate, water quality, and more. For this reason, expanded agricultural production must be performed in a sustainable manner so that the earth continues to provide for future generations.

The next three chapters examine the complex systems of economic geography. Economic geography describes the location of economic activity, be it a wheat farm, a clothing factory, or a social media technology company. It aims to describe how economic activity is driven by the local and global connections of people, ideas, raw materials, and finished products that flow between places. Economic activity can be broken down into three broad categories: the primary, secondary, and tertiary sectors. This chapter focuses on the primary sector of the economy, which involves the extraction of raw materials from the earth. Primary sector activities include farming, fishing, mining, and logging. The following two chapters focus on the secondary sector, which consists of manufacturing, and the tertiary sector, which involves the provision of services.

Economic geography is one of the most relevant aspects of human geography in that it shapes the lives of all people and the natural world. It helps us understand where and why there is satisfying employment in some places and underemployment or unemployment in other places. It helps describe the spatial distribution of poverty and wealth globally and locally. It gives insight into environmental problems (and solutions). And it helps us explain the overall quality of life for humans in our cities and countryside.

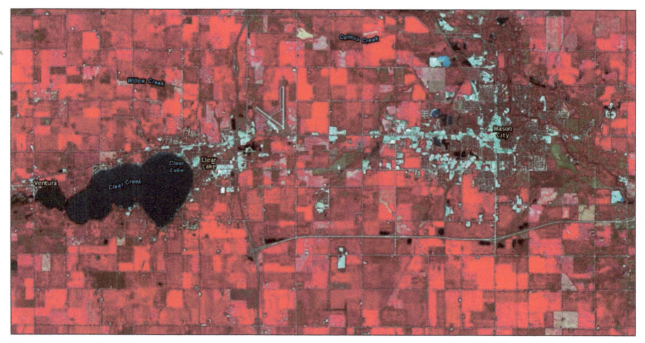

Figure 6.1. Agricultural land in Iowa. This image shows endless fields of corn in color infrared taken from the Landsat 8 satellite. The challenge for agricultural producers is to produce a quantity, quality, and diversity of food to nourish an expanding population. Explore this map at https://arcg.is/1HabvX. Data sources: Esri, USGS, AWS, NASA.

Origin and diffusion of agriculture

The first agricultural revolution

For most of human history, hunting and gathering was the sole means of survival for human groups. But sometime around 9000 BCE, some groups became sedentary and began to domesticate seeds for farming, resulting in the *first agricultural revolution*. In all likelihood, the process of plant domestication was a gradual one. People selected the larger seeds and fruits when gathering, unconsciously choosing specimens that would produce additional larger offspring. Some of these seeds inevitably fell to the ground and grew in places where humans spent more time. This artificial selection of larger seeds and fruits ultimately led to domesticated seeds that differed substantially from their native origins. Planted food would have supplemented hunting and gathering initially, but over time, active farming came to supersede it and became the dominant source of food. Somewhat later, people

began to domesticate animals as well, selectively breeding them for docility and to provide food, clothing, transportation, and labor.

Agriculture developed as an independent innovation in various locations around the world (figure 6.2). Possibly the oldest origin point lies in the Middle East, specifically the Fertile Crescent around 9000 BCE. Here, early agriculturalists grew various types of wheat and raised domesticated sheep, goats, and cattle. Chinese agriculture originated about 8000 BCE, with wheat dominating a northern core while rice dominated a southern core. Around the same time agriculture originated in Mesoamerica and South America. In West Africa and the Ethiopian highlands, agriculture formed around 4000 to 3000 BCE. Eastern North America saw the development of agriculture around 2000 BCE.

From these points of origin, agriculture then spread through *relocation* and *contagious diffusion*. Via contagious diffusion, hunter-gatherer societies

Agriculture **developed independently** in several locations.

The oldest points of origin are likely the **Middle East**, followed by **China**.

Agriculture spread via **contagious** and **relocation diffusion**.

Agricultural surplus led to **growing populations**, occupational **specialization**, and ultimately **urbanization**.

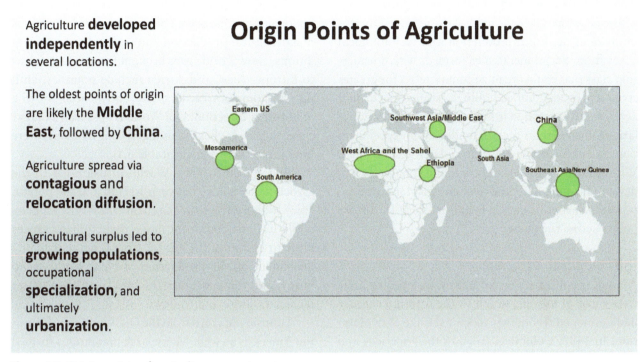

Origin Points of Agriculture

Figure 6.2. Origin points of agriculture. Image by author.

near agricultural societies were exposed to farming techniques and soon adopted them. At the same time, the movement of people to new locations spread agriculture via relocation diffusion. Crops and farming techniques from the Middle Eastern Fertile Crescent diffused westward through Europe and around the Mediterranean Sea and eastward into South Asia. Likewise, crops and techniques from China diffused west. Before too long, seeds and farming technology had diffused throughout the Eurasian landmass.

Of course, there were initial limits to the diffusion of agriculture that slowed its spread to all parts of the world. The techniques and crops that were developed in agricultural points of origin faced intervening obstacles as people were deterred from moving across unfamiliar environments. For instance, people of the Mediterranean were unlikely to migrate across the Saharan desert into sub-Saharan Africa, thus slowing diffusion southward. Similarly, Andean llamas and potatoes were slow to diffuse from their highland environments in South America northward across the

tropical lowlands of Central America. And naturally, large bodies of water, such as the Pacific and Atlantic oceans, slowed diffusion for many millennia.

The results of the first agricultural revolution were profound. Prior to the development of agriculture, hunter-gatherer societies were limited to relatively small bands of people whose numbers were limited by availability of food. But with the advent of farming, food supplies increased and populations grew. This led to an increase in the number of people as well as an increase in population density. Whereas bands of people were once spread thinly over the landscape, they came to be settled in more densely populated sedentary farming communities.

As farming techniques improved, agricultural communities reached levels of efficiency where some people could perform tasks other than farming. This meant that there were growing settlements where some members of the community could specialize in nonagricultural labor. Some could specialize as blacksmiths or carpenters, developing new tools for farming and defense. Others formed a priestly or scientific class, whose observations of the natural world helped further improve agricultural production and technological innovation. Social hierarchies formed, with a leadership class that could organize communities for greater economic production and military strength. The first agricultural revolution led to larger and denser populations, then surplus food that allowed for specialization, and ultimately, innovation and new technologies. With sufficient numbers of people and enough time to develop farming and other technological innovation, powerful urban societies began to form in diffuse points across the globe.

The Columbian exchange

Among the most powerful urban societies to arise were those of Europe, which by the fifteenth century had developed technology to spur the age of exploration. In 1492, Columbus sailed to the Americas and initiated an exchange of crops, disease, people, and ideas that reshaped global political and economic geography (figure 6.3). This process is referred to as the *Columbian exchange*, which overcame the intervening obstacles of the great oceans, finally connecting the Eastern and Western Hemispheres.

Figure 6.3. *The Landing of Columbus, October 11, 1492*, painting by Currier & Ives, 1846. Contact between the Old and New Worlds initiated an exchange of crops and animals between the hemispheres and a massive reconfiguration of global agricultural production. Image by Everett Historical. Stock illustration ID: 252139165. Shutterstock.

The Columbian exchange is one of the most important processes contributing to current patterns of agricultural production and regional culinary traditions. New World crops brought from the Americas to Europe, Asia, and Africa include potatoes, chili peppers, tomatoes, cacao, maize, and tobacco. Moving in the other direction, Old World crops, such as sugar cane, coffee, soybeans, oranges, and bananas, were brought to the Americas (table 6.1).

Old World to New World diffusion led to the association we have between Ireland and the potato. Spicy Indian curries are based on Old World peppers, while Italians could not make their tomato sauces prior to exchange. In Africa, Cote d'Ivoire is highly dependent on the production of New World cacao beans, and top consumers of Mexican-origin maize include Lesotho and Malawi.

Diffusion of crops from the Old World transformed the Americas as well. Many large plantations formed to grow sugar cane, coffee, and bananas for export to markets in the Old World, while soybeans consumed large tracts of land in Brazil and the United States. Not only did these crops alter the agricultural landscape, they dramatically transformed the demographic landscape as well. Large numbers of slaves from Africa and indentured servants from Europe and Asia were brought to the Americas to provide labor for these export enterprises.

From its inception, the Columbian exchange brought more land under cultivation globally, contributing to an increase in caloric output and nutritional diversity. This fed global population growth worldwide and was the first major step in development of our system of international agricultural trade.

The second agricultural revolution

From the beginning of the first agricultural revolution, technological and economic systems changed slowly. Stone and wood tools were gradually developed to help till the land, and by the Bronze Age (circa 2000 BCE), metal tools and animal-drawn plows were put into use. While agricultural products were traded to some degree, most agriculture was for subsistence

Sample Crops from the Old and New Worlds

New World Crops	Old World Crops
cacao beans	apples
cassava	bananas
chilies/peppers	barley
eggplants	coconuts
maize	coffee
natural rubber	grapes
pineapples	oats
potatoes	olives
sunflower seeds	onions
sweet potatoes	oranges
tobacco	palm oil
tomatoes	rice
	rye
	sorghum
	soybeans
	sugar beets
	sugar cane
	watermelons
	wheat
	yams

Table 6.1. Sample crops from the Old and New Worlds. Data source: Nunn and Qian, 2010.

purposes, intended to feed local populations and, at best, nearby expanding cities.

But by the mid-eighteenth century, dramatic changes in agricultural production occurred. As the Industrial Revolution accelerated in Europe, especially in Great Britain, technological and economic change in agriculture took hold. These dramatic changes, from roughly 1750 through 1880, are known as the *second agricultural revolution*. This was a time when agriculture in Europe transformed from a predominately subsistence model to a factory farm model.

These changes were driven by several factors. Whereas two-field crop rotation, which helps maintain soil nutrients, had been in use for hundreds of years, advances in agricultural science led to widespread adaptation of four-course crop rotation in eighteenth-century Great Britain. At the same time, enclosure of agricultural fields took hold. Enclosure is the process whereby individual famers enclose land and take ownership. This practice contrasted with medieval customs of commonly farmed land with communal or feudal ownership. With better crop rotation methods and economic incentives to increase investment and maximize output on privately held farms, agricultural intensity rose and production rapidly increased. Agricultural output further increased as farming took a more industrial approach to production. Farmers increasingly purchased fertilizers and seeds rather than producing their own. They also invested in new machinery, such as mowers, reapers, threshers, and tractors. Landowners specialized in specific crops to maximize efficiency.

Through changes wrought by the second agricultural revolution, food output rose substantially, producing enough food for a growing urban industrial workforce.

The green revolution

Another major transformation in agriculture began in the second half of the twentieth century. This transformation is known as the *green revolution* and is sometimes called the *third agricultural revolution*. The green revolution involved advances in agricultural science that allow for the development of high-yield crops by intentionally breeding seeds that produce more output per acre of land. Crops were developed that increased yields, were resistant to diseases, and responded better to fertilizers. Wheat and rice were the first major crop types to benefit from this technology, and their use spread throughout Asia and Latin America. Later, crops such as maize, cassava, and others were developed for use in Africa.

The green revolution was successful at increasing agricultural output globally. It is estimated that without this technology, prices in the year 2000 would have been 66 percent higher and 14 to 19 percent more land would have been devoted to agriculture in developing countries. Higher food prices would have led to lower caloric intake in developing countries, higher levels of malnourishment, and greater levels of infant and child mortality. Likewise, more land under cultivation would have had detrimental environmental impacts, reducing natural habitat and impacting water and other natural resources. Without the green revolution, the world may not have faced a complete Malthusian catastrophe, but it would have been heading in that direction.

More recently, yield increases have slowed, so new technology is being used to continue the benefits of the green revolution. Whereas early green revolution technology involved cross-breeding plants for desired characteristics, the process is much more precise today. Scientists now manipulate plants at the genetic level by altering specific segments of DNA. These GMOs have been created to resist pests, thus reducing the need for pesticides, and to resist herbicides, allowing farmers to quickly and easily spray their fields to kill weeds but not the crops. Other genetic modifications include rice and other crops that are drought tolerant or flood tolerant, allowing them to be grown on land that was formerly unsuitable for production. As climate change accelerates, increasing droughts in some places and flooding in others, crops with greater environmental

resistance will be even more essential for feeding the world population.

While GMO crops can increase yields, there are criticisms, some valid and some not, related to their use. One criticism is that GMO crops are "unnatural" and can cause adverse health effects in humans. To date, this is unproven, with millions of people consuming GMO crops and no documented negative health impacts. Other criticisms have more value. While less pesticide is used with GMOs, herbicide use has increased, so overall chemical use has not declined as promised (figure 6.4). Another concern is that engineered genes may spread into wild plant populations. There is mixed evidence as to the extent and impact of this spreading.

More broadly, critics of green revolution technology point out that it has benefited large landowners much more than small subsistence farmers. Small-scale farmers often cannot afford to purchase the materials needed for successful implementation of these technologies. Green revolution crops require the purchase of fertilizers and herbicides. In the case of GMOs, seeds often must be bought from large multinational corporations. Likewise, successful green revolution farming often requires water pumps, tractors, and other

Figure 6.4. Crop duster spraying chemicals in Idaho. The green revolution has been a key factor in expanding global food production, but it often requires substantial chemical inputs, such as herbicides. Photo by B. Brown. Stock photo ID: 203957230. Shutterstock.

equipment. Frequently, farmers face debt to purchase the required inputs and may struggle to turn a profit. Finally, these technologies rely on greater scientific knowledge rather than local farming knowledge, which many small-scale farmers do not possess.

While the green revolution has resulted in greater food output, many small landowners, unable to complete with larger farm enterprises, have become low-wage landless agricultural laborers or have migrated to urban areas. Yet overall, the green revolution has helped increase food output for a growing world population. As the world population continues to grow, innovations in technology and farm operations must continue to advance.

Go to ArcGIS Online to complete exercise 6.1: "The Columbian exchange: Origins of agricultural globalization."

The spatial distribution of agricultural activity

Two essential factors drive the spatial distribution of agricultural activity: the natural environment and economics. That the natural environment influences which crops are grown in which locations is relatively self-evident. Grapes need plenty of warmth and sunshine, sugarcane and bananas need ample amounts of water, green beans can thrive in short growing seasons, while citrus needs a long one. Thus, which types of crops grow best in which locations is strongly influenced by temperature, precipitation, and length of the growing season. These factors vary greatly by latitude and elevation. For instance, tropical rainforests along the equator are warm and wet year round; midlatitude Mediterranean climates have hot, dry summers with cooler, wetter winters; subarctic climates have long, cold winters and short, cool summers. These and other climate types shape the spatial distribution of crops.

But while the natural environment restricts where different crops can or cannot grow, economic forces ultimately drive the spatial distribution of agricultural activity. After the first agricultural revolution, humans migrated to new locations and traded with other societies, spreading crops with them. For example, wheat diffused from the Fertile Crescent westward into Europe and North Africa and eastward to the Indus Valley of South Asia. Barriers to diffusion prevented crops from spreading globally at first. So, while wheat grows well in southern Africa, barriers such as the Saharan desert and tropical rainforests inhibited migration and trading with that region, preventing its diffusion. With sea navigation and the Columbian exchange after 1492, barriers to diffusion broke down. Ships allowed people to migrate and trade at a global scale, sailing from Europe to Asia around the southern tip of Africa and across oceans to the Americas and Australia.

Globalization and comparative advantage in agricultural production

In the scramble for dominance, European powers settled and colonized lands across the globe, and agricultural systems were developed to create wealth for the mother countries. Old and New World crops were transplanted around the world, dramatically altering agricultural landscapes. By the nineteenth century, the Industrial Revolution and the second agricultural revolution were in full swing, feeding (literally and figuratively) the modern age of globalization.

The globalization of agriculture is based on the idea of *comparative advantage*. Comparative advantage describes the fact that some places have an advantage at producing certain crops compared to other places. This advantage comes from efficient use of what economic geographers call the factors of production: land, labor, and capital. In the case of agriculture, land describes the natural environment, such as climate and soil conditions. Labor refers to the people who work in agriculture, including their quantities and skill sets. Capital is the machinery and buildings available for production. Each place has a unique combination of these factors, with differing environmental conditions; a varying number of people willing and able to work in agriculture; and differing amounts of farm machinery, storage, facilities, and transportation networks. Regional specialization occurs as places produce agricultural products in which they have a comparative advantage, then trade with other places that have comparative advantages in different crops.

We can see comparative advantage at work in agricultural patterns of the United States (figure 6.5). California is the nation's top producer of almonds, while potatoes are grown primarily in Idaho and Washington. The environment of California is well suited for a wide range of crops, and a substantial amount of potatoes could grow there if farmers chose to plant them. However, California can produce almonds more efficiently than potatoes, so it makes sense to specialize in almonds instead. Of course, that does not mean that the people of California cannot get potatoes and the people of Idaho cannot get almonds. Instead, the two

states trade, allowing Californians and Idahoans to get lower-cost almonds and lower-cost potatoes.

The same process of regional specialization and comparative advantage takes place at a global scale. You may have had a cup of coffee this morning, but it is unlikely that it was produced near you. Coffee grows best in the equatorial zone, so in North America, this means that it is produced only in Hawaii. Globally, tropical Brazil and Vietnam are the largest producers. Figure 6.6 shows the top agricultural exports by value for selected countries in 2013. If your breakfast included a banana, it may have come from Ecuador or Panama. A salad at lunch could have included olive oil from Greece and tomatoes from Morocco. Dinner could consist of rice from India and meat from Ireland. As each country specializes in what it can produce most efficiently compared to other countries, comparative advantage leads to a greater diversity of foods and a lower price than if everything was sourced locally.

Comparative Advantage: Almonds and Potatoes

Potatoes are more efficiently produced in Idaho and Washington.

Almonds are more efficiently produced in California.

Through **comparative advantage**, **regional specialization** and trade produce more goods at a lower price.

Figure 6.5. Comparative advantage: Almonds and potatoes. Data source: Agricultural Census, 2012. US Department of Agriculture, National Agricultural Statistics Service.

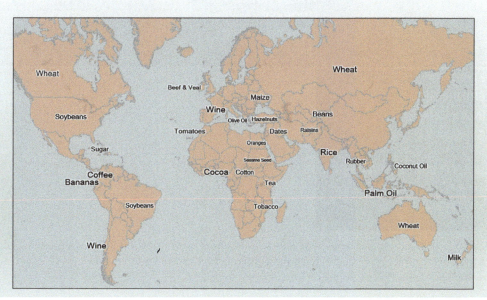

Figure 6.6. Comparative advantage at a global scale. Data source: Food and Agriculture Organization of the United Nations.

Agriculture and level of development

As countries develop and become richer, the role that agriculture plays in their economy changes. Increasing development corresponds with rising levels of productivity. More machinery is used in production, transportation networks connect more farms with markets and trade hubs, better storage facilities reduce losses of product, and more chemical inputs increase yields. This means that a great deal of agricultural output can be produced per worker. As fewer workers are needed to work in agriculture, they are freed up to work in urban occupations in industry (the secondary sector) and services (the tertiary sector). With more workers in the secondary and tertiary sectors of the economy, agriculture becomes a less significant portion of economic output.

Table 6.2 shows how several agriculture-industry variables change as GDP per capita increases. Ethiopia

is a very poor country, with a GDP per capita of only US $603 per year. Being a poor country, capital machinery is very limited, so many workers are needed to produce food. Because food production is very labor intensive, each worker produces only $303 worth of agricultural goods per year. Also, since many people must work in agriculture, fewer are available to work in the secondary or tertiary sectors of the economy. Thus, nearly 38 percent of Ethiopia's GDP comes from agriculture.

Other countries with low GDP per capita are in similar situations, though not to the same degree as Ethiopia. For instance, the Philippines and Indonesia also have relatively low agricultural value added per worker, a substantial portion of their workforces in agriculture, and larger GDP contributions from agriculture.

On the other hand, developed countries with high GDP per capita, such as Germany, the United States, and Switzerland, have very productive agricultural

Agriculture and Level of Development

COUNTRY	AGRICULTURE VALUE ADDED PER WORKER (CONSTANT 2005 US$)	EMPLOYMENT IN AGRICULTURE (%)	GDP FROM AGRICULTURE (%)	GDP PER CAPITA (US$)
ETHIOPIA	303	NA	37.9	603
PHILIPPINES	1,315	30.4	10.3	2,904
INDONESIA	1,281	34.3	13.5	3,346
GERMANY	35,477	1.4	0.6	41,686
UNITED STATES OF AMERICA	68,872	1.5	1.0	56,054
SWITZERLAND	23,225	7.1	0.7	80,831

Table 6.2. Agriculture and level of development. Lower-income countries typically have a larger proportion of workers in the agricultural sector due to lower levels of mechanization and thus lower levels of productivity. Data source: Food and Agriculture Organization of the United Nations.

sectors. Plentiful capital for machinery and other inputs means that fewer workers are needed in the agricultural sector. The value added in agriculture per worker in each of these countries is over $20,000, making them tens to hundreds of times more productive than in poorer countries. High agricultural productivity frees up workers to produce higher-value goods and services in urban industrial and services jobs. Therefore, very small proportions of workers are in agriculture, and national GDP relies less on agriculture and more on secondary and tertiary goods and services.

Types of agricultural activity

The level of integration into the global economy varies widely depending on the factors of production, especially in relation to the availability of capital. The most globally connected agricultural producers rely on large amounts of capital for purchasing wide tracts of high-quality land, machinery, storage and processing facilities, and means of transportation. Those least connected rely on human labor rather than capital investment to produce food for the family or small, local community.

Commercial agriculture describes production with the intent to sell for cash. The degree of labor and capital involved can vary, depending on the type of crop and the cost of human labor versus that of capital machinery. When labor is abundant and cheap, large numbers of people often work the land. When labor is scarce or expensive, capital machinery is often used in its place.

Subsistence agriculture describes systems where food is produced for consumption by the household. The objective of production is to sustain the household, not to earn cash. It tends to be labor intensive, requiring ongoing work by household members, but not capital intensive. Thus, machinery and chemical inputs tend to be limited.

While agriculture is often categorized as commercial or subsistence, production, in reality, falls along a continuum. Some commercial producers sell all of their output for cash, and some subsistence producers do not sell any. However, especially in developing countries, many farmers consume some of their output but sell some for cash.

Commercial and subsistence agriculture are often described as being *intensive* or *extensive*. Intensive agriculture involves activity that produces a larger yield per acre of land. This is achieved with large amounts of labor and/or capital. With more people and/or machinery, land can be intensively worked for optimal output. In contrast, extensive agriculture involves lower yields per acre. Less labor and capital inputs are used, and extensive agriculture can be profitable only with large quantities of land. For this reason, it functions only where land is relatively cheap and population density is low.

Large commercial agribusiness

At the top of the commercial-subsistence continuum are large agribusinesses that sell to national and international markets. These companies can be owned by families or larger groups of investors and rely on substantial amounts of capital and/or labor. Their goal is to maximize efficiency by using the latest agricultural technologies, such as hybrid seeds; fertilizers and pesticides; and production, processing, and distribution machinery. Comparative advantage drives the types of crops and animals produced by large agribusinesses. Based on the unique combinations of land, labor, and capital, production is geared toward export to distant markets. These companies supply large, vertically integrated food companies, such as Dole, Chiquita, and Archer Daniels Midland Company, that can provide everything from processing and packing of crops to retail distribution.

Commercial plantation agriculture consists of extensive, large-scale, single-crop production in the tropics and subtropics that is exported for consumption in the global market. It is very labor intensive, requiring large numbers of lower-skilled workers to cultivate and harvest crops (figure 6.7). At the same time, large amounts of capital are required for purchasing large tracts of land, for processing and storage facilities, and for transportation to export facilities in seaports and airports. Common crops include rice, pineapple, coffee, tea, rubber trees, sugarcane, oil palms, cacao, and bananas.

Plantation crops often dominate the agricultural economy of less-developed tropical and subtropical countries, mainly in Africa (figure 6.8). For instance, over half of revenue earned from crop and livestock exports in 2013 came from tobacco in Zimbabwe. Likewise, rubber in Liberia, cotton in the Central African Republic, coffee in Burundi, and cocoa in Ghana reflected over half of revenue from crop and livestock exports.

Commercial grain farming involves the extensive cultivation of wheat, barley, oats, corn, rice, and other cereals where ample land and capital machinery is available (figure 6.9). Such cereal crops are found in inland regions where summers are short and winters are cold, while rice is found in more humid and warmer environments. With a very small number of workers, commercial grain farming produces large amounts of food for the world population. This is done with modern equipment, such as combine harvesters, storage facilities, and rail and ship transportation. Some of these crops have been modified as part of the green revolution to produce higher yields and to resist pests. Nevertheless, herbicides and pesticides are often still important capital inputs. The largest producers

Figure 6.7. Female tea picker on a tea plantation in Maskeliya, Sri Lanka. Plantations require large numbers of workers to harvest crops. Photo by Melis. Stock photo ID: 554643850. Shutterstock.

Plantation Crops: Over 50% of Agricultural Exports

Plantation crops dominate agricultural exports in many African countries.

69%

Sugar's share of agricultural export value in **Algeria**.

Coffee is **51%** of agricultural export value in **Burundi**.

In Sierra Leone, cocoa represents **75%** of agricultural export earnings.

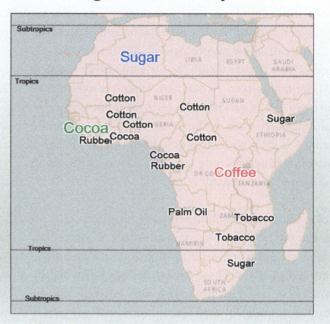

Figure 6.8. Plantation crops: Over 50 percent of agricultural exports. Data source: Food and Agriculture Organization of the United Nations.

of these crops are China, the United States, India, Russia, and Brazil, where farms are often thousands of acres in size.

Figure 6.9. Commercial grain harvesting. Extensive and capital-intensive production requires large amounts of land and machinery. Photo by Sasa Prudkov. Stock photo ID: 148705532. Shutterstock.

Large commercial livestock ranching requires few workers, but capital is needed to purchase large numbers of animals and for permanent water wells and facilities for shearing, holding animals, and transporting them to market. Naturally, it is an extensive operation, which requires sizable amounts of ranchland for grazing. Livestock ranching tends to take place in arid to semi-arid locations where a lack of water makes crops production less feasible. Large ranchlands can be found in places such as the prairies of Canada and the western United States, the Pampas of South America, and the Australian Outback (figure 6.10).

With a growing global middle class, more people now demand meat in their diet, increasing demand for livestock ranching. This contributes to environmental pressure, as more land is required for raising livestock. In South America, for instance, ranchers are clearing rainforest to create pastureland for cattle, threatening the habitat and native species of that landscape.

Figure 6.10. Commercial livestock ranching, Australia. This extensive form of agriculture requires capital for equipment and buildings to support large-scale production. Photo by Anne Greenwood. Stock photo ID: 310552712. Shutterstock.

Large commercial dairy farms are common in developed countries with plentiful capital for purchasing land, animals, and milking equipment. Whereas dairy products were once sold only locally due to being highly perishable, they are now part of a global market that is facilitated by refrigeration and rapid transport. Global competition has put price pressure on farms in places such as the United States, forcing consolidation into larger, more efficient commercial enterprises (figure 6.11). In 1997, there were 252,079 US farms with milk cows, but by 2012, their number had fallen to 130,208 as small producers were pushed out of the market.

Figure 6.11. Commercial dairy farming. Economic pressures are forcing dairy farms to consolidate into large, capital-intensive operations. Photo by Pavel L. Stock photo ID: 156577604. Shutterstock.

Commercial smallholder agriculture

The primary difference between large commercial agribusiness and commercial *smallholder farms* is the size of the enterprise, not the types of crops that are produced. In developed countries, such as the United States, these farms sell their output for cash, just as large agribusinesses do. They can produce crops such as grains, fruits, and vegetables and raise animals for meat and dairy. In 2012, about 65 percent of farms in the United States were less than 180 acres in size, with only about 9 percent of farms over 1,000 acres. Likewise, about three-quarters of farms had sales under $50,000 per year, while just under 20 percent had annual sales over $100,000. Smallholder farms may sell in local markets, but they also supply large national and international food companies with goods.

In developing countries, farms are substantially smaller than in the developed world. Table 6.3 shows the median farm size in several countries in sub-Saharan Africa, Asia, and Latin America. As the table shows, most farms are operated by smallholders. This is due to a scarcity of capital to purchase larger plots of land and limited machinery to efficiently work the land. But while many farms are small, most operate largely or partly as commercial enterprises where output is sold for cash. Hired labor can be employed during peak planting and harvesting time, and some producers have access to mechanized equipment. Despite their small size and relative inefficiency compared to larger

Smallholder Agriculture Acreage

		NATIONAL MEDIAN, IN ACRES
SUB-SAHARAN AFRICA	Kenya, 2005	2.1
	Ethiopia, 2012	4.5
	Malawi, 2011	1.8
	Niger, 2011	10.2
	Nigeria, 2010	2.5
	Tanzania, 2009	3.7
	Uganda, 2012	2.8
ASIA	Bangladesh, 2005	1.0
	Nepal, 2003	2.3
	Vietnam, 2002	1.6
LATIN AMERICA AND THE CARIBBEAN	Bolivia, 2005	3.7
	Guatemala, 2006	2.6
	Nicaragua, 2005	23.5

Table 6.3. Smallholder agriculture. In much of the world, farms are relatively small and operated by individual families. Data source: Food and Agriculture Organization of the United Nations.

commercial agribusiness, many of these producers supply large food companies that distribute at national and global scales. Output is also sold frequently in local markets.

One common type of production found in regions with smallholder farms is *mixed farming*. Mixed farming regions have a greater diversity of agricultural output than regions that specialize primarily in a limited number of crops or animals. Depending on the environment, crops can include grains, fruits, vegetables, and animal products such as meat and dairy. The advantage of mixed farming is that it can reduce economic and environmental risk for smallholders. By producing different types of crops and livestock, farmers are less exposed to price fluctuations of any single crop. Likewise, diversity can reduce risk from pests or weather fluctuations that impact some crops and animals more than others.

Mixed farming can take place on a single farm, referred to as *on-farm mixing*, where different crops and animals are raised, or within a region where

individual farms specialize, known as *between-farm mixing*, where products are exchanged. On-farm mixing typically uses an integrated approach, where waste products are recycled on the farm. For instance, chicken droppings are used to promote algae growth in fish ponds, water from fish ponds is used to irrigate vegetables, and vegetable residues are used to feed livestock. Livestock can also graze under fruit trees, fertilizing them with their manure (figure 6.12). Between-farm mixing works in a similar way but through the exchange of waste products between different farms. For instance, cattle from one farm can provide manure fertilizer, while grain from another farm can provide livestock feed.

In areas with Mediterranean climates, many smallholder farmers work in *Mediterranean agriculture*. Mediterranean climates are those with characteristics similar to the areas around the Mediterranean Sea. These regions have mild, wet winters and hot, dry summers, and they lie roughly between 30 and 45 degrees latitude (figure 6.13). Outside of the Mediterranean

Figure 6.12. Mixed farming in the Caucasus region. Mixed farming can include grazing cows within orchards and using their manure as fertilizer. Photo by Ibrahim Buraganov. Stock photo ID: 299739680. Shutterstock.

Sea region, Mediterranean climates are located on the western or southwestern coasts of continents, which include California, central Chile, and the southwestern portions of South Africa and Australia.

Mediterranean agriculture, as with agriculture in all regions, is influenced by environmental characteristics. With only seasonal rainfall, it is not suitable for water-intensive crops such as rice unless irrigation from other regions can be brought in. However, ample year-round sunshine and short frost periods allow for a wide range of crops and livestock. In these regions, barley, oats, wheat, olives, citrus and deciduous fruits, cotton, and other goods are produced (figure 6.14) either in labor-intensive or in capital-intensive manners, depending on the availability and cost of each.

Based on 2013 export values of crops and livestock, Spain, Italy, and Chile all earned the largest portion of their revenue from wine made of locally grown grapes. Greece earned substantial revenue from olive oil, while California produced nuts, dairy, and wine. When capital and labor are available in sufficient amounts, productivity in Mediterranean agriculture regions can be high. California alone provides over one-third of the United States' vegetables and two-thirds of its fruits and nuts. Likewise, Spain and Italy

Mediterranean Agriculture

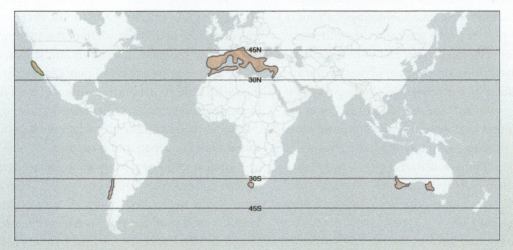

Hot, dry summers and wet, mild winters make Mediterranean regions highly productive and important suppliers of fruits, vegetables, and nuts.

Figure 6.13. Mediterranean agriculture regions. Image by author.

Figure 6.14. Mediterranean agriculture. Olive harvesting in Caltavellotta, Italy. Mediterranean agriculture often provides a substantial amount of fruits and vegetables to surrounding regions. Photo by Yulia Grigoryeva. Stock photo ID: 336578609. Shutterstock.

are the largest producers of fresh fruits, especially grapes and citrus, in the European Union. Although Mediterranean agriculture traditionally has been produced by smallholder farmers, it too is consolidating into larger commercial enterprises as global competition creates pressure to increase efficiency, pushing small producers out of the market.

Subsistence agriculture

While the difference between large commercial agribusiness and commercial smallholder agriculture is based largely on size of the enterprise, the difference between commercial smallholder and subsistence agriculture is based more on the proportion of output that is consumed by the household. Today, very few farmers are so isolated that they do not sell or trade any of their output. However, many consume most of what they produce, selling only small amounts of surplus each year. Subsistence farming is largely labor intensive, with limited capital available for purchasing land, livestock, or machinery that would allow greater commercial production.

Nomadic herding involves the raising of animals such as cattle, goats, sheep, yaks, camels, musk oxen, and more, without having a permanent home base. Nomadic herders live in family or tribal groups and move from place to place in search of forage and water for their animals. They are typically found in regions

that are arid or semi-arid and thus not well suited for agriculture. Given the arid nature of the landscape and the large area needed to support nomadic herding, it is considered a form of *extensive agriculture*. While historically nomadic herding is considered subsistence agriculture, today herders may work for wages or sell or trade some of their animals for other goods and services. Nomadic herding requires minimal to no capital machinery but rather relies on a small number of herders to move flocks.

The Sahel region of Africa, an environmental transition region that lies south of the Saharan desert and north of more fertile savanna landscapes, has one of the largest populations of nomadic herders today (figure 6.15). In recent years, conflict between nomadic herdsmen and farmers has increased as drought conditions dry up grazing lands and growing populations convert land to urban and farm uses.

Another traditional form of subsistence agriculture is shifting cultivation, also known as *swidden agriculture*. This is the process by which land is cleared for farming, used until soil nutrients are exhausted, then abandoned and left to regenerate. In many cases, vegetation is cut and burned so that ash enriches the soil. This process is known as *slash-and-burn agriculture* (figure 6.16). Shifting agriculture is commonly practiced in low-latitude tropical regions of Latin America, Africa, and Asia, where heavy rainfall leaches away

Figure 6.15. Nomadic herding in Djibouti. While nomadic herding is being replaced by stationary commercial agriculture, it can still be found in much of the Sahel region of Africa. Photo by Kertu. Stock photo ID: 422647387. Shutterstock.

soil nutrients in a relatively short time once the natural vegetation is removed. A plot of land can typically be used for a few years before the soil is exhausted and the land abandoned. Shifting cultivation uses human labor more than capital, as people clear land with basic tools and lack capital inputs such as fertilizers to replace soil nutrients.

This form of agriculture has been used sustainably for thousands of years, as villagers cleared land, farmed it, then let it regenerate. However, in places where populations are increasing and land becomes more valuable, forests are increasingly being cut permanently and not allowed to regenerate, leading to increasing deforestation.

Throughout much of South Asia, Southeast Asia, and East Asia, intensive subsistence rice cultivation has been an important form of agriculture. In poorer communities, it has maintained relatively large populations for centuries. Rice requires large amounts of water, so paddies can be developed only in irrigable deltas, floodplains, coastal plains, and terraces. Production relies on large amounts of labor but limited capital machinery (figure 6.17). Nearly all work is performed by hand, with animals assisting in the plowing of fields. Given that this type of agriculture is labor intensive, output per capita is relatively low, and poverty levels are high.

Local patterns of agricultural production

Zooming in to a larger scale, it is also possible to see the spatial distribution of agriculture in local areas. Naturally, the environment plays a role in the specific

Figure 6.16. Slash-and-burn agriculture in Uganda. This form of agriculture has been sustainable for thousands of years, but as populations grow and land becomes scarce, it becomes more difficult to leave land fallow long enough to regenerate. Photo by 360b. Stock photo ID: 160929149. Shutterstock.

Figure 6.17. Intensive subsistence rice cultivation in India. This labor-intensive form of agriculture has supported large populations in much of Asia for centuries. Photo by Arti Arun. Stock photo ID: 532069486. Shutterstock.

types of crops grown in a local area. But more important, economic factors, especially as they relate to distance, shape agricultural production patterns.

Johann Heinrich von Thunen's work from 1826 is one of the most influential in agricultural economic geography. Von Thunen was one of the pioneers in describing the role of accessibility in land use. Land that is close to settled areas has a higher value in that it is more accessible to people. People will bid higher prices for land that is easily accessible, since transportation costs in terms of time and money are lower. As one moves farther from settled areas, transportation costs increase, thus making peripheral land less accessible and less valuable.

Based on the idea of accessibility, von Thunen developed a generalized model of agricultural location based on three key variables: *revenue per unit area*, *land cost*, and *transportation cost*. Based on these variables, the von Thunen model illustrates how agricultural land uses will form rings around urban consumer markets (figure 6.18). As we saw in chapter 5

with urban land uses, the cost of land is greatest near the center of a city or town. Near the center, transportation costs to reach market are low. With distance from town, transportation costs increase, thus driving down land costs.

Per the von Thunen model, agricultural rings form as follows:

- **Market gardening and dairy farming.** Close-to-town agriculture is intensive, with high levels of output per unit area. Vegetables and milk bring higher revenues per unit area of land and are difficult to transport due to their perishability. Farmers produce these products in the first ring, delivering them to market daily.
- **Silviculture**. This ring consists of forest land for fuelwood and construction timber. While these products are not perishable, they are heavy and bulky, so transportation costs are high. This makes their production feasible only in areas relatively close to market.

Von Thunen Model of Agricultural Land Use

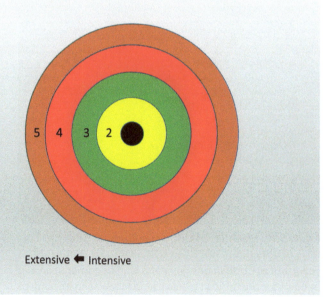

Land values are strongly influenced by **accessibility** to urban markets.

Types of agricultural products vary by distance based on their **revenue per acre**, **land value**, and **transportation cost**.

Land Use Zones:
1. Village or City
2. Market gardening and dairy farming
3. Silviculture
4. Crop farming
5. Livestock grazing

Extensive ← Intensive

Figure 6.18. Von Thunen model of agricultural land use. Image by author.

- **Crop farming**. As one moves farther from urban markets, agricultural production becomes more extensive. In this ring are found crops that earn lower revenue per unit area of land and are easy to transport. Wheat, corn, and other grain crops need more land for their production and can be transported relatively easily to urban markets without risk of perishing in route.
- **Livestock grazing.** The outer ring consists of extensive areas of land used for grazing by cows, sheep, goats, and other animals. Very large tracts of land are required for grazing, resulting in low revenue per unit area. While meat is perishable, animals can be transported live to slaughterhouses close to urban markets.

The von Thunen model was developed in the early nineteenth century, a time when refrigeration was close to nonexistent and agricultural products were brought to market by farmers in horse-drawn wagons. Refrigeration technology has now become widespread, and transportation connections now link lands worldwide, thus reducing the applicability of the model. As discussed in the previous section, comparative advantage and trade allow for specialization over much larger regions.

Nevertheless, the von Thunen model can still describe agricultural patterns in some cases. In poorer, less-developed places, where refrigeration and transportation are limited, the model is more likely to apply. Furthermore, the von Thunen model applies to modern urban farmer's markets in developed countries. The idea behind urban farmer's markets is that consumers can purchase locally grown fruits and vegetables that have traveled short distances from nearby farms. The intended benefits are that the produce has not been frozen or refrigerated for long journeys and therefore is fresher and more flavorful.

Local versus global agriculture

Although much food is produced globally based on the concept of comparative advantage, some farmers still grow crops, especially higher-value fruits and vegetables, close to urban areas, as per the Von Thunen

model. So, which should a consumer choose? Some people emphatically defend a *locavore* approach, arguing that buying locally produced agricultural goods offers a wide range of environmental, economic, and social benefits. Other people prefer the efficiency and convenience of buying globally produced food.

Those who promote buying locally produced foods focus on several perceived benefits. From an environmental standpoint, it is argued that locally produced food creates less greenhouse emissions, since food is transported tens of miles rather than hundreds or thousands of miles. Transporting a local crop forty miles to an urban market requires less fuel and therefore has less of an environmental impact than transporting a globally produced crop 1,500 miles.

Another potential benefit of locally produced food is that is it fresher and more flavorful. It does not have to be frozen or preserved with synthetic chemicals for shipment to another country or continent. Correspondingly, different produce is consumed according to the season in a manner that is more in tune with the natural environment.

Another stated benefit of locally produced food is social. By purchasing locally produced foods at a farmer's market, it is possible to have a personal relationship with the farmer. This may reflect a nostalgic desire for a more "analog" life, with interaction at a human scale, in a time when it is easier to simply click a computer screen and have groceries delivered directly to your home (figure 6.19).

Lastly, promoters of locally produced food make an economic argument: that when people buy locally, money stays within the community. This can help with local economic development by circulating the money locally.

Those who promote a global approach to food production say that the case in favor of locally produced food is not so clear. As the underlying goal of comparative advantage is to produce efficiently, goods can typically be raised at a lower cost and sold to consumers at a lower price. This is a very important issue for low-income households around the world. However, if consumers purchase locally grown foods in places with large agricultural economies during peak

Figure 6.19. Farmer's market, Zagreb, Croatia. There are benefits to both locally produced and globally produced food. Photo by Don Mammoser. Stock photo ID: 138644645. Shutterstock.

season, there may be some cost savings as compared to purchasing globally grown foods.

The environmental case for locally produced food is also less clear. While shorter distance transportation can reduce energy use and carbon emissions, much more energy is used in agricultural production than in transportation. One study has shown that 83 percent of emissions occur during food production, making the argument in favor of production efficiency even more salient. Just consider how much less energy can be used to produce tomatoes in Spain for shipment to Sweden as compared to growing tomatoes locally in Sweden. In fact, what one consumes can have a greater impact on carbon emission than how far food travels. By switching from red meat and dairy to chicken, fish, and vegetables, greenhouse gas emissions can be reduced significantly because cows consume large amounts of grain and because they produce substantial amounts of methane, a greenhouse gas.

While locally produced food may indeed be fresher, the global agricultural economy allows for a greater diversity in diets. Southern hemisphere countries can supply the north with fruits and vegetables in January, while the opposite is true in August. In much of the world, diets would be much more limited if the only produce available was from locally grown sources.

Lastly, global trade in agricultural goods can help farmers in developing countries earn more money and expand their economies. As developing countries grow wealthier and global aggregate demand increases, they have more money to purchase goods and services from developed countries, including different agricultural goods, more sophisticated manufactured goods, and consumer and professional services (figure 6.20).

Ultimately, it is likely that people will buy some combination of locally and globally grown foods, depending on what is produced locally, the season, and personal preference. For many people in developed countries, a quick stop by the grocery store can be convenient, but a nice stroll through a farmer's market may be more pleasant. In developing countries, many will choose the lower costs of locally grown seasonal foods, with only occasional purchases of essentials that cannot be provided locally.

Go to ArcGIS Online to complete exercise 6.2: "Agriculture and development," and exercise 6.3: "Farmer's markets: Consumption patterns of buying locally."

Figure 6.20. Agricultural exports in Thailand. One benefit of globally produced agricultural products is that it can help grow economies in the developing world. This can ultimately benefit developed countries as global incomes and consumption increase. Photo by Sergey Edentod. Stock photo ID: 664265806. Shutterstock.

Agriculture and sense of place

As you'll recall from chapter 1, the unique combination of physical and human features can form a strong sense of place. In many parts of the world, agriculture and agricultural products are key components that contribute to the unique "personality" of places. For instance, what would Napa Valley be without wine? Georgia is known as the Peach State. When many people think of Mexico, tequila comes to mind, while coffee is associated with Colombia. The association between agricultural products and specific places invokes an important essence of a place, be it a hot cup of Colombian coffee at a café in Bogotá or a cool margarita on a beach in Cancún.

Typically, the association between agricultural goods and places is based on centuries, if not millennia, of traditional knowledge that formed from the combination of unique environmental characteristics and cultural practices. Authentic parmesan cheese comes from several northern provinces of Italy, where the cows are fed with locally grown fodder and production processes have been traced back at least to the fourteenth century. Likewise, wine production in Champagne, France, goes back to the ancient Roman empire 2,000 years ago. The climate and soils of this region, combined with specialized techniques for growing grapes and processing and fermenting them, make it world famous for its sparkling wine. Colombian coffee gets its uniqueness from the Andes' volcanic soils and climate patterns that cross between the Amazon basin and the Pacific and Atlantic oceans.

Because of the close relationship between environment and cultural traditions, many producers argue that products cannot be created in just any place. They are not simply a recipe to be copied but something that can be produced only in a specific location in a specific manner. In an attempt to protect place-based agricultural products from inauthentic imitators, legal systems have been put in place in many countries and regions. In essence, these legal protections function similarly to copyrights, but they are tied to specific places rather than specific companies.

For instance, the European Union protects hundreds of products on the basis of their place of origin. These include feta cheese that can be produced only in Greece and champagne and other wines that must come from specific regions in Europe. Some but not all of these protections apply in the United States. For example, parmesan cheese sold in the European Union must be from specific regions of northern Italy, but the rule does not apply to parmesan sold in the United States. The same holds true for champagne, where sparkling wines produced in California can carry the same name. Within the North American Free Trade Agreement (NAFTA), tequila and its coarser cousin, mescal, can be produced only in Mexico, while Canadian whiskey can come only from Canada and bourbon whiskey from the United States (figure 6.21). Other places-of-origin protections include Colombian coffee and Napa Valley wine.

These protections have two major goals. One is to protect traditional culture. Since most protected agricultural products have long histories in their place of production, legal protection can help preserve traditional production methods and prevent mass-produced copies from places without the same environment or indigenous knowledge. In short, it can help preserve

Figure 6.21. Agriculture and sense of place. Blue agave for tequila production, in Tequila, Jalisco, Mexico. Some agricultural products are inextricably linked to specific places and can even be protected by law. Photo by T photography. Stock photo ID: 291156659. Shutterstock.

and enhance the sense of place of many rural communities. The other goal is simply economic. By legally protecting an agricultural product, producers in the region face less competition from firms in other places. Marketing campaigns can be developed to promote exports and attract tourists, contributing to rural economic development.

Hunger and food security

Ultimately, the most important reason for understanding the geographic patterns of agriculture is that it is what sustains human life on our planet. The global population continues to grow, although not as quickly as in previous decades, meaning that more food must be produced to sustain it. As you'll recall from chapter 2, a Malthusian catastrophe has been avoided so far, as world food supply has grown faster than population (figure 6.22). More food has meant less

undernourishment. Figure 6.23 shows that the average proportion of undernourished people in a sample of about 100 developing countries has fallen by nearly ten percentage points since 1989. Data is not available for earlier years, but it can be assumed that it has fallen even more so since the 1960s, when the green revolution was beginning to increase crop yields.

So, the good news is that overall food production has increased relative to population growth, resulting in less global hunger. The benefits of the green revolution and the efficiencies gained through comparative advantage have saved millions from hunger and starvation. But with that said, there is still much work to be done.

Before moving on, it is important to define the terms *hunger* and *food insecurity*. Hunger was traditionally used by geographers and others to describe a prolonged lack of food. Typically, places with high rates of hunger would have high rates of stunted growth and underweight people. Food security or insecurity

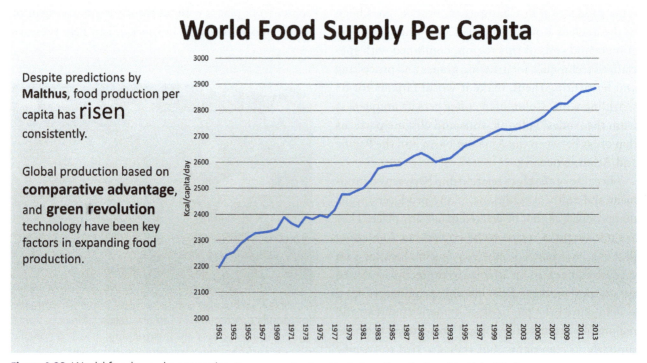

Figure 6.22. World food supply per capita. Data source: Food and Agriculture Organization of the United Nations.

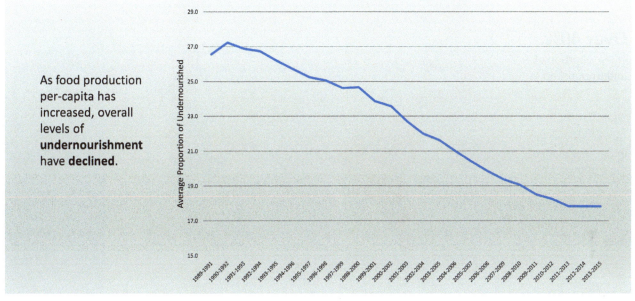

Average Proportion of Undernourished, Selected Developing Countries

As food production per-capita has increased, overall levels of **undernourishment** have **declined**.

Figure 6.23. Average proportion of undernourished, selected developing countries. Data source: Food and Agriculture Organization of the United Nations.

is now preferred as a broader concept. Food security includes not only access to food but also quality and diversity necessary for a healthy diet. Hunger is one result of food insecurity.

Hunger and food insecurity are still grave problems in too many parts of the world. Figure 6.24 shows the wide variation by region in undernourishment. In the region of Middle Africa, over 40 percent of the people are undernourished, while East Africa is just slightly better off at over 30 percent. In the third-highest region, South Asia, over 15 percent of the population is undernourished.

Further evidence of hunger is shown in figure 6.25 that describes a sample of countries where over 20 percent of children under age 5 are underweight. Low weight for children is associated with increased risk of mortality as well as a child's growth potential. It can lead to lifelong cognitive and physical impacts, which can limit economic opportunities for the individual and, collectively, for a country.

The economic foundation of food insecurity

Based on the world food supply per capita in figure 6.22, it is clear that there is more food than ever, yet hunger persists in some places. But availability of food must also be accompanied by accessibility and quality. Most hunger is due to a lack of access to food, while in some cases, it is caused by the wrong types of foods. Poverty is one of the most powerful causes of hunger, and it is typically the poorest of the poor within a country who have insufficient food. When a country is poor, such as in much of Africa and parts of Asia, higher numbers of people struggle to find adequate nutrition on a regular basis.

At the individual level through the societal level, poverty that results in food insecurity often leads to a vicious cycle of ongoing poverty and hunger. When people are hungry or undernourished, they are sicker and less economically productive, making it difficult to rise out of poverty and attain food security. This

Proportion Undernourished by Region

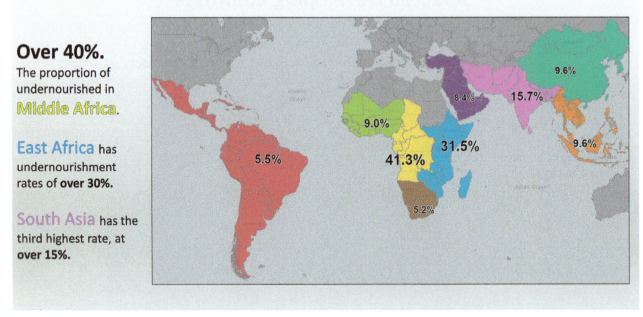

Over 40%.
The proportion of undernourished in **Middle Africa**.

East Africa has undernourishment rates of **over 30%**.

South Asia has the third highest rate, at **over 15%**.

Figure 6.24. Proportion undernourished by region. Data source: Food and Agriculture Organization of the United Nations.

Children Under Age 5, Underweight

Lack of food and nourishment leads to high levels of underweight children in some countries.

This can have long-term negative impacts on individuals, and overall economic development.

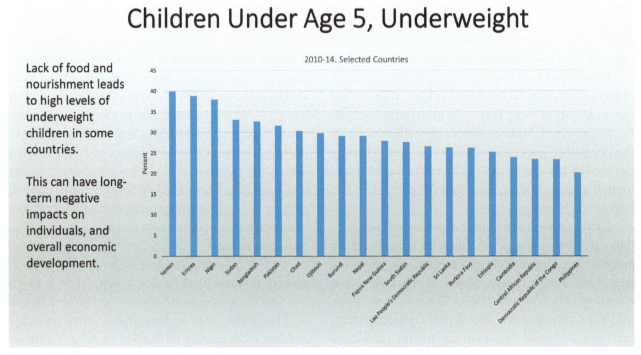

Figure 6.25. Children under age five, underweight. Data Source: Food and Agriculture Organization of the United Nations.

is obviously bad for individuals and households, but in aggregate, high rates of undernourishment can also hold back overall development of a country. For instance, the sub-Saharan country of Malawi loses about US $600 million annually due to child undernutrition. Poor health from undernourishment increases costs in medical care and lost income from family members who must care for the ill. From an educational standpoint, losses come from grade repetition and lower levels of school retention. Extra time spent repeating classes has a direct cost on schools, while children who drop out have lower levels of human-capital skills. Ultimately, a sicker and less-educated population lowers overall productivity of the workforce, thus reducing economic potential.

Economic growth is the surest way to reduce poverty and, in turn, food insecurity. In recent years, China's economic growth has pulled millions of people out of poverty. As seen in figure 6.26, as GDP per capita has increased, the proportion of undernourished has decreased correspondingly. In 1991, nearly 24 percent

of China's population was undernourished, but by 2014, the proportion had fallen to under 10 percent. This represents 160 million fewer undernourished people between 1991 and 2014.

However, when economic growth is not inclusive, that is, when a substantial segment of the population does not benefit from a growing economy, then poverty and undernourishment can remain high. This was the case in the United Republic of Tanzania through much of the 1990s and 2000s (figure 6.27). In this case, economic liberalization unleashed economic growth, but many people, such as smallholder farmers without access to capital and export markets, were unable benefit from growth. Despite economic growth, the rate of undernourishment is higher than in the early 1990s.

To reduce food insecurity, countries need to ensure that smallholder farmers can increase productivity through training on land, soil, and water management and access to capital inputs such as fertilizer and improved seeds. Better access to markets, both domestic and international, can also help increase

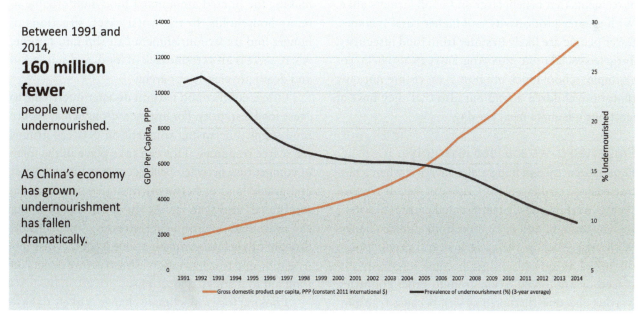

Figure 6.26. China: Economic growth and undernourishment. Data source: Food and Agriculture Organization of the United Nations.

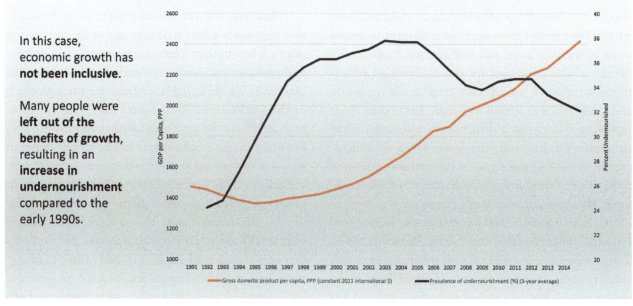

Figure 6.27. United Republic of Tanzania: Economic growth and undernourishment. Data source: Food and Agriculture Organization of the United Nations.

rural incomes. This requires improvements in transportation infrastructure as well as international trade agreements that benefit large and small farmers alike. With greater productivity and higher rural incomes, fewer people are likely to suffer from food insecurity. Targeted social protection programs can also help. For example, school lunch programs can ensure not only that more children attend but also that they have at least one adequate meal per day.

Protracted crises and food insecurity

Food insecurity and hunger also increase when a protracted crisis strikes a place. Crises can include conflict, such as warfare and civil unrest, and natural disasters, such as drought and disruption from climate change. With crises such as these, agricultural production is disrupted as agriculturalists become internally displaced or are forced to emigrate to other countries. Markets become disrupted as well because transportation routes become dangerous to travel and roads

and bridges are destroyed. With a decline in food production and restricted market activity, prices rise, making the limited amounts of food unaffordable to many households. In the worst cases, situations of hunger and undernourishment can slip into famine, where there is an extreme lack of food, and starvation and death are clearly apparent.

Often, conflict and natural disaster intertwine to cause food insecurity. For instance, drought may reduce food supply, increasing competition between groups for scarce resources. This can take place in the form of competition between producers over limited water and arable land or between consumers facing rising food prices and scarcity. At the same time, conflict can exacerbate environmental stresses. A manageable drought can become unmanageable if conflict disrupts irrigation and transportation infrastructure or cuts off food aid to distressed locations.

Looking forward, climate change is likely to have an increasing impact on food insecurity. Sub-Saharan

Africa will see lower yields of all-important cereal crops such as maize, as the frequency of extremely dry and extremely wet years increase. In Asia, higher temperatures will reduce yields of the region's most important staple, rice. Likewise, wheat and maize yields will decline in North Africa and the Near East. Poor regions such as these that already have food insecurity will be less able to adapt to changes as economic resources to help agricultural producers adapt are limited.

In 2017, nearly five million people faced food crises, with 100,000 facing famine, in the young country of South Sudan. With its independence from Sudan in 2011, there was great hope that South Sudan would move in a direction of greater economic and social development. However, civil war based on political and ethnic divisions soon ripped the new country apart (figure 6.28). This led to millions of internally displaced people and refugees. Following the typical pattern, crop production fell, food could not get to markets, foreign aid supplies were interrupted, and prices skyrocketed.

Similarly, conflict in northern Nigeria in 2017 led to food crises. The terrorist organization Boko Haram, which affiliated itself with Islamic State, used violence and kidnapping in an attempt to carve out

an independent state in this region, resulting in over 20,000. Combined with poverty and environmental degradation, the resulting large-scale displacement of people reduced crop production and markets.

In the Western Hemisphere, the poverty-stricken country of Haiti has faced severe food insecurity due to environmental shocks. A major earthquake devastated Port-au-Prince in 2010 (figure 6.29). Before it could fully recover, a major drought reduced food output in 2015, while a devastating hurricane struck in 2016. In a country where three-quarters of the population lives on less than $2 per day, the ability to absorb environmental shocks is severely limited. Current models predict that tropical regions of Latin American and the Caribbean, such as where Haiti is located, will face lower agricultural productivity from climate change–induced heat stress and drought.

Food insecurity in the developed world: Food deserts

Whereas food insecurity in developing countries such as South Sudan, Nigeria, and Haiti often manifests as a lack of calories, in the developed world, food insecurity more often consists of inadequate nutrition. In the United States in 2015, 5 percent of households

Figure 6.28. Conflict and food insecurity. A South Sudanese soldier. Conflict can create or exacerbate food insecurity. Photo by Punghi. Stock photo ID: 287374589. Shutterstock.

Figure 6.29. Natural disaster and food insecurity. Massive damage from the 2010 earthquake in Port-au-Prince, Haiti. Damage from natural disasters can push poor countries into situations of severe food insecurity. Photo by Arindambanerjee. Stock photo ID: 60773563. Shutterstock.

had very low food security, where at least one person had disrupted food intake during the year. Another 7.7 percent of households faced low food security, where less diverse food was consumed or food was obtained from federal food assistance programs or community food pantries.

From a geographic standpoint, malnutrition can be seen as a combination of low income and low access to nutritious food. Food deserts are places where low levels of income and access act as barriers to food security. Some urban neighborhoods and rural communities are not served by supermarkets or grocery stores that provide healthy and affordable food. When this is combined with poverty, where people lack vehicles and adequate public transit, people are often forced to rely on local convenience stores or fast-food restaurants for meals. These typically offer fewer nutritious options, such as a large produce section, and charge higher prices than large, efficiently run supermarkets. Beer and candy may be readily accessible in local convenience stores, but fresh fruits and vegetables are less so.

In the United States, food deserts have been found in a number of urban and rural areas. Figure 6.30 shows food deserts in New Orleans. These places have been identified as census tracts where at least 100 households are low income, lack a supermarket within a half mile, and do not have a vehicle. With a substantial number of low-income households without vehicles, people in these areas are more likely to purchase food at nearby corner convenience stores or possibly fast-food restaurants. As seen in the figure, not all low-income census tracts are food deserts according to this definition. In many low-income communities, the vast majority of households have access to a vehicle and can drive to supermarkets that are outside of the immediate neighborhood. Likewise, there are supermarkets found in some low-income neighborhoods. It is the combination of poverty, lack of access, and lack of a vehicle that limits food options for people.

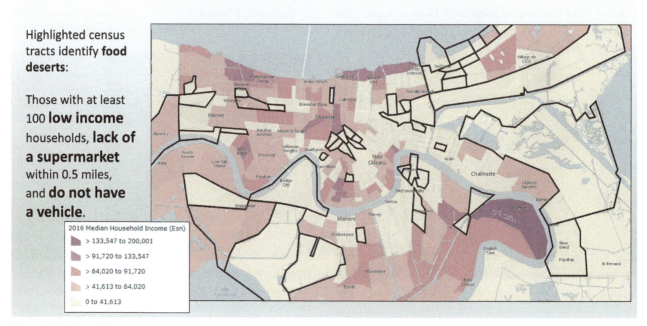

Figure 6.30. Food deserts. Data source: Economic Research Service (ERS), US Department of Agriculture (USDA).

It should be noted, however, that access to nutritious foods does not guarantee that people will consume them. Some research shows that even when supermarkets with healthy options are opened in places that were formerly food deserts, people's diets do not change. Two factors may explain why some people do not improve their diet when options become available. First, "junk food" with lots of calories per gram are substantially cheaper than healthier fruits and vegetables. Thus, for low-income households, it is cheaper to satiate hunger with less-healthy foods. Second, poor eating habits are closely associated with education levels, especially of mothers. Merely adding a grocery store to a community will have limited impact on eating habits when dietary knowledge is limited and unhealthy habits are ingrained.

The challenge of food loss and food waste

As a final note on hunger and food security, it is important to understand the role that food loss and food waste play in our global food supply. On a positive note, food production is up and malnutrition is down. Yet, populations continue to grow, and pockets of food insecurity persist. To reduce food insecurity and feed a growing world population, as much food as possible must get from the fields to people's mouths. Yet, astonishingly, around one-third of the food produced for humans gets lost or wasted each year. Food loss occurs between production and arrival at a retail market, while food waste is when food spoils or is discarded by retailers or consumers (figure 6.31). Generally, there is more food loss in developing countries, while food waste is a greater problem in developed countries.

In Africa, a place with relatively high levels of food insecurity, food is lost to mold, insects, and rodents due to inadequate storage facilities. Without refrigeration, dairy and fish perish, while vegetables such as tomatoes rot and get crushed in transit on bumpy, slow roads. In India, another place with high levels of food insecurity, the same is found. But waste is not exclusive to poor developing countries and regions. In rich countries, crops are left in the field or turned into compost merely because they are not the right size, shape, or color.

Food waste is a greater concern in richer developed countries. There, consumers waste nearly as much food as that which is produced in sub-Saharan Africa on an annual basis. Waste occurs when consumers throw away food at restaurants and at home and when businesses throw away unsold product when new deliveries arrive.

In Europe alone, food waste could feed 200 million people, while food loss in Africa could feed an additional 300 million. To reduce food loss, better storage and transportation infrastructure is needed in developing countries, a great challenge given economic constraints. Getting people in developed countries to waste less may be just as difficult. Possible solutions include smaller restaurant portions and better descriptions of production "use by" dates so that people do not immediately toss items out. There are even businesses now that sell "ugly" or "imperfect" fruits and vegetables, creating a market for items that would have previously been thrown away by farmers or grocers.

Go to ArcGIS Online to complete exercise 6.4: "Food deserts and racial disparities," and exercise 6.5: "Undernourishment and development."

Figure 6.31. Food waste in Bangkok, Thailand. Food waste is a serious problem in developed countries but is also common in developing countries. Photo by Hadkhanong. Stock photo ID: 159058547. Shutterstock.

References

Boone-Heinonen, J., P. Gordon-Larsen, C. I. Kiefe, J. M. Shikany, C. E. Lewis, and B. M. Popkin. 2011. "Fast Food Restaurants and Food Stores: Longitudinal Associations with Diet in Young to Middle-aged Adults: The CARDIA Study." *Archives in Internal Medicine* 171, no 13: 1162–70. doi: https://doi.org/10.1001/archinternmed.2011.283.

Bornstein, D. 2012. "Time to Revisit Food Deserts." *New York Times*, April 25, 2012. https://opinionator.blogs.nytimes.com/2012/04/25/time-to-revisit-food-deserts/?_r=0.

Brouwer, C., and M. Heibloem. 1986. *Irrigation Water Management: Irrigation Water Needs.* Rome: Food and Agriculture Organization of the United Nations. http://www.fao.org/docrep/s2022e/s2022e00.htm.

California Department of Food and Agriculture (CDFA). 2015. *2015 Crop Year Report.* Sacramento, CA: CDFA. https://www.cdfa.ca.gov/statistics.

Centre for the Promotion of Imports (CBI). 2015. *CBI Trade Statistics: Fresh Fruit and Vegetables in Europe.* The Hague, Netherlands: CBI Ministry of Foreign Affairs. https://www.cbi.eu/sites/default/files/market_information/researches/trade-statistics-europe-fresh-fruit-vegetables-2015.pdf.

Cho, R. 2012. "How Green Is Local Food?" *State of the Planet.* Earth Institute, Columbia University blog comments. http://blogs.ei.columbia.edu/2012/09/04/how-green-is-local-food.

Dallman, P. 1998. *Plant Life in the World's Mediterranean Climates: California, Chile, South Africa, Australia, and the Mediterranean Basin.* Berkeley, CA: University of California Press.

Dean, A. 2007. "Local Produce vs. Global Trade." *Global Policy Forum.* Carnegie Council for Ethics in International Affairs. https://www.globalpolicy.org/component/content/article/220/47372.html.

DeWeerdt, S. 2009. "Is Local Food Better?" *World Watch Magazine* 22, no. 3 (May-June). http://www.worldwatch.org/node/6064.

Diamond, J. M. 2017. *Guns, Germs, and Steel: The Fates of Human Societies.* New York: W.W. Norton.

Economic Research Service, US Department of Agriculture (USDA). 2017. *Food Access Research Atlas.* Washington, DC: USDA. https://www.ers.usda.gov/data-products/food-access-research-atlas.

The Economist. 2014. "The New Green Revolution: A Bigger Rice Bowl." *The Economist*, May 10, 2014. http://www.economist.com/news/briefing/21601815-another-green-revolution-stirring-worlds-paddy-fields-bigger-rice-bowl.

Evenson, R. E. 2003. "Assessing the Impact of the Green Revolution, 1960 to 2000." *Science* 300, no. 5620: 758–62. doi: https://doi.org/10.1126/science.1078710.

Federación Nacional de cafeteros. "Colombian Coffee Is Unique." https://www.federaciondecafeteros.org/particulares/en/nuestro_cafe/el_cafe_de_colombia.

Folger, T. 2014. "The Next Green Revolution." *National Geographic,* October, 2014. http://www.nationalgeographic.com/foodfeatures/green-revolution.

Food and Agriculture Organization of the United Nations (FAO). 2015. *The State of Food Insecurity in the World (SOFI) 2015.* FAO Agriculture and Economic Development Analysis Division. http://www.fao.org/3/a-i4646e.pdf.

———. "Climate Change, Agriculture, and Food Security." *The State of Food and Agriculture 2016.* FAO. http://www.fao.org/publications/sofa/2016/en.

Food Security Information Network (FSIN). 2017. *Global Report on Food Crises 2017.* Rome: FSIN. http://www.fao.org/3/a-br323e.pdf.

Fuller, D. Q., T. Denham, M. Arroyo-Kalin, L. Lucas, C. J. Stevens, L. Qin, R. G. Allaby, and M. D. Purugganan. 2014. "Convergent

Evolution and Parallelism in Plant Domestication Revealed by an Expanding Archaeological Record." *Proceedings of the National Academy of Sciences* 111, no. 17: 6147–52. doi: https://doi.org/10.1073/pnas.1308937110.

Hakim, D. 2016. "Doubts About the Promised Bounty of Genetically Modified Crops." *New York Times,* October 29, 2016. https://www.nytimes.com/2016/10/30/business/gmo-promise-falls-short.html?_r=0.

International Coffee Organization. 2016. "Trade Statistics Tables." http://www.ico.org/trade_statistics.asp?section=Statistics.

Jennings, R. 2013. "What Makes Champagne Special? A Brief History." *The Huffington Post,* December 6, 2013. https://www.huffpost.com/entry/what-makes-champagne-spec_b_4278904.

MacDonald, J. M., P. Korb, and R. A. Hoppe. 2013. *Farm Size and the Organization of U.S. Crop Farming.* Washington, DC: US Department of Agriculture, Economic Research Service.

Mailin, J. 2017. "The 100-Year-Old Loophole That Makes California Champagne Legal." *VinePair* blog. https://vinepair.com/wine-blog/loophole-california-champagne-legal.

Martin, K., and J. Sauerborn. 2013. "Origin and Development of Agriculture." In *Agroecology.* Netherlands: Springer.

McMillan, T. 2016. "Shift to 'Food Insecurity' Creates Startling New Picture of Hunger in America." *National Geographic,* July 2016. http://news.nationalgeographic.com/news/2014/07/140716-hunger-america-food-poverty-nutrition-diet.

Nair, K., G. W. Crawford, K. Mellanby, W. D. Rasmussen, G. Ordish, and A. W. Gray. et al. 2017. "Origins of Agriculture." *Encyclopædia Britannica.* https://www.britannica.com/topic/agriculture#toc10760.

Nunn, N., and N. Qian. 2010. "The Columbian Exchange: A History of Disease, Food, and Ideas." *Journal of Economic Perspectives* 24, no.2: 163–88. doi: https://doi.org/10.1257/jep.24.2.163.

National Coffee Association of USA. "Coffee Around the World." http://www.ncausa.org/About-Coffee/Coffee-Around-the-World.

National Geographic Society. 2012. "Ranching." National Geographic Society Resource Library. https://www.nationalgeographic.org/encyclopedia/ranching.

Olmsted, L. 2016. "Most Parmesan Cheeses in America Are Fake, Here's Why." *Forbes,* November 19, 2016. https://www.forbes.com/sites/larryolmsted/2012/11/19/the-dark-side-of-parmesan-cheese-what-you-dont-know-might-hurt-you/#4b00374e4645.

Parker-Pope, T. 2007. "A High Price for Healthy Food." *New York Times,* December 5, 2007. https://well.blogs.nytimes.com/2007/12/05/a--high-price-for-healthy-food/?_r=0.

Pirog, R, and McCann, N. 2009. "Is Local Food More Expensive? A Consumer Price Perspective on Local and Non-Local Foods Purchased in Iowa." *Leopold Center for Pubs and Papers* 63. Iowa State University. https://lib.dr.iastate.edu/leopold_pubspapers/63/ l.

Price, T., and Bar-Yosef, O. 2011. "The Origins of Agriculture: New Data, New Ideas: An Introduction to Supplement 4." *Current Anthropology* 52, no. S4: S163–74. doi: https://doi.org/10.1086/659964.

Reardon, T. 2016. "Growing Food for Growing Cities: Transforming Food Systems in an Urbanizing World." Chicago Council on Global Affairs. https://www.thechicagocouncil.org/publication/growing-food-growing-cities-transforming-food-systems-urbanizing-world.

Royte, E. 2014. "One-Third of Food Is Lost or Wasted: What Can Be Done." *National Geographic,* October 13, 2014. http://news.nationalgeographic.com/news/2014/10/141013-food-waste-national-security-environment-science-ngfood.

Schiere, H., and L. Kater. 2001. *Mixed Crop-Livestock Farming: A Review of Traditional Technologies Based on Literature and Field Experience.* Rome: Food and Agriculture Organization of the United Nations, Animal Production and Health Division.

Simon, J. 2017. "Clashes Over Grazing Land in Nigeria Threaten Nomadic Herding." *NPR,* April 23, 2017. http://www.npr.org/sections/goatsandsoda/2017/04/23/525117431/clashes-over-grazing-land-in-nigeria-threaten-nomadic-herding.

Thompson, F. 1968. "The Second Agricultural Revolution, 1815–1880." *The Economic History Review,* 21, no. 1, new series: 62-77. doi: https://doi.org/10.2307/2592204.

Thornton, P. K., and M. Herrero. 2015. "Adapting to Climate Change in the Mixed Crop and Livestock Farming Systems in Sub-Saharan Africa." *Nature Climate Change 5*, no. 9: 830–36. doi: https://doi.org/10.1038/NCLIMATE2754.

United Nations World Food Programme. 2015. *The Cost of Hunger in Malawi.* http://www.wfp.org/content/cost-hunger-malawi.

US Department of Agriculture (USDA). 2012. "2012 Census of Agriculture." USDA National Agricultural Statistics Service. https://www.agcensus.usda.gov/Publications/2012.

———. 2015. "A Data Portrait of Smallholder Farmers." USDA fact sheet. http://www.fao.org/family-farming/detail/en/c/385074.

Utting, P. 2015. *Revisiting Sustainable Development.* Geneva: United Nations Research Institute for Social Development.

Whittlesey, D. 1936. "Major Agricultural Regions of the Earth." *Annals of the Association of American Geographers* 26, no. 4: 199–240.

World Health Organization. 2014. "Frequently Asked Questions on Genetically Modified Food." http://www.who.int/foodsafety/areas_work/food-technology/faq-genetically-modified-food/en.

World Intellectual Property Association. 2015. "The King of Cheese and Its IP Crown." http://www.wipo.int/ipadvantage/en/details.jsp?id=3664.

Zwerdling, D. 2009. "'Green Revolution' Trapping India's Farmers in Debt." *NPR,* April 14, 2009. http://www.npr.org/2009/04/14/102944731/green-revolution-trapping-indias-farmers-in-debt.

Chapter 7
Manufacturing

As countries get richer, fewer people are needed in the primary sector of the economy. By 2014, only 1.5 percent of Americans and 1.8 percent of Canadians worked in agriculture, while in Mexico, which is poorer than its North American neighbors, the figure was just 13.4 percent. Machinery and technology have rendered most agricultural workers useless, but clearly this does not mean that most people are now without work. Rather, agricultural mechanization freed people up to pursue other types of work. Thus, new jobs in the secondary and tertiary sectors have absorbed displaced agricultural workers.

As you'll recall, the secondary sector of the economy involves manufacturing. The transformation from agriculture to manufacturing has had profound impacts on the world. Demand for industrial workers exerted a powerful gravitational pull toward urban areas, feeding urbanization first in Europe and North America and later in Asia, Latin America, and Africa. Industrialization led to the development of an urban working class, which eventually organized into unions and claimed solid wages and benefits. The new urban working classes could consume manufactured products like never before, ultimately enjoying household appliances, consumer electronics, automobiles, travel by ship and airplane, and more. But a dark side has also accompanied industrialization. Where regulations are weak, worker exploitation and pollution have caused great suffering to people and the planet. Likewise, competitive economic pressures drive manufacturers to seek efficiencies, which sometimes means resisting wage increases or relocating factories to new locations.

Because of the benefits and costs associated with manufacturing, it is important to understand the geography of industrial production. As industrial production develops in specific places and relocates to others, it impacts local and national economies. When a factory opens, it creates new jobs, often offering employment opportunities superior to those that were available previously. For much of the nineteenth and twentieth centuries, this benefit was seen in North America and Europe, while the benefit shifted to other continents in the twenty-first century. But the flip side of this benefit is that when a factory closes, it often devastates communities, leaving people with few employment opportunities. These issues have had important political ramifications in recent years, as declining industrial centers in Europe, North America, and elsewhere face unemployment and despair in declining urban regions. Looking toward the future, manufacturing will continue to impact people and economies, especially as increasing automation means that fewer workers will be needed on the factory floor.

The Industrial Revolution

For several thousand years, the most successful agricultural societies accumulated power and influence over others. Agricultural surplus allowed part of the population to specialize in nonagricultural pursuits, developing technology, tools, and administrative structures that enhanced power. During this time, the fabrication of tools, clothing, and other goods was performed in small workshops of urban guilds or individual homes in rural districts. By the late 1700s in Western Europe, urban merchants were contracting with hundreds of thousands of rural residents to produce cloth, thread, and other items. These cottage industries were small scale but formed the nucleus of an industrial working class. With invention of the steam engine and "spinning jenny" for cloth production, these workers were soon brought together into large textile mills. Thus began the *Industrial Revolution*, first in the United Kingdom and later in other parts of Western Europe and beyond.

The Industrial Revolution led to a new form of political and economic dominance, one held by countries that increasingly produced manufactured goods rather than just agricultural ones (figure 7.1). Following the production of basic goods, such as clothing, came the development of steam-powered locomotives and ships, which allowed goods and materials to move quickly to new places. Canals and rail lines connected places, facilitating trade and production based on comparative advantage. Cities grew as more labor left the countryside for work in urban factories (figure 7.2).

In the early twentieth century, the *Second Industrial Revolution* began in the United States. This revolution is often associated with Henry Ford, who helped develop and implement mass-production

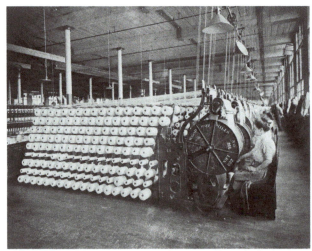

Figure 7.2. Women working at textile machines, at the American Woolen Company, Boston. Circa. 1910. Photo by Everett Historical. Stock photo ID: 244389922, Shutterstock.

Figure 7.1. Manufacturing has played an essential role in the development of modern economies. Photo by 279photo Studio. Stock photo ID: 372863239, Shutterstock.

manufacturing in his automobile plant in Michigan. This form of standardized production employed an assembly line where machines and less-skilled workers performed repetitive tasks, greatly increasing production efficiency and reducing costs. Along with decent wages for his workers, "Fordism" led to an era of industrial mass consumption, where broad segments of the population could purchase an ever-increasing number of mass-produced consumer goods.

The spatial distribution of manufacturing

Industrialization began in the United Kingdom, diffused through Western Europe, and hopped the Atlantic to North America. As with agriculture, forces driving the spatial distribution of industrial development played out locally and globally. At a local scale, industry often located near coal deposits initially, which were used to power manufacturing plants. Globally, European and North American powers used their industrial strength to extract raw materials from colonies in Latin America, Asia, and Africa, forging a global system of manufacturing based on the concept of comparative advantage. Countries in Europe and North America had advantages in manufacturing, with capital for machinery and urban workers for labor. Poorer countries had available land, with natural resources such as metals, cotton for clothing, and agricultural goods to feed urban workers.

Figure 7.3 illustrates how industrial power was concentrated in Europe and North America through the first half of the twentieth century. In 1900, nearly half of all exported manufactured goods came out of two

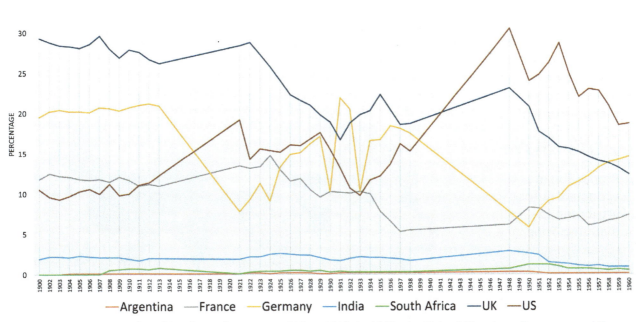

Figure 7.3. Percentage of world industrial exports: 1900 to 1960. In 1900, the United Kingdom, Germany, and France accounted for nearly half of all manufactured exports. By 1960, the United States was the dominant industrial power. (Data points are missing around much of the European war years: 1914–20 and 1939–47). Data source: International Trade Statistics 1900-1960, Statistical Office of the United Nations.

countries, the United Kingdom and Germany. France and the United States also produced a substantial amount of industrial export goods. But countries such as India, South Africa, and Argentina in other continents manufactured very little. By 1960, the relative strength of European and North American industrial power had shifted, with the United States becoming the largest industrial exporter. The United Kingdom remained a significant player but lost its dominance. Germany's industrial strength was damaged during the two world wars but leaped ahead of the United Kingdom by 1960. Despite these shifts, countries in Latin America, Africa, and Asia remained insignificant producers.

Spatial distribution in the twenty-first century

In recent decades, improved data collection has provided varied means of measuring the spatial distribution of industrial activity. Two measurements provide an indication as to how significant manufacturing is to a place and its people. Looking at the proportion of a population employed in the manufacturing sector shows how important it is in supporting households. In some places, a manufacturing job is a ticket to a stable income and an improvement over agricultural employment or no job at all. Likewise, by looking at the proportion of a place's GDP that consists of manufacturing output, one can see where it is a driver of a country's economic development. A third measure, manufacturing value added, illustrates the dollar amount of industrial goods produced in a place. Unlike the other two measurements, it does not show the relative size of the manufacturing sector but rather the straight dollar value of output. This helps identify places that produce a large amount of industrial goods even if other sectors of the economy are larger.

Global distribution

As of 2013, manufacturing was a strong component of GDP in parts of East Asia and Central and Eastern Europe (figure 7.4). Roughly 30 percent of GDP in South Korea and China and nearly 25 percent of GDP

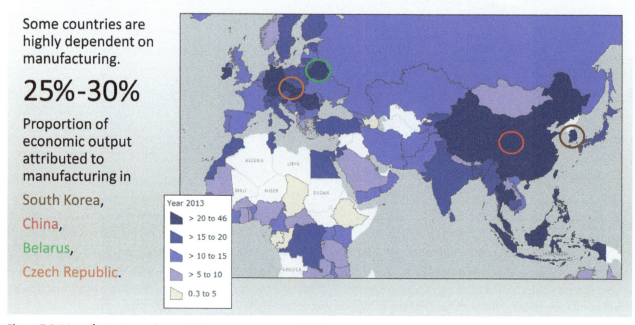

Figure 7.4. Manufacturing value added as percentage of GDP. Data source: World Bank.

in Belarus and the Czech Republic came from manufacturing. Of the major Western European and North American countries that led the Industrial Revolution, only Germany still earned a significant percentage of GDP from manufacturing. In addition to being an important component of economic activity, industry also employed a substantial number of people in these countries, with about 30 percent or more of the workforce in the manufacturing sector.

That China and Korea rely heavily on their manufacturing sectors is probably not a surprise. While few Chinese brand names are known in the West, people worldwide are familiar with the "Made in China" label. Since the 1990s, the country became a major player in the manufacture of toys, clothing, consumer electronics, and more. For instance, China is the primary producer of iPhones, and in Zhengzhou, known as "iPhone City," 500,000 iPhones can be produced in a single day.

In recent years, China has been rising in the manufacturing value chain, moving away from low-value items and assembly of products for Western technology companies to more sophisticated design and manufacturing of goods. China now produces millions of automobiles each year, some for major US companies such as Ford and General Motors but others domestically developed by companies such as Great Wall and Jiangling (figure 7.5). It has also moved into advanced semiconductor design and manufacturing and now builds a commercial jetliner. These more sophisticated manufactured goods supply the domestic market, and a handful are exported globally, but with time they should challenge large global brands as they push to expand exports further into new foreign markets.

As with China, South Korea has become a major industrial power, but with more globally recognized brands. Global automobile producers such as Hyundai and Kia are sold worldwide, as are appliances and electronics produced by LG and Samsung.

In Central and Eastern Europe, both Belarus and the Czech Republic are reliant on industry for a substantial portion of GDP, but for different reasons. Belarus, a former republic of the Soviet Union, has an industrial sector that is large but outdated and inefficient. About 80 percent of industry is controlled by the state, with exports such as machine tools, work vehicles, household appliances, and fertilizers that favor Russian markets. The Czech Republic is a more open market that is heavily dependent on automobile production for export to countries of the European Union. It manufactures a domestic automobile for sale in emerging markets, as well as Peugeot, Hyundai, and other foreign-owned brands.

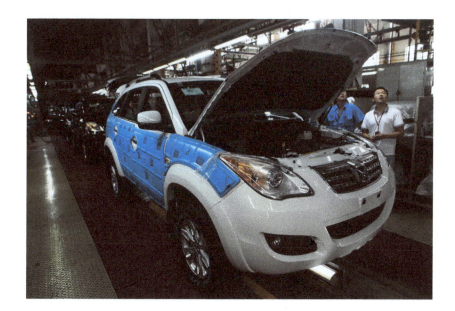

Figure 7.5. Jiangling Motor Group Company in Nanchang, China. China is moving up the manufacturing value chain to higher-end products, such as automobiles. Photo by Humphery. Stock photo ID: 509065885, Shutterstock.

While some countries are heavily dependent on manufacturing for employment and economic growth, their influence in the global industrial economy is not necessarily great. Figure 7.6 shows industrial production by value added. This represents the total value of industrial output regardless of number of workers or proportion of GDP. Here the map looks substantially different and may surprise many who think China is the world's only factory floor. China produced nearly one-quarter of the value of all manufactured goods in 2013, but that was followed by the United States at nearly 17 percent. Four countries alone account for over half of all global industrial output: China, the United States, Japan, and Germany. These are the true industrial powerhouses that supply everything from the dollar-store goods to multimillion-dollar aircraft and industrial machinery.

One reason people may be unaware of the United States' industrial strength is that many consumer goods, from say, Walmart, often are made abroad. Yet, many products consumed within the United States and exported to places around the world, still carry the "Made in the USA" label. The largest category of US-manufactured goods is food and beverage and tobacco products. This includes a wide range of items purchased at grocery stores, such as processed meat and poultry, soft drinks, dairy, and tobacco. While agricultural production is considered part of the primary sector, it typically must be processed and packaged in manufacturing plants for consumption.

Each time residents of the United States fill their car with gas, they are "buying American." This is because, after food and tobacco manufacturing, the United States produces petroleum products, primarily converting oil into fuel for cars, trucks, and aircraft. Likewise, when people in the US are sick, their medicine was probably produced domestically as part of the chemical manufacturing sector. This sector also includes plastics, soaps and detergents, paints, pesticides, and much more. And despite fierce global competition, the fourth-largest manufacturing sector in the US is motor vehicles and parts. Given the global nature of the motor

Manufacturing Value Added, 2013

Despite the perception many have, the **United States** is the second-largest manufacturing country after **China**.

Just **four** countries produce **55%** of manufacturing value added.

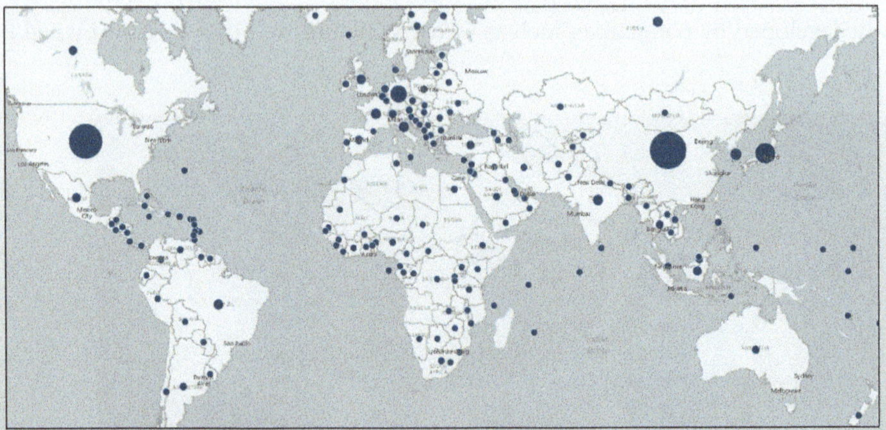

	Current $US	% World
China	$2,856,981,212,483	23.6%
United States	$2,005,988,000,000	16.6%
Japan	$1,002,078,388,793	8.3%
Germany	$759,714,564,717	6.3%
World	$12,092,318,763,830	100.0%

Figure 7.6. Manufacturing value added, 2013. Data source: World Bank.

vehicle industry, an automaker's corporate headquarters does not necessarily reflect where a vehicle is made. Many Ford and General Motors vehicles sold in the US are manufactured in, or contain a substantial number of parts from, Canada or Mexico. At the same time, Japanese companies such as Honda and Toyota, as well as German brands such as BMW, are often produced in the United States. In fact, four of the top ten most "American made" cars produced in the US included Japan's Honda and Acura brands in 2017.

Local distribution

Moving from a global view to a larger, local scale, more fine-grained industrial regions can be identified. The European industrial core for many decades was described as a "Blue Banana" that stretched from the southern United Kingdom southward to Milan, Italy (figure 7.7). But as discussed in the previous section, Europe's industrial core is shifting eastward, encompassing an area from southern Germany south and east to Romania.

China, while an industrial giant, has seen its manufacturing spatially concentrated along its eastern coast (figure 7.8). Just five provinces (Jiangsu, Shandong, Guangdong, Henan, and Zhejiang) account for over half of all revenue earned from manufacturing. Over time, however, industrial production is spreading west, slowly producing a more dispersed pattern of industrial production.

In the United States, the traditional industrial heartland stretches from eastern New York and Pennsylvania westward into Illinois and Wisconsin (figure 7.9). But while this region still holds sway in the minds of many Americans, its industrial dominance has declined in recent decades. Now, other regions of the United States, such as the Southwest, Pacific Northwest, and Eastern Seaboard, also produce a substantial amount of industrial output. Small pockets of industrial production can also be found in much of the South.

Within the traditional industrial heartland, manufacturing still employs a substantial proportion of the

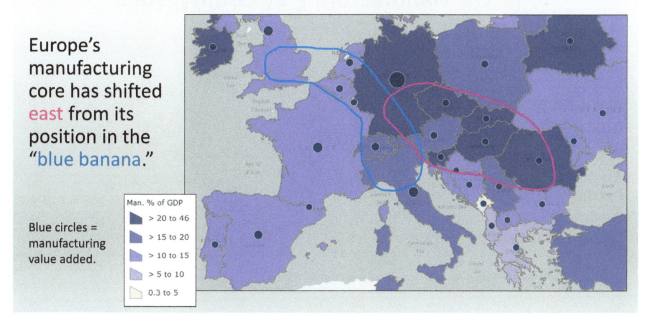

Figure 7.7. European manufacturing regions. Data source: World Bank.

China: Manufacturing Revenue by Province

Just **five** provinces account for **over 50%** of all revenue earned from manufacturing: Jiangsu, Shandong, Guangdong, Henan, and Zhejiang.

But industrial production is diffusing **west** over time.

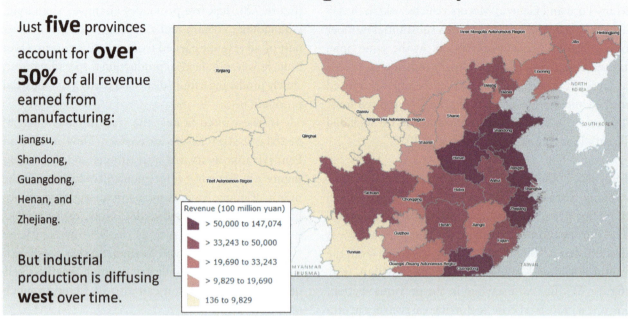

Figure 7.8. China: Manufacturing revenue by province. Data source: National Bureau of Statistics of China.

US Manufacturing by County, 2012

Manufacturing is found in many counties outside of the traditional **manufacturing heartland**.

These include the **Pacific Northwest**, the **Southwest**, **Eastern Seaboard**, and scattered **Southern counties**.

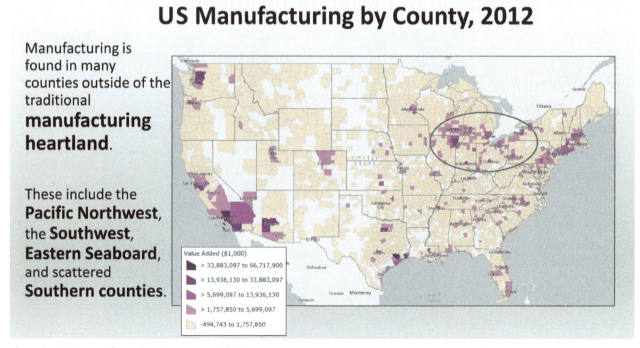

Figure 7.9. US manufacturing by county, 2012. Data source: 2012 Economic Census of the United States.

workforce, although less than in the past. Michigan, the home of Ford and General Motors, continues to be known for its motor vehicle production, while much of Iowa's manufacturing is linked to its agricultural production. Not only does Iowa manufacture food and beverage products, it is the home to large agricultural machinery companies, such as John Deere.

Manufacturing also encompasses a large proportion of employment in southern states, such as Alabama and Kentucky, both of which now have important motor vehicle and parts industries. Likewise, food and tobacco processing employ a substantial proportion of workers in Arkansas, Mississippi, and Kansas.

Although rarely talked about, the largest numbers of manufacturing workers are located in California and Texas. California's largest manufacturing sector is in computer and electronic products, which include semiconductors, aircraft instruments, navigation equipment, and other sophisticated electronics. Texas also produces computer and electronic products, but not surprisingly, its primary manufacturing output lies in chemical and petroleum products.

Factors driving industrial location

Over time, new industrial regions grow while others go into decline. The industrial dominance of Europe and North America has waned as new competitors from Asia and elsewhere have gained market share. At the same time, spatial patterns of industrial production within regions and countries continue to shift. Chinese manufacturing is gradually shifting from the coast to interior provinces, European manufacturing is moving from west to east, and America's industrial heartland has faced decline as competitors in other states and countries outcompete it.

To explain how and why these patterns shift, we can turn back to several concepts already introduced in previous chapters. In most cases, industrial location is tied to market forces. Investors want to find the best location for maximizing profit, considering the factors of production: land, labor, and capital.

A factory needs physical space and natural resources as inputs to production. It also needs a sufficient number of workers with the right types of skills. Finally, capital machinery and facilities are required for production and transportation.

The factors of production are in turn related to the concepts of site and situation. As discussed in chapter 3, site refers to the local characteristics of a place, while situation refers to the relative location of a place. For example, a good location for a factory may require site characteristics that include a large workforce with at least a high school education. It may also require a site where machinery and tools are available for production. Likewise, it may need to be relatively close to raw materials, such as coal and iron.

Based on site and situation and their influence on the factors of production, each place has advantageous characteristics for different types of production. Some places have low-cost energy for petrochemical manufacturing, others have large numbers of low-skill workers for textile fabrication, and yet others have high-tech facilities and an educated workforce for aerospace production. On the basis of comparative advantage, firms will concentrate in specific geographic locations where they can most efficiently produce their products. In essence, they will look for a location with characteristics that give it an advantage over producing in other locations.

Weber and early industrial location theory

One of the most influential figures in the theory of industrial location was Alfred Weber, who in 1929 studied the roles that transportation, labor, and agglomeration play in the spatial patterns of manufacturing. Given the nature of manufacturing in the early twentieth century, Weber argued that transportation was the most important factor influencing where factories located. Manufacturing facilities would locate at the least-cost location, or the place where transportation costs would be lowest, based on the location between market and sources of raw materials, taking weight into consideration. Manufacturing at that time often involved large amounts of raw materials and fuel. Think of steel plants with their requirements for heavy iron ore and coal fuel. In this case, massive amounts of raw iron ore would have to be shipped to the factory, where smelting would extract pure iron. At the same time, large volumes of coal would be burned to fuel the furnaces. Thus, the weight of raw material inputs (raw iron ore and coal) were much greater than the output (pure iron or steel). These weight-losing industries would locate close to iron ore and coal deposits, since transporting the raw materials (mostly via rail) was an expensive endeavor (figure 7.10). Factories with fewer heavy inputs, such as textile factories, could locate closer to markets. It is for this reason that the industrial heartland of the United States developed near the coal mines of Appalachia and Illinois, while in Europe, factories concentrated near coal regions in northern and central United Kingdom and Germany's Ruhr Valley.

But industrial location is not based on transportation costs alone. Labor, or the workforce in a place, also plays an important role. The cost of labor varies from place to place, depending on a wide range of variables, such as the strength of the local economy, the supply of workers, their skills and efficiency, local labor laws, unionization, and much more. With industrial growth through the nineteenth and early twentieth centuries, a large and low-cost workforce of rural-to-urban migrants, along with limited regulation and unionization, fed manufacturing in North American and Europe.

The third factor influencing industrial location, according to Weber, is *agglomeration*. Agglomeration is the process whereby individual manufacturers benefit by clustering in the same location. This clustering benefits firms in several ways. First, smaller factories can combine into larger entities, increasing efficiency through economies of scale. Second, agglomeration means that specialty suppliers can more efficiently support manufacturers. If many factories locate in the same area, then specialty suppliers can offer a wide range of business-to-business goods and services. For instance, machinery maintenance services, transportation and logistics specialty firms, and technical and managerial training centers will be found in larger industrial clusters. These will be harder and more expensive to source if a factory is in a remote site. Third, clusters will have an experienced workforce, reducing costs in training new workers. Fourth, an industrial agglomeration will have shared infrastructure. Roads, port and airport facilities, electricity, and water can be developed and maintained for the benefit of all firms in the area.

Figure 7.10. Loading iron ore on a train. Weight-losing manufacturing, such as steel production, tends to be located near raw materials. In this case, the weight of raw materials is much greater than the weight of final output. Photo by Kaband. Stock photo ID: 76088089, Shutterstock.

But agglomeration benefits function only up to a certain size. When industrial clusters become too concentrated, *deglomerative* forces come into effect. These can most importantly include high rents but can also include congestion, which inhibits transportation, and upwardly spiraling wages.

Contemporary industrial location

Weber's work helped explain industrial location through the early twentieth century. He pointed out the importance of how site characteristics influence where manufacturers locate. These characteristics include labor costs and skill sets in particular places and the shared benefits of agglomeration, such as shared specialty subcontractors, a trained workforce, and shared infrastructure. He also showed how situation is relevant in terms of transportation costs between the factory, raw material sources, and markets. Today, these factors are still important, but manufacturing and the political and economic environment in which it operates is now influenced by additional concerns.

One of the greatest changes to industrial location factors in the twentieth and twenty-first centuries has been the declining cost of transportation. As transportation costs have fallen, the relative distance between places has decreased. Goods now travel faster via road, rail, air, and sea, essentially "shrinking" the distance between places. This has opened most of the world as potential locations for industrial production. As transportation costs have become less important, more manufacturers have become footloose and can locate in just about any place that offers a good return on investment.

With worldwide options for locating factories, the role of governments has become an important variable in industrial location decisions. Government policy at the international, national, and local scales alter the site and situation characteristics of places. Governments make trade agreements and influence investment in transportation linkages, impacting the relative distance between sources of raw materials and markets. Governments also influence the cost of labor via rules and regulations. The formation of industrial agglomerations is also influenced by government policy in some cases, with governments actively offering incentives to attract companies to specific locations. Environmental regulations and taxes imposed by governments further influence decisions on where to place factories.

With these changes to the site and situation characteristics of places, companies can now incorporate in just about any jurisdiction or can set up parts of their production chain in different places. In many cases, companies outsource some (or even all) of their manufacturing to subcontractors in different locations. With outsourcing, the corporate entity does not own production facilities. Rather, it contracts with other companies to produce components or assemble products. When this is done in another country, it is called *offshoring*.

The role of transportation

The cost of moving raw materials to factories and finished goods to markets is a central component of industrial location. But for many industries, its significance has declined. The cost of road transportation fell by nearly 40 percent over a thirty-year time period leading up to the turn of the twenty-first century, as more and better roads were built and fuel efficiency improved. Likewise, air freight costs declined substantially with proliferation of jet engine aircraft after the 1950s. Rail transport, the focus of Weber's work, saw prices fall by 87 percent between 1890 and 2000.

Ocean shipping costs have not fallen significantly; however, technological changes have cut shipping times, thus reducing overall costs. With the invention of the shipping container in the 1950s, goods could be loaded into a container just once, at the factory, and unloaded just once, at the retail distribution center or retail outlet (figure 7.11). Prior to the shipping container, goods were loaded and unloaded separately each time there was a change in mode of transportation. These break-of-bulk points, where goods are transferred between truck, ship, aircraft, or train, now move goods at a pace unheard of prior to containerized shipping.

Figure 7.11. Container terminal. The shipping container helped reduce the time and cost involved with shipping goods, making transportation much less important in manufacturing location decisions. Photo by tcly. Stock photo ID: 151227356, Shutterstock.

As transportation costs as a proportion of total value of goods have fallen, more places now serve as potential production locations. Manufacturers can now search globally, and within wide swaths of countries, for the most profitable place to build factories.

The role of government policy

When the Industrial Revolution began, manufacturing was minimally regulated by government. Smokestacks belched out toxic black smoke, and effluent poisoned lakes and rivers; worker safety was an afterthought and industrial accidents common; wages were as low as the market would bear. Robber barons who owned industry became immensely rich, while common workers were viewed as mere interchangeable parts, to be replaced when needed just as any other machine. But as awareness of worker and environmental exploitation grew, government regulation increased.

Today, the role that government plays in industrial location is a significant factor. Government policies now regulate wages, safety, and benefits for workers and control industrial pollution. Governments also set different tax rates and trade and investment policies. These policies vary widely from place to place. Government regulation varies substantially between countries, as well as between US states. In countries with less unitary, or centralized, forms of government, regulations can even vary from city to city.

Trade

Industrial production requires inputs of raw materials and manufactured components and produces outputs for sale to consumers. In a global economy, rarely are all inputs and consumers located near the manufacturing plant. More commonly, inputs and consumers are found in many parts of the world. For this reason, trade laws are an important factor that manufacturers must consider when locating a facility.

Data from the World Bank shows that many European countries are open to free trade, where goods can cross borders relatively quickly and cheaply. Goods can move freely within the European Union but also between EU countries and many other markets around the world. This means that goods can easily move between countries without facing heavy tariffs (import taxes), import quotas, or other restrictions. Thus, goods produced in places such as Germany and the Czech Republic can be shipped to trade partners such as Canada, Mexico, South Korea, and South Africa. At the same time, German and Czech manufacturers can easily import raw materials and other inputs from these countries.

Other countries with relatively open trade policies are found in Asia, such as South Korea, Singapore, and Japan. On the other end of the ease-of-trade spectrum lies much of sub-Saharan Africa and the Middle East. The Democratic Republic of Congo, Iraq, and others have slow and expensive customs procedures that restrict trade, making them much less desirable locations for industrial production.

Labor

Government regulation of labor is another consideration that manufacturers must account for. Data from the Heritage Foundation, an organization that promotes free enterprise and small government, scores countries on the basis of labor flexibility. Labor flexibility is the degree to which employment is based on market mechanisms of supply and demand rather than on government regulation. Flexible labor markets have limited regulation in terms of minimum wages, mandated taxes and worker benefits, and job security. Places with more labor flexibility include the United States, where the federal minimum wage is low; health, vacation, and holiday benefits are not mandated by the government; and "at-will" employment contracts make it easy to hire and fire workers. Other places with labor flexibility are geographically dispersed. The top ten most flexible countries include affluent Singapore and Denmark as well as poorer countries such as Uganda and Somalia.

Labor laws in places like the United States can vary substantially as well. State minimum wages in 2017 varied from $12.50 in Washington, DC, to the federally mandated minimum of $7.25 in Alabama and Mississippi. Some cities mandate even higher minimum wages. Seattle's minimum wage was $15 in 2017, and San Francisco's was $14.

Taxes

It goes without saying that tax burden is another factor that manufacturers consider when search for the most profitable location. Based on Heritage Foundation data, the countries with the lowest tax burdens are spatially clustered in the oil states of the Middle East. Other low-tax countries include small "tax haven" places such as Liechtenstein in Europe and the Bahamas in the Caribbean.

Overall business environment

Determining the overall business environment of places is highly subjective and depends on which variables are deemed important. The World Bank ranks countries according to "business friendliness," which includes costs of and time associated with obtaining permits to start a business, getting construction permits and electrical hookups, obtaining financing, paying taxes, importing and exporting goods, and legal protections for investors and contract enforcement. The Heritage Foundation looks at the rule of law; the size of government, such as tax burden; the overall regulatory environment, including labor laws; and the openness of markets to investment and trade. While each organization uses slightly different criteria, there is substantial overlap in their country rankings.

Thirteen of the top twenty best business environments are included in both rankings (table 7.1). These include places many people would expect to rank highly, such as Singapore, Hong Kong, the United States, and the United Kingdom. Also included are some Eastern European countries, such as Estonia and Latvia, that liberalized their economies after the fall of the Soviet Union. Possibly more surprising are the Scandinavian countries of Denmark and Sweden that many people associate with strong social benefits more than free enterprise.

The worst regulatory environments also overlap considerably (table 7.2). These are the places that do not support private investment and make starting and growing an industry extremely difficult.

It is interesting to note that the countries at the top of these rankings not only offer regulatory environments that are conducive to conducting business but also are affluent with high standards of living. Those at the bottom of the list, by contrast, are poor and underdeveloped. This should give pause to those who see corporations as greater sources of evil than good. Heavy-handed government regulation can stifle

Best Regulatory Environments

	WORK BANK RANKING	HERITAGE RANKING
AUSTRALIA	15	5
DENMARK	3	18
ESTONIA	12	6
GEORGIA	16	13
HONG KONG SAR, CHINA	4	1
IRELAND	18	9
LATVIA	14	20
NEW ZEALAND	1	3
SINGAPORE	2	2
SWEDEN	9	19
TAIWAN, CHINA	11	11
UNITED KINGDOM	7	12
UNITED STATES	8	17

Table 7.1. Best regulatory environments. Data source: The World Bank and the Heritage Foundation.

Worst Regulatory Environments

Afghanistan
Angola
Chad
Congo, Republic of
Djibouti
Equatorial Guinea
Liberia
Timor-Leste
Venezuela

Table 7.2. Worst regulatory environments. Data source: The World Bank and the Heritage Foundation.

business and wealth creation. Regulations are necessary to protect workers and the environment, but they must be well targeted and managed so that benefits outweigh costs. Strong economic growth allows for sufficient tax revenue to provide essential quality-of-life services.

Government incentives

In addition to generalized regulations, governments sometimes attempt to guide industrial development to specific geographic locations though various forms of business incentives. These programs can involve a wide range of incentives, from tax breaks to subsidized land, labor and import laws that are more liberal than those outside of the zone, and modern and reliable infrastructure. Typically, incentives are designed to attract industries into areas that may not be good locations based purely on market-based criteria. The aim of governments is to essentially subsidize industrial development until it can stand on its own without special incentives.

One of the most important industrial incentive programs is China's *Special Economic Zones* (SEZs). The first SEZs were established in the coastal provinces of Guangdong and Fujian in 1980. From an original four SEZs, various types of business-friendly zones were established through the 1990s. For the most part, they were located along the coast, but over time, some were established in interior provinces as well. Industry was attracted to these areas by new infrastructure, such as roads and reliable water, electricity, sewer, and telecommunications. Regulations were streamlined to facilitate the easy import of industrial inputs and export of finished products. Business services were provided to assist companies with legal, marketing, and management issues. Inexpensive land with secure leases was provided, as was a large labor force with flexible regulations for hiring and firing.

Based on these incentives, *foreign direct investment* (FDI) flooded into China. Economic output growth in SEZs and other special areas far outpaced the overall economy of China. As foreign companies set up facilities, local firms grew to supply needed inputs of parts and raw materials, eventually leading to intricate webs of thousands of industrial firms.

Incentives from the Chinese government continue to drive industrial development. Below-market-rate loans and other incentives are being used to attract high-tech manufacturing such as advanced semiconductors. China's iPhone City did not just appear in Zhengzhou by chance. Rather, manufacturers in the area have received over US $1.5 billion from the Chinese government to help build facilities. The local government also pays for new roads for moving goods and new power plants to supply energy to the factories. It also assists with recruiting workers and offers bonuses to industry when export goals are met.

Mexico also used geographically targeted industrial incentives to spur its manufacturing economy. In 1965, it established the *Border Industrialization Program*, a type of export-processing zone that had been used in other countries for promoting export-oriented industrial development. This program mandated that benefits applied only within 20 kilometers (12.5 miles) of the US-Mexico border and in officially designated industrial parks. Within this zone, raw material and machinery could be imported duty free. Skilled foreign managers and technicians could also legally work at the plants. Components were imported, assembled in Mexico, then exported for sale abroad. When US-made components were used in assembly, the US applied tariffs only to the value added in Mexico. This arrangement had great advantages for foreign manufacturers. Close proximity to the US market reduced transportation costs, while low Mexican wages kept labor costs down.

The Border Industrialization Program initiated what came to be known as the *maquiladora industry*, whereby goods are assembled in Mexico for export to foreign markets. Ultimately, hundreds of thousands of workers were employed in these factories, and geographic restrictions were loosened so that manufacturers could operate deeper south into Mexico.

In the United States, state and local governments frequently offer incentives to attract companies. In essence, governments compete with each other to attract companies in hopes of increased tax revenue and jobs. Incentives can include reduced property and business taxes for a certain number of years, below-market-rate loans, grants, and other financial benefits. More geographically targeted programs include *enterprise zones*, which are economically distressed areas of a state or city where businesses can take advantage of a wide variety of incentives (figure 7.12).

Chobani yogurt, for instance, opened a factory in Twin Falls, Idaho, in part because of incentives from federal, state, and local governments. The city spent over $6 million to upgrade its wastewater treatment plant and waived various sewer and building fees. The state offered reimbursements for worker training, while the federal government helped with grant money. In total, it is estimated the Chobani plant received nearly $55 million in government assistance to locate in Twin Falls. Of course, the expectation is that, over time, the economic stimulus provided by the Chobani plant will outweigh the amount spent on incentives, as employment grows, local spending increases, and resulting tax receipts rise.

The role of labor

Manufacturing, like all business, is highly dependent on labor, because ultimately, it is people who make products, even if they work closely alongside machines. But the type of labor required varies greatly by type of manufacturing. Therefore, different types of manufacturing locate in different types of places.

Some manufacturing requires skilled workers, while other industries require unskilled labor. *Skilled labor* refers to workers with specialized training and skills. Training is typically postsecondary and can include technical certification, community college, or university-level degrees. In manufacturing, examples include CNC machine operators, industrial

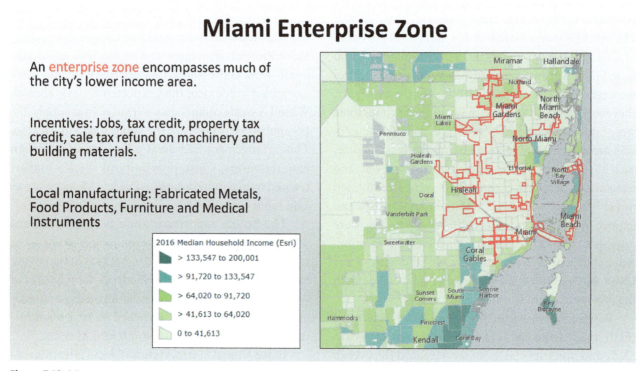

Figure 7.12. Miami enterprise zone. Data sources: Enterprise zone data from Miami-Dade County GIS Open Data. Median Household Income from Esri, U.S. Census Bureau.

and mechanical engineers, machinery installation and maintenance, and much more (figure 7.13). As manufacturing processes become more automated, these skills are increasingly required in the manufacturing sector. While automation means a lower proportion of jobs in manufacturing, it also means that higher-skilled, and thus higher-paid, workers are needed.

Manufacturing that requires skilled labor can be found in developed countries in North America, Europe, and Asia. Sophisticated Boeing aircraft are built throughout the United States, while Europe-based Airbus aircraft are made in France, Germany, Spain, China, the United States, and the United Kingdom. Likewise, handmade Swiss watches require high levels of skill and are still made in Switzerland.

In contrast, unskilled labor does not require any specialized knowledge or training. Workers are largely interchangeable and generally earn low wages. Simple, repetitive assembly of goods and basic monitoring of production fall into this category (figure 7.14). Other positions that require less than a high school education include hand packers and packagers, meat cutters, and sewing machine operators.

Manufacturing that requires unskilled labor has largely left the developed world and is now more commonly found in developing countries of Asia, Latin America, and Africa. To mention China's iPhone City again, the reason cell phones and other electronic devices are assembled in that country is that China offers a massive labor force of lower-wage workers. The same holds true for textile manufacturing. Wages are simply too high in developed countries for most clothing to be made there.

When companies with unskilled labor needs do produce in richer countries, they often struggle. For instance, American Apparel, which made clothing in Los Angeles, filed for bankruptcy in 2016. Meat and dairy processing plants also struggle to find workers. In most cases, unskilled labor in richer countries consists of immigrants with limited skills and native-language ability.

The role of agglomeration

Transportation, government regulation, and labor play more obvious roles in industrial location decision making. But as identified by Weber, agglomeration is an important, if less obvious, factor driving where manufacturers locate.

Figure 7.13. Skilled labor in the production of electronic components at a high-tech factory in Moscow, Russia. Skilled labor is required for much modern manufacturing. These types of jobs typically require postsecondary education and are higher paying than unskilled positions. Photo by Nikitabuida. Stock photo ID: 326369231, Shutterstock.

Figure 7.14. Unskilled labor at a clock factory in Shenzhen, China. Unlike skilled labor used to make high-end Swiss watches, unskilled workers make basic clocks for mass consumption. Photo by Bartlomiej Magierowski. Stock photo ID: 57598960, Shutterstock.

Recall that agglomeration involves clustering of industries in the same location, which offers benefits that would not be available if a factory were in an isolated place. Agglomeration facilitates consolidation of smaller firms into larger and more efficient ones, it allows for small specialized subcontracting firms to offer needed services, it provides a larger pool of experienced workers, and it offers shared urban infrastructure.

In the past twenty or so years, many industrial agglomerations have formed in China, entrenching its role as the world's factory floor. In Zhejiang province, a socks manufacturing cluster has formed, which includes an interconnected supply chain of 3,000 small and medium-sized firms. The Shenzhen province produces iPhones and iPads for Apple, offering not only assembly but a network of enterprises that can provide design, manufacturing production expertise, glass, memory chips, plastic buttons, and more. Agglomerations such as these function as complex ecosystems and are difficult to move or replicate in other places.

The Seattle, Washington, region is home to Boeing and hosts a large agglomeration of aerospace firms. Over 1,300 firms in the region supply parts and services to the aerospace industry and employ over 135,000 people. By being physically located in the same region, these firms can quickly share and exploit new ideas and technology, drawing on a large and experienced workforce.

The same holds true for pharmaceutical manufacturing in New Jersey. The state is the home of fourteen of the world's twenty largest pharmaceutical firms. These firms are tightly integrated with universities that provide a steady supply of skilled biotechnology workers as well as teaching hospitals and research institutions.

Agglomeration benefits firms and broader economic regions much more than would be possible with isolated manufacturing plants. Challengers to China's electronics sector, Washington's aerospace industry, or New Jersey's pharmaceutical cluster face tough competition from integrated networks that are difficult to replicate.

Go to ArcGIS Online to complete exercise 7.1: "Determining industrial location for a textile factory," and exercise 7.2: "Industrial site selection in the United States."

Shifting patterns of industrial production

Transportation costs, government rules and regulations, labor considerations, and agglomeration effects shape important site and situation characteristics and drive industrial location decisions. Transportation costs influence the connectivity between places and their spatial situation in relation to one another. Sites will also vary, in part, on the basis of government actions that promote or hinder industrial development and the characteristics and costs of workers. Sites will also offer varying levels of agglomeration benefits, or in some cases deglomeration costs, that influence where manufacturing is located. As these variables change, the desirability of places for manufacturing can increase or decrease.

United States

Manufacturing in the United States has gone through major transformations since the late 1800s. It was one of the industrial powerhouses in the early twentieth century and became even more influential with the postwar years of midcentury. But as global competition and technology changed, so too did the nation's industrial landscape.

Total manufacturing employment fell after the 1970s in real numbers (figure 7.15), in what is often called *deindustrialization*. Today, less than 8.5 percent of US nonfarm workers are in the manufacturing sector. Yet, as discussed earlier in this chapter, the United States is still a major industrial producer, and despite a decline in industrial workers, the total value of US manufacturing continues to increase, being outpaced only by China (figure 7.16).

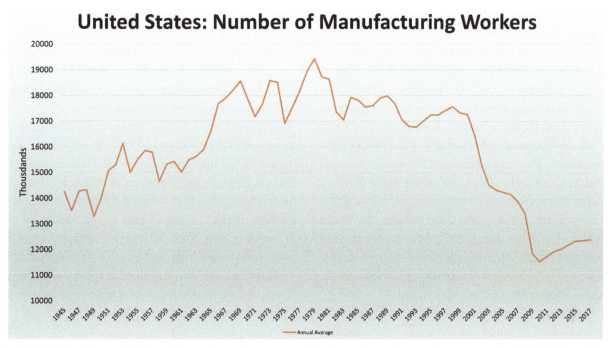

Figure 7.15. United States: Number of manufacturing workers. Total manufacturing employment peaked in the United States in the late 1970s. Since then, the number of manufacturing workers has fallen substantially. Data source: US Bureau of Labor Statistics.

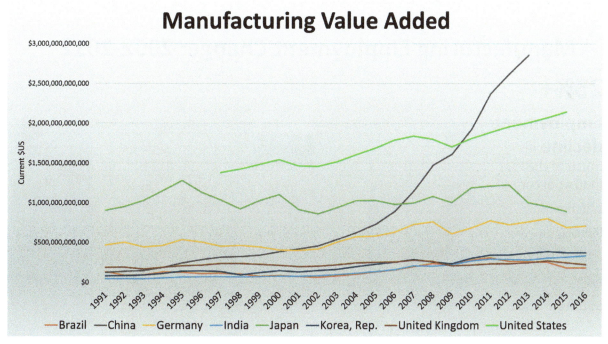

Figure 7.16. Manufacturing value added. Although US manufacturing employment has fallen, the total value of manufactured goods produced in the US continues to increase. Manufacturing is now done with a smaller number of, but more highly skilled, workers. The US is the second-largest manufacturing economy after China. Data source: World Bank.

Broadly, manufacturing in the US has been impacted by changes in technology and globalization. First, advances in robotics and automation technology now means that more goods can be produced with fewer workers. Thus, the apparent contradiction between declining manufacturing jobs and growing value of manufacturing output is due in large part to the use of industrial robots and a skilled workforce to produce higher-end manufactured goods. Second, global competition has increased substantially in recent decades. Improvements in road, port, and air facilities in many parts of the world have shrunk the world in terms of relative distance. Increased connectivity allows manufacturers to search for the most profitable site based on government regulations and incentives, labor requirements, and agglomeration.

Labor-intensive, lower-skilled jobs, such as in the textile and apparel industries and electronics assembly, have moved to lower-cost countries. Places in Latin America and Asia can offer lower wages and less government regulation than in the US. Furthermore, some capital-intensive higher-end goods, be it cars from Japan or machinery from Germany, compete with US-produced vehicles and machines. Consumers buy goods that offer the best value in terms of quality and price, so manufacturers respond by searching for sites that optimize efficient production.

Nearly every US state lost manufacturing jobs, as measured in the 2002 and 2012 US Economic Censuses (figure 7.17). Only Wyoming, North Dakota, and South Dakota gained a small amount of manufacturing. The most dramatic decline occurred throughout the eastern portion of the country. Much of the industrial core of western New York, Michigan, Illinois, and Ohio saw declines of 25 percent or more in their industrial workforces. The same held true for many southern states, where textile mills were shuttered and moved to lower-cost manufacturing nations.

Competition between states functions in the same way as international competition. For example, while the automobile industry struggled in northern cities, there has been growth in the South. Alabama,

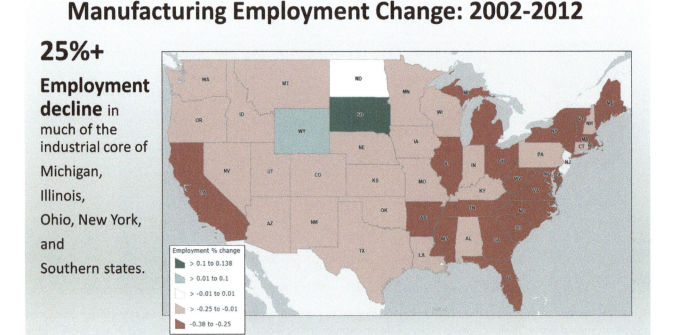

Figure 7.17. Manufacturing employment change: 2002–2012. Data source: 2002 and 2012 Economic Census of the United States.

Kentucky, and Tennessee now produce for major auto manufacturers, such as Ford, General Motors, Toyota, Nissan, and Volkswagen. Southern states have advantages for manufacturers over states such as Michigan and Ohio in that wages are lower and unions are weaker. These advantages, combined with state tax credits and other incentives, have attracted domestic and foreign auto manufacturers to the South.

Zooming to an even larger scale, it becomes apparent that manufacturing clusters in specialized urban agglomerations. Urban areas offer experienced workers, specialized suppliers and business customers, and the sharing of ideas through face-to-face interaction. This specialized clustering entails complex linkages between multiple firms, workers, educational and research centers, and more that boost efficiency and reduce overall costs. On the downside, however, when a city or region is highly specialized, competition from another country, state, or city can quickly lead to serious economic damage. Figure 7.18 shows the location quotient for textile manufacturing employment. Cotton production in the South means that a large cluster of towns dependent on textile mills are found in the same region. However, with lower wages abroad, many cities that are dependent on the textile industry have suffered tremendously from foreign competition.

Europe

Just as in the United States, European manufacturing has seen dramatic changes over time. Its global industrial dominance has waned with foreign competition, and manufacturing that remains is being transformed by automation and a shift to lower-cost regions of the continent. These changes have led to an overall decline in manufacturing employment in the region, while manufacturing value added has remained constant or increased.

The shift from the Blue Banana eastward has been driven by political and economic trends. Politically, many Eastern European countries, such as the Czech Republic, Poland, and Romania, joined the European Union in the early 2000s. Their participation meant

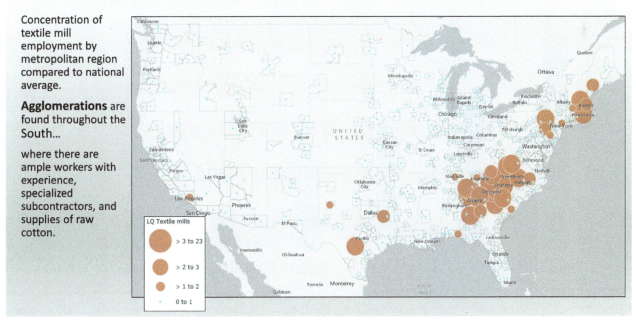

Figure 7.18. Location quotient: Textile mill employment, 2015. Data source: US Census Bureau.

that goods produced in Eastern European EU-member countries could be shipped throughout most of Europe without customs and border barriers. Capital machinery could be imported to eastern countries and finished products exported to western countries with ease. The greatest benefit for manufacturers was lower wage levels in the east, which can be one-third to one-fifth or less than the cost in Western Europe (figure 7.19).

Asia

The biggest impact on global manufacturing in recent decades has clearly come from Asia. Over one-third of the entire global population lives in China and India alone, offering not only a vast pool of industrial workers but an increasingly affluent group of consumers. Other countries in the region are gaining significance as well. Vietnam, Bangladesh, Indonesia, Pakistan, and other countries are producing and buying goods at growing rates. Thus, the demographic and economic

weight of the world is moving from its traditional centers in Europe and North America.

China has become the industrial behemoth of the region. Beginning with its SEZs, China began to attract foreign direct investment and expanded its role in manufacturing for foreign markets. In 2001, it joined the World Trade Organization, cementing its place in the global economic system. Low wages and government support attracted manufacturers to China, and massive industrial agglomerations began to form. Looking back at figure 7.16, it is evident how the value of China's manufactured goods increased since the 1990s, making it the largest manufacturing country in the world.

China, too, is seeing a transformation of its manufacturing sector. Whereas it once attracted vast amounts of foreign direct investment with its cheap workforce, wages have increased substantially in recent years. Wages more than tripled from 2003 to 2015,

Figure 7.19. Hourly compensation costs in manufacturing. European manufacturing clusters have shifted east to lower-cost countries. Data sources: US Bureau of Labor Statistics, International Labor Comparisons.

significantly upsetting many factory owners' calculus when deciding where to produce their products (figure 7.20). These increases are attributed to traditional supply and demand. Demand for workers increased as China's manufacturing sector boomed, while the supply of workers slowed, in large part due to the decades-long one-child policy.

National wage data masks spatial differences, however (figure 7.21). Costs have increased the most in the booming coastal provinces of Guangdong, Shanghai, and Jiangsu. Moving inland, labor costs are lower, forcing some manufacturers to relocate or expand further from traditional manufacturing hubs.

Labor costs are also forcing companies to automate rather than rely on human workers. Just as in richer countries, automobile factories are using robots to weld, paint, and perform other tasks previously done by humans. And in factories that produce the iPhone and other electronics, one major company aims to have 30 percent of production done by robots.

Manufacturers that cannot relocate to lower-cost provinces in China or replace workers with robots are leaving the country altogether. Those that rely most on low-skilled labor, such as in apparel, shoes, and toys, are moving to the low-labor-cost countries of Vietnam, Bangladesh, Indonesia, and Cambodia (figure 7.22).

As an example, Bangladesh's massive ready-made-garments industry, which is responsible for over 80 percent of all export earnings, is based on ideal site and situation characteristics for global clothing retailers. Wages are much lower than in China, and a manufacturing agglomeration ensures experienced labor, deep supply chains for production inputs, and technical development facilities. The government has a major role as well, through relatively weak regulation and through trade agreements linking Bangladesh to global consumer markets in Europe, North America, and Japan.

As a final note on Asian manufacturing, the question of India comes up. Its population is nearly the size of China's and wages are generally low, theoretically

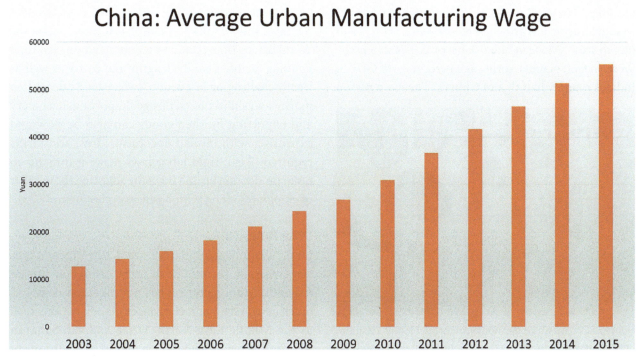

Figure 7.20. China: Average urban manufacturing wage. Chinese wages have steadily risen since the early 2000s, eroding its advantage in low-cost production. Data source: National Bureau of Statistics of China.

China: Manufacturing Wages by Province, 2015

Booming coastal provinces such as **Guangdong, Shanghai,** and **Jiangsu** have higher wages.

Some companies are now moving westward to lower wage provinces.

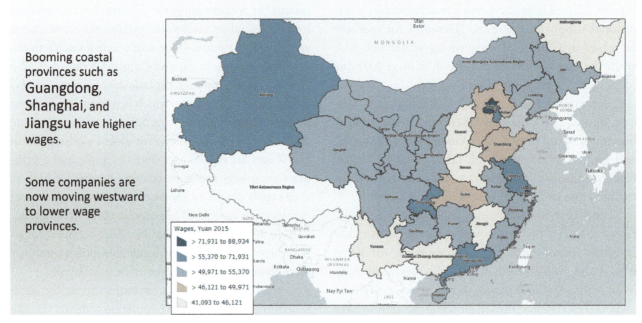

Wages, Yuan 2015
- > 71,931 to 88,934
- > 55,370 to 71,931
- > 49,971 to 55,370
- > 46,121 to 49,971
- 41,093 to 46,121

Figure 7.21. China: Manufacturing wages by province, 2015. Data source: National Bureau of Statistics of China.

making it an ideal place for global manufacturing. Manufacturing has grown in recent years, yet it is still rare to see a "Made in India" label on goods sold in the West. In fact, manufacturing as a percentage of GDP in India is roughly half that of China's. The main reason

Figure 7.22. Low-wage garment workers in Hanoi, Vietnam. Low-wage, labor-intensive manufacturing in Asia is shifting from China to lower-cost countries, such as Vietnam. Photo by Jimmy Tran. Stock photo ID: 428180893, Shutterstock.

for its weak manufacturing performance is government red tape. India ranks 130 out of 190 countries on the World Bank's Doing Business report. Acquiring land and building facilities can be lengthy and costly as well as legally challenging in a country where land title is often murky and subject to challenge. Importing machinery and exporting finished goods can also be slow and expensive, with myriad documents, fees, and tariffs required. Also, rigid labor laws make it difficult to ramp production up and down by adjusting the number of employees along with business cycles (figure 7.23).

Latin America

Of the major geographic regions, Latin America falls somewhere in the middle in terms of manufacturing strength. Manufacturing has not taken off nearly as much as in Asia, but it is stronger than in regions such as the Middle East/North Africa and sub-Saharan Africa. The region's lukewarm industrial environment is broadly tied to the factors of labor, infrastructure, and government institutions and regulation.

Figure 7.23. Brick factory workers in India. Workers make and move bricks by hand. India's infamous government bureaucracy has constrained manufacturing growth, keeping it low tech and labor intensive. Investment has been held back by uncertain land title, restrictions on importing machinery and other inputs, corruption, and more. Photo by Stanislav Beloglazov. Stock photo ID: 517935013, Shutterstock.

In terms of labor, wages are mostly too high to attract low-skilled manufacturers, such as textile mills, but skills are mostly too low to attract high-tech manufacturers. Figure 7.24 illustrates hourly compensation per manufacturing worker and human capital (skills and education) in the three largest economies of Latin America plus two Eastern European states and the United States. A manufacturer searching for relatively low labor costs but a much more skilled workforce will be better off in Poland or the Czech Republic than in Brazil or Argentina. Rather than Latin America, for the lowest wages, a firm may look to places such as Vietnam or Bangladesh, while a firm that needs high skill levels will search Europe and North America.

Another barrier to manufacturing in Latin America relates to infrastructure. Factories need reliable and affordable electricity and water as well as roads and ports to move their goods. Yet many Latin American countries still have inadequate and poorly maintained

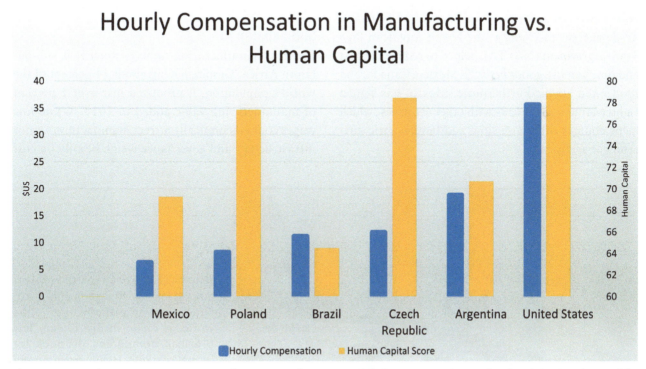

Figure 7.24. Hourly compensation in manufacturing vs. human capital. Countries such as Poland and the Czech Republic offer lower wages along with higher skills. Much of Latin America, on the other hand, offers only lower wages.

Data sources: US Bureau of Labor Statistics, Division of International Labor Comparisons. World Economic Forum.

road networks, slow and inefficient ports, and unreliable or high-cost electricity. Other countries, such as China and many in Eastern Europe, offer higher-quality infrastructure. When infrastructure quality is combined with labor costs and skills, once again, Latin America comes out behind.

A third barrier to manufacturing relates to government institutions and regulation. The three largest Latin American economies—Mexico, Brazil, and Argentina—score near the bottom of government institution competitiveness, according to the World Economic Forum. Too often, property rights are insecure, corruption is an ongoing concern, security is weak, and government efficiency is low.

As in all regions, there is substantial variation among Latin American countries. Mexico, for instance, while not offering the lowest cost or most highly skilled workforce, still manages to manufacture and export a significant quantity of goods. Its situation adjacent to the United States offers short transit times to move consumer goods and manufacturing components from border factories to customers in the massive US market. Trade agreements, such as the North American Free Trade Agreement (NAFTA), helped to reduce the cost and time to move goods between Mexico, Canada, and the United States. Furthermore, Mexico has signed eighteen trade agreements with other countries, which helps counter otherwise slow and bureaucratic government agencies.

Brazil and Argentina also have substantial manufacturing sectors, but they are traditionally hindered by heavy state influence. Both are in the bottom half of countries in terms of global competitiveness, according to the World Economic Forum. High tariffs restrict imports on machinery and other goods, while labor laws push wages higher than productivity. This often results in low-quality and high-priced goods. For example, in 2010, Argentina imposed nearly 40 percent tariffs on imported cell phones. High tariffs and government incentives convinced some phone manufacturers (but not Apple) to produce in Argentina, but by the time they reached consumers, they were twice as expensive as newer models in the United States. Likewise, cars produced in Brazil tend to have low levels of technology and are among the most expensive in the world for Brazilian consumers. The third-largest commercial aircraft manufacturer, Embraer, is based in Brazil, but otherwise, very few products outside of local markets are labeled "Made in Brazil" or "Made in Argentina" (figure 7.25).

Sub-Saharan Africa

The real manufacturing underperformer is sub-Saharan Africa. Despite having about 14 percent of the world's population, it produced just over 1 percent of manufacturing value added in 2015. While low wages could theoretically attract manufacturers, poor infrastructure and governance weigh heavily on cost

Figure 7.25. Brazilian-made Embraer aircraft. While Embraer, the third-largest aircraft manufacturer in the world, is made in Brazil, very few Brazilian companies produce for global markets. Heavy government influence reduces manufacturing competitiveness. Photo by Tupungato. Stock photo ID: 595726355, Shutterstock.

considerations. Only South Africa, which by no coincidence is the region's economic powerhouse, ranks in the top fifty most economically competitive countries. Most of the least competitive countries ranked by the World Economic Forum lie within sub-Saharan Africa.

With that said, manufacturing has been growing at least at the pace of overall GDP growth. As in Latin America, much production is targeted at local markets rather than being integrated into global supply chains. Low-cost cell phones for regional markets are being produced in South Africa, while much food and beverage production is local as well. Heavy-duty bicycles with puncture resistant tires and frames made for heavy loads are produced in a half-dozen African countries. Nevertheless, some manufactured goods are exported to global markets. Ethiopia is supplying stores such as H&M with clothing, and cocoa is being manufactured into chocolate bars in Madagascar.

With a young population, sub-Saharan Africa has the right conditions to use its demographic dividend for a manufacturing renaissance. The key challenge is to improve skills, infrastructure, and government effectiveness in order to harness its human potential.

Go to ArcGIS Online to complete exercise 7.3: "The changing geography of global manufacturing," and exercise 7.4: "The changing geography of US manufacturing."

Spatial relationships: The impacts of industrial change

As manufacturing shifts around the world, it has a profound impact on people and places. Places with growing manufacturing sectors often see improvements in employment opportunities, with more work and higher wages. This can be seen in developed countries such as the United States as well as in developing countries in Asia, Latin America, and Africa. However, when manufacturing declines, deindustrialization can have devastating impacts on local economies. As jobs disappear, unemployment rises and wages fall. Economic decline can ripple through a community, leading to the closure of additional businesses and further loss of jobs.

Manufacturing can also transform the social fabric of places. Especially in developing countries, manufacturing can upset traditional social norms, such as gender roles, as conservative rural economies transition to urban manufacturing ones. Yet again, when there is a decline in industrial employment, the social fabric of a place can be torn apart. This has been seen all too often in declining industrial towns, where drug and alcohol use spikes alongside declining opportunities for work.

Economic relationships

Manufacturing brought great wealth and development to much of Europe and the United States after the Industrial Revolution. Factories employed millions, productivity rose, and unions ensured good wages for substantial numbers for workers. Blue-collar, middle-class jobs sustained cities and towns across the developed world through the multiplier effect, as well-paid industrial workers fed local economies by spending their money on local goods and services. More recently, the same process has been happening in developing countries. As manufacturers search globally for places with ideal site and situation characteristics, such as lower-cost wages for labor-intensive production, employment opportunity has diffused.

China's phenomenal transformation to become the world's factory floor offers one of the most salient examples. In 1988, over 81 percent of Chinese lived on less than US $2 per day (in constant 2011 dollars), a World Bank measure of extreme poverty. By 2013, only 1.85 percent did, which represents hundreds of millions of people lifted out of poverty by China's manufacturing-led economic development. While factory working conditions included long hours at relatively low pay, workers voluntarily took the jobs, seeing them as a step forward from even lower wages and difficult working conditions in their rural hometowns.

More broadly, the relationship between manufacturing and places with better economic opportunities can be seen in the apparel industry. While the textile industry is often associated with sweatshops—factories with low wages and dangerous working conditions—in reality, it has been a powerful force in reducing poverty. Research has shown that textile workers earn more than they would in other unskilled jobs, and in factories tied to the global economy, working conditions are generally better. Places such as Honduras and Cambodia, among many others, have benefited from above-poverty-level wages and decreased poverty because of apparel manufacturing.

But places where industry brings economic opportunity are not confined to the developing world. In the United States, southern cities have seen their economies grow as manufacturing shifted from the northern industrial core. German and Japanese automobile companies have stimulated economies around Nashville, Tennessee, and Ford employs thousands in Louisville, Kentucky. Smaller rural places can see the most dramatic change when a factory opens. In 2013, in addition to the successful Chobani yogurt plant in Twin Falls, Idaho, Clif Bar & Co. also helped dramatically spur economic development in that city. Wages and benefits throughout Twin Falls were pushed up by competition from manufacturing employment, restaurants and other businesses were flourishing, and new homes were being built.

The flip side of manufacturing-stimulated economic growth is economic decline via deindustrialization. Global competition means that manufacturers can pack up and move to new locations when economic conditions change, leaving unemployed workers, falling wages, and declining economic activity. This process was seen Mexico in the early 2000s, as textile factories moved to China. In two years, Mexico lost 200,000 factory jobs, as many manufacturers relocated to China where production costs were lower. Incomes in many Mexican textile towns fell and unemployment increased. Yet ironically, by 2015, some manufacturing jobs were leaving China for Mexico. As Chinese wages increased, the difference between Chinese and Mexican labor costs narrowed. After considering shorter delivery times and easier monitoring of production with Mexico's proximity to the United States, overall costs were becoming lower by producing in Mexico.

In the United States, deindustrialization has a strong spatial relationship with economic suffering. Of US manufacturing workers who were displaced between 2013 and 2015, over one-third were not working at the beginning of 2016 (table 7.3). Of those who were working, only about 32 percent were making wages the same as or higher than at their previous job. The rest were working part time, were self-employed or caring for family members, or were working full time for lower wages.

Manufacturing is important not only for producing the goods we need but also because it is an important source of livelihood for people around the world. It has provided jobs and economic growth in places where it expands as well as economic pain in places it has left.

Social relationships

Outside of the manufacturing industry's impact on wages and employment, it can have a profound effect on the social fabric of places. When wages are high and economic opportunity is strong, communities tend to thrive. Workers have lower levels of stress when earning a steady paycheck. Paying rent and feeding a family become easier. In the United States, economic opportunity is associated with stronger families, including higher rates of marriage and lower rates of divorce. Yet, when economic opportunity declines, the social fabric can begin to tear. Job loss increases stress and family structures unravel. Unemployed workers, especially men, see greater social isolation and declining health.

In fact, middle-aged white (non-Hispanic) men with a high school education or less in the United States are seeing an increase in mortality despite declining mortality for all other age and racial/ethnic groups. This is largely due to increased deaths from drugs, alcohol, and suicide, so called "deaths of despair" that are consequences of diminishing economic opportunities for those with lower levels of education. The smaller map in figure 7.26 shows the death rate per 100,000 in 2015 caused by drug and alcohol use. Clear patterns are difficult to discern, as counties in many

Displaced Manufacturing Workers, 2013-15

Employment Status	
Employed	63.2%
Unemployed	15.0%
Not in Labor Force	21.7%
Full-Time or Part-Time	
Working Part-Time	10.5%
Working Full-Time	82.2%
Full-Time Worker Earnings	
Below previous wage	42.4%
Equal or above previous wage	44.3%
Self-employed and unpaid family workers	9.1%

Table 7.3. Displaced US manufacturing worker outcomes 2013–15. Displaced manufacturing workers have had a difficult time recovering financially. One-third are not working. Of those who are working, only about one-third are making wages the same as or higher than at their previous job. Data source: US Bureau of Labor Statistics, Division of Labor Force Statistics.

Death Rate by Drugs and Alcohol, 2015

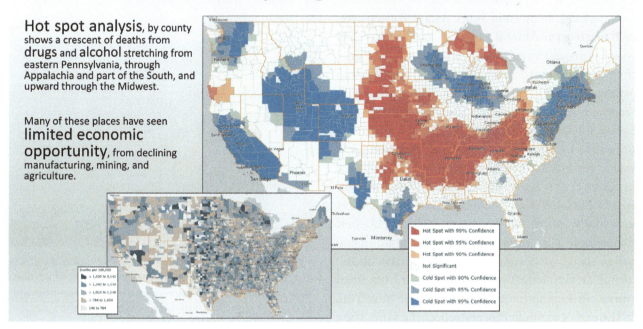

Hot spot analysis, by county shows a crescent of deaths from **drugs** and **alcohol** stretching from eastern Pennsylvania, through Appalachia and part of the South, and upward through the Midwest.

Many of these places have seen **limited economic opportunity,** from declining manufacturing, mining, and agriculture.

Figure 7.26. Death rate by drugs and alcohol, 2015. Data source: Centers for Disease Control and Prevention, National Center for Health Statistics.

states have high rates. However, a hot spot analysis in figure 7.26 helps clarify spatial patterns. A large cluster of counties with high levels of drug and alcohol deaths wraps around from eastern Pennsylvania through Appalachia and part of the South and upward through the Midwest. While not all these areas were former manufacturing centers, they do represent many places with loss of jobs and struggling economies, be it from declining manufacturing, mining, or agriculture.

One of the big ongoing stories in the US has been the struggle of laid-off factory workers. But in less-developed countries, the manufacturing story is one of social transformation in gender empowerment. Again, this story often contradicts common perceptions that manufacturing "sweatshops" exploit workers, often young, poorly educated women. Garment manufacturing relies heavily on female workers, typically employing a 70 to 90 percent female workforce. Hours are long and wages low. Workplace deaths and injuries are common, as in the 2013 Rana Plaza building in Bangladesh, where 1,120 textile workers died when nonpermitted floor additions collapsed like pancakes.

Yet women seek employment in the textile industry because it offers more opportunity than life in rural villages. In much of South Asia and elsewhere, conservative rural mores ensure that girls and young women are kept largely secluded in their homes until they can be passed on to another family through arranged marriage. Moving to the city to take a job is seen as risky and often dishonorable for their families. Still, a job in a textile factory gives them a freedom unheard of in the village. Much of their income is sent home to help their family, but girls also buy the occasional earring or bracelet as well as inexpensive cell phones to chat with new boyfriends. Most eventually return to their home village, but their exposure to paid work and somewhat more liberal urban social environments gradually shifts cultural norms. In fact, letters written by women from small towns working in the textile factories of New York in the early twentieth century reflect the same sense of excitement and liberation expressed by Indian and Bangladeshi women today.

Future trends in the geography of manufacturing

Manufacturers will continue to shift production to places that offer the most efficiency and profit. We have already seen that rising wages in China are pushing some manufacturing to other locations, be it low-wage Vietnam or close-to-the-US-market Mexico. Some manufacturing has even returned to rich places in the United States and Europe. As technology advances, location considerations will continue to change. The role of labor in the production process will decline as factories become even more automated (figure 7.27). The factory of the future will increasingly be filled with robots, not rows of human workers. More important will be the proximity to consumer markets and availability of skilled engineers and technicians to operate and maintain industrial machines. Government incentives and policies as well as agglomeration benefits will weigh more heavily.

Figure 7.27. Automated manufacturing: Industrial robots in a Skoda auto factory in the Czech Republic. The future of manufacturing includes more use of automation with fewer, but higher-skilled, human workers. Photo by Nataliya Hora. Stock photo ID: 368354981, Shutterstock.

This means that the demand for lower-skilled manufacturing labor will continue to decline. The days of going straight from high school to the factory floor in the United States or from the farm to a manufacturing job in China are coming to an end. Instead, manufacturing employees will have to possess higher-level skills, requiring postsecondary training of some sort. Jobs will involve operating and programming computer-controlled tools, electronics and mechanical repair, and process control. Manufacturing employment may increase in large consumer markets such as the United States, but it will not absorb millions of workers as in the past. Manufacturing employment as a percentage of all jobs will continue to decline.

Thus, products aimed at the US market will more likely be produced in the US, and goods for sale in China will be made there. The same will hold true for other regions as well, with locally made products designed with local tastes and preferences in mind.

Automation poses challenges to developing countries hoping to grow their economies and raise wages. No country has broken through to affluence without passing through a manufacturing stage. But this path to development may be cut off if low-cost labor ceases to be a significant factor of production. Even the sewing industry is moving toward automation, as advanced computer vision systems and agile robots are developed to replace human sewing factory workers.

Go to ArcGIS Online to complete exercise 7.5: "Trouble in the heartland: Manufacturing decline and deaths of despair."

References

Barboza, D. 2016. "How China Built 'iPhone City' with Billions in Perks for Apple's Partner." *New York Times,* December 29, 2016. https://www.nytimes.com/2016/12/29/technology/apple-iphone-china-foxconn.html?_r=0.

Barry, E. 2016. "Young Rural Women in India Chase Big-City Dreams." *New York Times,* September 24, 2016. https://www.nytimes.com/2016/09/25/world/asia/bangalore-india-women-factories.html.

Bradsher, K. 2017. "A Robot Revolution, This Time in China." *New York Times,* May 12, 2017. https://www.nytimes.com/2017/05/12/business/a-robot-revolution-this-time-in-china.html.

———. China's New Jetliner, the Comac C919, Takes Flight for First Time. *New York Times,* May 5, 2017. https://www.nytimes.com/2017/05/05/business/china-airplane-boeing-airbus.html.

Case, Anne, and Angus Deaton. 2015. "Rising Morbidity and Mortality in Midlife among White Non-Hispanic Americans in the 21st century." *Proceedings of the National Academy of Sciences of the United States of America* 112, no. 49: 15078–83. doi: https://doi.org/10.1073/pnas.1518393112.

Diagne, A. F. "Made in America: Computer and Electronic Products." US Department of Commerce, Economics and Statistics Administration. https://www.commerce.gov/sites/commerce.gov/files/migrated/reports/made-in-america-computer-and-electronic-products.pdf.

The Economic Policy Institute. 2017. "Minimum Wage Tracker." http://www.epi.org/minimum-wage-tracker/#/min_wage/California.

Economics and Statistics Administration, US Department of Commerce. 2015. "What Is Made in America?" http://www.esa.doc.gov/reports/what-made-america.

The Economist. 2012. "The Boomerang Effect." *The Economist,* April 21, 2012. http://www.economist.com/node/21552898.

———. 2014. "An Awakening Giant: Manufacturing in Africa." *The Economist,* February 8, 2014. https://www.economist.com/news/middle-east-and-africa/21595949-if-africas-economies-are-take-africans-will-have-start-making-lot.

Ernst & Young Global Limited. "The Central and Eastern European Automotive Market—Czech Republic." https://www.ey.com/en_gl/automotive-transportation.

European Commission Directorate-General for Trade. 2017. "Trade Agreements." European Commission. http://ec.europa.eu/trade/policy/countries-and-regions/agreements/#_other-countries.

The Heritage Foundation. 2017. "2017 Index of Economic Freedom. Promoting Economic Opportunity and Prosperity by Country." http://www.heritage.org/index.

The International Trade Administration, US Department of Commerce. 2016. "Bangladesh—Textiles and Textile Machinery and Equipment." https://www.export.gov/article?id=Bangladesh-Textiles-and-Textile-Machinery-and-Equipment.

———. 2016. "Brazil—Automotive Industry." https://www.export.gov/article?id=Brazil-Automotive-Industry.

Johnson, K. 2017. "An Idaho Town Bucks the Perception of Rural Struggle." *New York Times,* April 3, 2017. https://www.nytimes.com/2017/04/03/us/a-small-idaho-town-bucks-the-perception-of-rural-struggle.html.

Jordan, M. 2003. "Mexico Loses Jobs to China." *Wall Street Journal,* December 4, 2003. https://www.wsj.com/articles/SB10704760193998700.

Kotkin, J. 2016. "The Cities Leading a US Manufacturing Revival." *Forbes,* July 23, 2016. https://www.forbes.com/sites/joelkotkin/2015/07/23/the-cities-leading-a-u-s-manufacturing-revival/#2a961bf46113.

"Latin America: Why Are Manufacturing Exports Still Lackluster?" 2016. *Coface for Trade,* September 23, 2016. http://coface.com/News-Publications/Publications/Latin-america-why-are-Manufacturing-exports-still-Lackluster.

Lopez-Acevedo, G., and R. Robertson. 2012. "The Promise and Peril of Post-MFA Apparel Production." *Economic Premise* 84. World Bank, Washington, DC. https://openknowledge.worldbank.org/handle/10986/10043.

Lowe, J. 2017. "When Countries Think They Can Go D.I.Y." *New York Times,* May 5, 2017. https://www.nytimes.com/2017/05/05/magazine/when-countries-think-they-can-go-diy.html.

Mays, K. 2017. "The 2017 Cars.com American-Made Index." *News from Cars.com,* June 26, 2017. https://www.cars.com/articles/the-carscom-2017-american-made-index-1420695680673.

Mozur, P. 2017. "Plan for $10 Billion Chip Plant Shows China's Growing Pull." *New York Times,* February 10, 2017. https://www.nytimes.com/2017/02/10/business/china-computer-chips-globalfoundries-investment.html.

National Association of Manufacturers. "State Manufacturing Data." http://www.nam.org/Data-and-Reports/State-Manufacturing-Data.

National Conference of State Legislatures. 2017. "2017 Minimum Wage by State." http://www.ncsl.org/research/labor-and-employment/state-minimum-wage-chart.aspx.

New Jersey Business Action Center. 2017. "NJ Advantages for Pharmaceutical Companies." State of New Jersey Business Portal. https://www.nj.gov/njbusiness/industry/pharmaceutical.

Rascon, R. 2015. "China Acknowledges Loss of Manufacturing to Mexico." Offshore Group. https://insights.offshoregroup.com/china-acknowledges-loss-of-manufacturing-to-mexico.

Rivoli, P. 2015. *The Travels of a T-shirt in the Global Economy: An Economist Examines the Markets, Power, and Politics of World Trade.* Hoboken, NJ: Wiley.

Saunders, E. R. 2012. "Chobani Opens Twin Falls Yogurt Facility Today, But at What Cost to Taxpayers?" NPR, December 17, 2012. https://stateimpact.npr.org/idaho/2012/12/17/chobani-opens-twin-falls-yogurt-facility-today-but-at-what-cost-to-taxpayers.

Scott, R. E. 2015. *The Manufacturing Footprint and the Importance of US Manufacturing Jobs.* Briefing paper #388. Economic Policy Institute. http://www.epi.org/publication/the-manufacturing-footprint-and-the-importance-of-u-s-manufacturing-jobs/#epi-toc-17.

Statt, N. 2016. "iPhone Manufacturer Foxconn Plans to Replace Almost Every Human Worker with Robots." *The Verge,* December 30, 2016. https://www.theverge.com/2016/12/30/14128870/foxconn-robots-automation-apple-iphone-china-manufacturing.

Taylor, P. 2015. "No More Blue Banana, Europe's Industrial Heart Moves East." *Reuters,* March 15, 2015. http://www.reuters.com/article/us-eu-industry-analysis-idUSKBN0MB0AC20150315.

Taylor Hanson, L. 2003. "The Origins of the Maquila Industry in Mexico." *Comercio Exterior* 53, no. 11: 1–16.

Torpey, E. 2014. "Got Skills? Think Manufacturing." US Bureau of Labor Statistics. https://www.bls.gov/careeroutlook/2014/article/manufacturing.htm.

US Bureau of Labor Statistics. 2016. "Displaced Workers Summary." https://www.bls.gov/news.release/disp.nr0.htm.

———. 2017. "National Employment." https://www.bls.gov/iag/tgs/iagauto.htm.

Washington State Department of Commerce. 2016. *Proposed Strategic Plan for Washington State Aerospace Sector, 2017–2019.* http://www.commerce.wa.gov/wp-content/uploads/2017/01/Sector-Leads-Aerospace-Strategy-2016.pdf.

Weber, A., and C. J. Friedrich. 1929. *Theory of the Location of Industrie*s. Chicago: University Press.

Wial, H. 2016. "Locating American Manufacturing." Brookings. https://www.brookings.edu/interactives/interactive-locating-american-manufacturing.

World Bank. 2009. *World Development Report 2009: Reshaping Economic Geography.* Washington, DC: World Bank.

World Economic Forum. 2017. *Human Capital Report 2016.* http://reports.weforum.org/human-capital-report-2016.

———. *The Global Competitiveness Report 2015–2016.* http://reports.weforum.org/global-competitiveness-report-2015-2016.

World Trade Organization (WTO). 2008. *World Trade Report 2008.* WTO Economic Research and Analysis Gateway. https://www.wto.org/english/res_e/reser_e/wtr08_e.htm.

Zeng, D. Z. 2015. "Global Experiences with Special Economic Zones." *Open Knowledge Repository.* https://openknowledge.worldbank.org/handle/10986/21854.

Chapter 8
Services

After agriculture and manufacturing, the third major component of economic geography is the tertiary, or service, sector. In all likelihood, you and your classmates will not be working on a farm or in a factory. Rather, you will be working in an office, store, school, hospital, or some other location where you are paid not to produce tangible goods such as apples or cell phones but to provide intangible goods (figure 8.1). The intangible goods produced by the service sector are wide ranging. They include wholesale and retail services, such as selling cars, food, electronics, and furniture. They also include transportation and warehousing services that employ workers such as truck drivers, stocking clerks, package delivery personnel, and airline pilots. Utility workers are employed in the service sector as well, providing the generation and transmission of power as well as water and sewer services. Information production is another large component of this sector and includes film and recording industries, internet publishing, data processing, and telecommunications. In finance, stockbrokers, insurance agents, and realtors are part of the service sector. Education and health care are two massive parts of the service sector, as are the arts and tourism.

As in other sectors of the economy, service providers search for the most profitable location. Companies need places with good site characteristics and access to the right types of land, labor, and capital. Government regulations, infrastructure, and agglomeration also influence the desirability of sites. *Situation*, or the relative location of service providers to other places, rarely depends on proximity to raw material inputs, but proximity to customers can play an important role. However, with digital technology, the service sector is now global, creating a spatial distribution where some companies rely on workers on one side of the world to provide services to customers on the other. Globalization has led to clusters in some places. Finance, entertainment, tourism, information technology (IT), arts, and more congregate in certain cities, driving economic growth in those cities. At a local scale, spatial patterns are also driven by site and situation characteristics. Banking and insurance may cluster in one part of a city, warehousing and transportation in another. Retail establishments also locate in areas on the basis of neighborhood demographics and the presence of competitors.

Figure 8.1. Service sector workers in an office. Photo by Monkey Business Images. Stock photo ID: 656511280. Shutterstock.com.

While the future of most workers lies with the service sector, it is important to note the challenges that this brings. First of all, wages in the tertiary sector vary substantially. Whereas agricultural and manufacturing work tends to have a narrower range of wages, services range from the very low to the very high. Many fast-food and home-health-care workers earn near minimum wage, while surgeons and corporate executives can earn annual incomes in the hundreds of thousands of dollars or more.

At the same time, outsourcing and offshoring of work to temporary contract and foreign employers threaten the wages and job security of many service workers. The so called gig-economy means that many service workers lack regular paychecks and the health and retirement benefits that often accompany them. Likewise, many companies have found that workers in poorer countries can perform service jobs at a fraction of the cost of workers in higher-income places. Everything from lower-skilled customer service call centers to higher-end IT jobs have been sent offshore, providing new opportunities for residents of developing countries but threatening prospects for workers in developed ones.

Finally, automation, as in agriculture and manufacturing, threatens many service jobs. The proportion of people working as bank tellers and secretaries has declined with the diffusion of the automatic teller machine and office computers. The composition of service jobs will continue to change as self-driving vehicles and warehouse-stocking robots replace many transportation and wholesale positions. Even higher-skilled jobs are being threatened. Machines that can analyze x-ray and magnetic resonance images may replace radiologists, while investment advice and data analytics are being done with artificial intelligence.

These changes can lead to social conflict and present economic and political challenges that are increasingly difficult to ignore. Retraining displaced service workers and ensuring adequate wages will take on greater political significance. As in all sectors of the economy, technological change and globalization offer great benefits but also significant challenges.

Types of services

The US Bureau of Labor Statistics breaks service-providing industries into seven *supersectors*, which are further divided into more detailed sectors (table 8.1). Sectors are then broken down into even more detailed industries. In essence, the service industries include all businesses outside of manufacturing, construction, and natural resources and mining.

The service industry is often broken up into two broad categories: producer services and consumer services. Producer services, also known as business-to-business services, are those that sell to businesses and the government. This sector of the economy has seen steady growth as more companies focus on their core competencies and outsource other functions to subcontractors whose nature is to sell services to other businesses. In contrast, consumer

Trade, Transportation, and Utilities
• Wholesale Trade
• Retail Trade
• Transportation and Warehousing
• Utilities
Information
Financial Activities
• Finance and Insurance
• Real Estate and Rental and Leasing
Professional and Business Services
• Professional, Scientific, and Technical Services
• Management of Companies and Enterprises
• Administrative and Support and Waste Management and Remediation Services
Education and Health Services
• Educational Services
• Health Care and Social Assistance
Leisure and Hospitality
• Arts, Entertainment, and Recreation
• Accommodation and Food Services
Other Services (except Public Administration)

Table 8.1. Service supersectors and selected sectors. Data source: US Bureau of Labor Statistics.

	Producer Service	Consumer Services
Trade, Transportation, and Utilities	FedEx delivers packages for Amazon.	Lyft transports you to a party.
Information	A web hosting company manages the website of the New York Times.	You subscribe directly to the online edition of the New York Times.
Financial Activities	A home appraiser determines the value of your home for a mortgage lender.	A real estate agent helps you purchase a new house.
Professional and Business Services	An accounting firm audits your college or university's finances.	A tax specialist does your income taxes.
Education and Health Services	A "travel nurse" company provides hospitals with temporary nurses.	A hospital nurse cares for you.
Leisure and Hospitality	An art restoration company repairs works for a museum.	A museum opens its collection to the public.

Table 8.2. Producer vs. consumer services. Table by author.

services sell directly to consumers. Table 8.2 shows that producer and consumer services are often found within the same service sectors. The difference is not in what service is being provided but rather who the customer is.

Services can also be divided into *tradable* and *nontradable* industries. Tradable services can be consumed in a place other than where they are produced. A corporate lawyer in London can provide services for a client in Mexico City, and a computer programmer in Mexico City can provide a service for a software company in London. In both cases, where the work is performed and where the work is consumed are different. In contrast, nontradable services are those that must be consumed in the same place they are produced. Even if they offer a better service for a better price, a barber in New York cannot cut the hair of a customer in Los Angles, and a plumber in Poland cannot fix the pipes of a customer in Chicago.

Spatial distribution of services

Global distributions
Services and development
Whereas agricultural strength conveyed power to ancient societies, as did industrial might to countries of the eighteenth through twentieth centuries, all affluent and powerful countries now draw their economic strength from services. In fact, of the rich countries identified as developed by the United Nations, the service sector dominates GDP production in all of them. In the United States, France, and the United Kingdom, nearly 80 percent of GDP came from services in 2015. Germany and Japan, two countries with sizable manufacturing sectors, still earned around 70 percent of GDP from services. At a broader scale, high-income countries earned about 75 percent of GDP from services, while low- and middle-income countries earned less than

60 percent, and the least developed countries earned under 50 percent of GDP from services. Figure 8.2 shows how developed Europe and North America rely heavily on the service sector, developing Latin American is somewhat mixed, and less-developed Africa has a weaker service sector.

Richer and more developed countries earn more from services because of high levels of automation and productivity. When people have higher levels of education and skills, and when machinery helps perform many tasks, more goods can be produced with fewer workers. Providing food, clothing, and other material necessities no longer requires a large segment of the working population, so people are freed up to work in other areas. Time and energy can be devoted to everything from potentially life-saving research in cancer treatments to less-essential yet enjoyable services

such as developing social media apps, dog walking, and personal training.

In developing countries, such as many in Africa and parts of Latin American and Asia, large segments of the population still work in low-productivity subsistence agriculture. Nevertheless, some improvements in productivity, especially among capital-intensive commercial agriculture producers, now mean that more people are leaving the agricultural sector. Manufacturing has not absorbed many of these workers, given the problems discussed in chapter 7, leaving most to find some type of work in the service sector. However, the service jobs they take on are not higher-skilled positions in business services and technology. Rather, they are in low-skilled consumer services. These can include motorcycle taxi drivers, small shopkeepers, street vendors, house cleaners, and sellers of prepaid

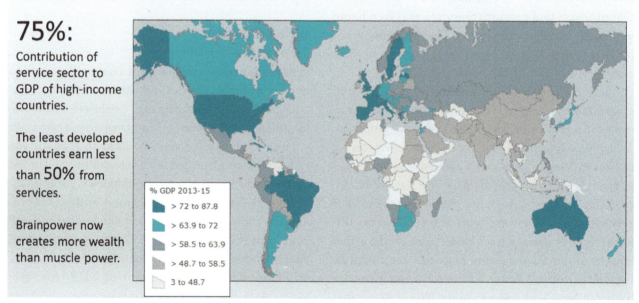

Figure 8.2. Services as percentage of GDP, 2013–15. Explore this map at https://arcg.is/OiS9f. Data source: World Bank.

Figure 8.3. Small shop in Cape Town, South Africa. In less-developed countries, many service workers are in low-productivity jobs, such as at small retail establishments in the informal economy. Photo by Daniel S Edwards. Stock photo ID: 235537198. Shutterstock.com.

cell-phone minutes. Most of these jobs are in the informal sector, with inconsistent incomes and no benefits (figure 8.3). So, while maps of service employment in developing countries may show some growth in that sector, it is important to remember that the shift does not always represent movement toward well-paid jobs in new sectors of the economy.

Figure 8.4 shows the proportion of GDP from services for selected countries that have data from the 1960s to the present. France, the richest and most developed of this group, earns a much higher proportion of GDP from services. France still produces agricultural and industrial goods, but over time, automation on farms and in factories has allowed the French to devote their time to other pursuits. This has allowed the service sector to grow and contribute a greater share to the French economy. Mexico's service sector has grown somewhat, but a greater transformation since the 1960s has been a shift from agriculture to manufacturing via the maquiladora industry. South Korea's service sector has grown substantially. Manufacturing is still an important component of its economy, but new sectors have grown as well, including entertainment such as K-pop (figure 8.5). Kenya and India sill remain largely agricultural. In the case of Kenya, tourism services are

important, although they have declined in recent years with fears of terrorism from nearby Somalia. India has seen rapid growth in its service sector, especially in back-office works such as call centers and data processing as well as information technology.

Services, cities, and the world economy

Some cities have a disproportionate influence in global services. By studying the location of corporate headquarters and the revenue earned, certain places stand out on the map as leaders in areas such as finance, consumer goods, information technology, and more. Even when a corporation's primary function is manufacturing, corporate headquarters provide the services that oversee it, such as management, accounting, marketing, and research.

The reason for this concentration of services lies again in the benefits of agglomeration. As similar services cluster in one place, the strength of the whole becomes greater than any single part. The most skilled and talented workers will be attracted by ample employment prospects. Also, despite ease of communication brought about by the internet, physical proximity leads to casual conversations at coffee shops, bars, and on the street, which contribute to new ideas

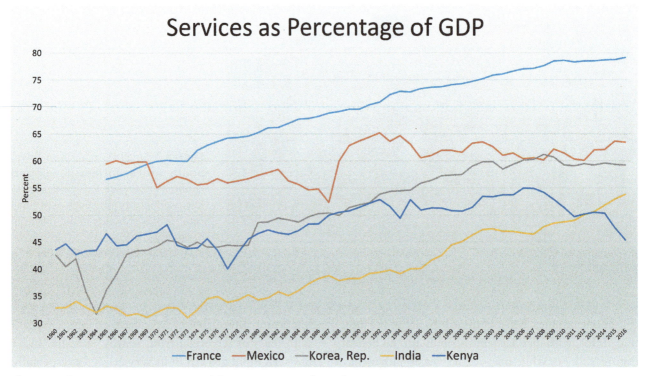

Figure 8.4. Services as a percentage of GDP over time, selected countries. Data source: World Bank.

Figure 8.5. K-pop group Blush. Korea now exports a wide range of services, such as music, which is consumed around the world. Photo by Randy Miramontez. Stock photo ID: 96074576.

as well as career tips. Specialized subcontractors are also drawn to the agglomeration, offering services outside of a corporation's core competencies. These can include human resources functions, IT network maintenance, payroll, wellness, and other services that help corporate headquarters run smoothly.

Figure 8.6 shows cities by revenue generated in different economic sectors in 2012. The financial capital of the world is New York, followed by Paris, London, and Tokyo (figure 8.7). These cities are where many financial institutions have their headquarters and earn vast amounts of revenue. Large banks such as JPMorgan Chase, Goldman Sachs, HSBC Holdings, BNP Paribas, Mitsubishi UFJ Financial, and many more are found in these cities. In these cities, flows of investment capital are managed, corporate merger and acquisitions deals are done, and lending to companies

and individuals is overseen. From this handful of urban nodes, major economic decisions are made that impact people around the world.

Consumer staple goods are also concentrated in a handful of powerful cities. These include large corporate food markets and providers of household and personal products. London tops the list in terms of revenue, followed by Paris, Tokyo, and New York. Food companies such as Tesco (groceries and general merchandise) and Carrefour (similar to Costco in the US) have their headquarters in London and Paris respectively. Colgate-Palmolive from New York oversees a massive market of personal hygiene and cleaning products. Beauty products come from L'Oréal in Paris and Estee Lauder in New York. Thus, many of the types of food, personal care, and beauty products are developed and marketed from a few influential cities.

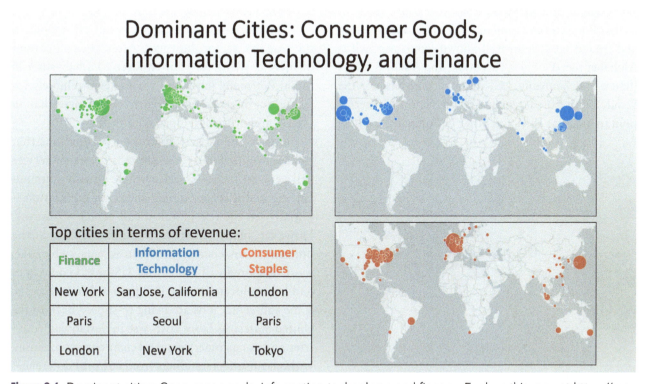

Figure 8.6. Dominant cities: Consumer goods, information technology, and finance. Explore this map at https://arcg .is/1aurPy. Data source: The data was produced by G. Csomós and constitutes Data Set 26 of the Globalization and World Cities (GaWC) Research Network (http://www.lboro.ac.uk/gawc) publication of inter-city data.

Figure 8.7. Office workers in the financial district of Tokyo, Japan. Many corporate headquarters locate in Tokyo, making it a major center of global services, including consumer staples, finance, information technology, and more. Photo by Kobby Dagan. Stock photo ID: 365261141. Shutterstock.com.

Information technology, both software and hardware, also concentrates in a handful of places. Not surprisingly, Silicon Valley's San Jose tops the list in terms of both revenue and number of headquarters. In and around San Jose are well-known companies such as Alphabet Inc. (Google), Facebook, and Adobe Systems (figure 8.8). Samsung, LG, and others are based in the Seoul, South Korea, region, and the New York metropolitan area has IBM and other IT companies. These

Figure 8.8. Alphabet (Google) office in Mountain View, California. Photo by Pozdeyev Vitaly. Stock photo ID: 383777227. Shutterstock.com.

cities shape a large part of our daily lives, providing tools for work and entertainment that many people use on a near-constant basis.

Services and offshoring

In a global economy, companies search for the most profitable location for production. This holds true for the service sector just as it does for manufacturing and agriculture. While rail, roads, and shipping containers facilitate global trade in agricultural and manufactured goods, advances in communication technology have facilitated the global service industry. As communications have evolved from letters to the telegraph to the phone to the internet, many service companies have been able to take site and situation factors into account in their location decisions, the most important of these being labor. Today, with nearly instant communication between places, skill sets, language, wages, and government labor laws play a big role in where service providers locate.

The globalization of services only functions in the tradable sector, however. After all, you cannot offshore day care for your children to a babysitter in India. But many jobs can be completed remotely for customers by workers who earn lower wages than in the US and other developed countries.

One of the most well-known occupations that has seen significant spatial shifts is customer-service call centers. India—a place where British colonialism created a large English-speaking population—initially led the revolution in call-center offshoring. However, many Americans complained about the difficulty of understanding Indian accents, so by 2011, most offshored call centers for US-based companies had been relocated to the Philippines. More recently, some call-center jobs are being re-shored back to the United States. As labor costs rise in developing countries and as customers demand clear communication, without misunderstandings from cultural differences and English-language idioms, some companies are deciding that it makes better economic sense to use workers in the United States.

The IT sector has also seen shifts in its global distribution. The nature of information technology is that digital data can be nearly instantly moved from place to place. In theory, this makes it among the most footloose of industries. India, Eastern Europe, and other lower-cost countries have benefitted from the offshoring of IT work, such as computer programming, network maintenance, and research and development. Major US companies, including Disney, Verizon, and the University of California, San Francisco, have moved or considered moving IT jobs abroad to reduce costs. Yet this shift is not one way. Ironically, Infosys, an Indian outsourcing company, is also moving jobs to the United States. While one attraction of India as an IT outsourcing destination is its large number of engineering graduates, Infosys found that only 5 percent of them could write code correctly.

So, the geography of IT services continues to fluctuate. Some companies are now outsourcing not to foreign countries but to other cities and states within the United States. With exorbitant housing prices in tech centers such as Silicon Valley, where six-figure starting salaries are the only way to attract employees, some technology firms have found that workers in, say, Michigan, can do the same high-quality work but for lower wages than a worker in the San Francisco Bay Area. Other changes in IT employment are spurring some companies to bring workers back into the corporate office. It turns out that much IT work is best performed when people are physically in the same location as coworkers and customers. New ideas and solutions often come out of serendipitous conversations in offices and around the water cooler, which cannot happen as easily when workers are isolated in remote workplaces. In recent years, large technology firms such as Yahoo! and IBM have called workers back to the office from telecommuting.

Thus, the global distribution of service employment faces some of the same issues as manufacturing. Sites with lower wages appeal to companies in search of lowering costs, yet once worker skill sets and product quality are taken into account, it often makes sense to locate closer to customers.

Local distributions

Zooming in to a larger scale, we can look at the spatial pattern of services and how they order themselves on the landscape regionally and within cities. Many of these patterns you may have noticed already. For instance, if you live in a smaller town, you probably know that you have to drive into the big city for a pop concert or (if you win the lottery) to buy an Armani suit or handbag. Likewise, you may be fine seeing a local doctor in town if you have the flu, but if she finds something more serious, you may have to see a specialist in an urban regional medical center.

Another pattern you may have noticed is that similar types of stores seem to cluster in the same location. Clothing retailers, even those with very similar lines, are typically in the same part of town, as are restaurants and other commercial services. Given that they compete with each other, this seems like an unlikely arraignment, yet clustering with competitors is much more common than dispersed relationships.

Clearly, the spatial distribution of services is not random. Some types of services are found in smaller towns, and others are located only in larger cities. Likewise, stores and other services are not evenly spread around town but are clustered in specific places.

Central place theory

To understand why services differ between small towns and big cities, we return to Christaller's central place theory, first described in chapter 5. Central place theory is based on the idea of market areas, which are determined by the concepts of threshold and range (figure 8.9). Each type of business has a threshold, or minimum number of customers needed to support it. A small convenience store has a small threshold, being able to make a profit from a relatively small number of customers in a local neighborhood. A high-end Armani store will not survive exclusively with customers from the nearby community. Since few people can afford Armani products, and those who can do not purchase them on a regular basis, it

Market Area: Central Place Theory

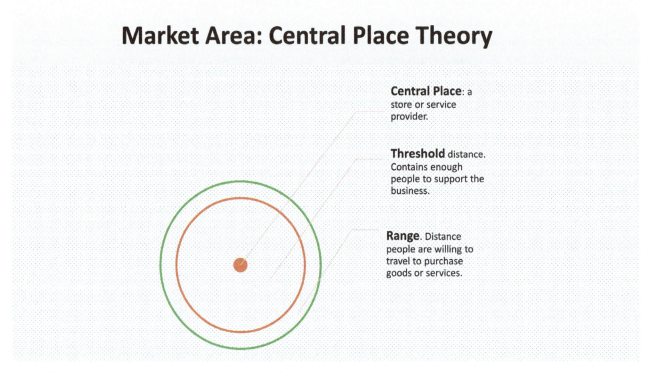

Central Place: a store or service provider.

Threshold distance. Contains enough people to support the business.

Range. Distance people are willing to travel to purchase goods or services.

Figure 8.9. Market area according to central place theory. Image by author.

must draw from a much larger population. For this reason, it has a larger threshold. Market areas are also determined by range, or the distance people are willing to travel to purchase goods or services. People will generally travel only a short distance to purchase a soda from a convenience store, but they will travel a greater distance to purchase an Armani bag or suit.

Businesses that have large market areas, with large threshold requirements and large ranges, must cluster in high-order urban areas. A large city and its surrounding urban areas will have a sufficient customer base to support specialized services. Smaller urban areas will have businesses with smaller market areas, as a limited population size supports only businesses with small threshold requirements.

Thus, low-order small towns have businesses with very small market area requirements. These can include convenience stores, gas stations, auto repair shops, hair salons, and other small businesses. Medium-order cities that are somewhat bigger can have primary care physicians, maybe a car dealership, larger grocery stores, and midlevel clothing stores. Large, high-order cities have the most specialized businesses, such as exotic car dealers, specialist health-care facilities, high-end clothing stores, and major league sports teams.

Figure 8.10 shows cancer specialty hospitals in the southwestern United States, which require highly specialized medical staff and equipment. Because of the level of specialization and cost involved in this type of medical care, these hospitals have very large thresholds. Thankfully, cancer is less common than ailments such as the flu, so a large population is required to support this type of facility. At the same time, cancer is a very serious disease, meaning that people are willing to travel greater distances to get treatment. For this reason, these facilities also have a large range. It is evident that specialty hospitals are located primarily in larger urban counties and metropolitan areas. Smaller (micropolitan) urban areas and rural regions, because of their small populations, have very few facilities of this type.

High-Order Service: Cancer Specialty Hospitals

Specialty hospitals, such as those that treat cancer, are found in larger urban areas rather than smaller urban and rural communities.

This is because they have a large **threshold**: a large population is needed in order for them to stay in business.

They also have a large **range**: people are willing to travel a larger distance for their specialized services.

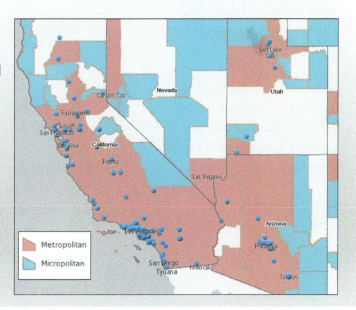

Figure 8.10. High-order service: Cancer specialty hospitals. Data sources: ArcGIS Business Analyst; US Census.

Figure 8.11 emphasizes the significance of large urban areas in hosting highly specialized services such as cancer hospitals. The densely populated Los Angeles–Long Beach–Santa Ana statistical area, defined by the US Census Bureau, has sixty-nine such hospitals, serving a population of nearly thirteen million people. Assuming there is a rough balance between the supply of cancer hospitals and the demand by cancer patients, then each of the sixty-nine hospitals operates with a threshold of about 188,000 people (13 million/69 = ~188,000).

Convenience stores, unlike specialty hospitals, have lower costs and are used more frequently by the nearby population (figure 8.12), so they are much more scattered throughout small, medium, and large cities. The city of Madera, California, population just over 60,000, does not have any specialty hospitals. However, as would be expected, it does have convenience stores. Thirteen of these establishments (not including combination gas stations and convenience stores) serve an average of less than 5,000 people each.

Hotelling's location model

While the central place theory helps explain where different types of services locate—specialized high-order services in larger cities and generic low-order services in smaller towns as well as larger ones—it does not explain where the same types of services locate. Often, when you and friends plan to meet for food or drinks, you pick a section of town where there are multiple restaurants. Likewise, when you go shopping, there is typically a neighborhood with multiple clothing stores. Auto dealers also frequently cluster in several blocks. On the surface, this seems illogical in that nearby businesses must compete with one another for business. But in 1929, Harold Hotelling, a mathematical statistician and economic theorist, came up with a theory that shows that the natural result of competition will draw businesses together in the same location.

High-Order Service in Large Urban Areas

Large urban areas such as Southern California have sufficient population to support services with high thresholds.

Thus, specialty cancer hospitals cluster there.

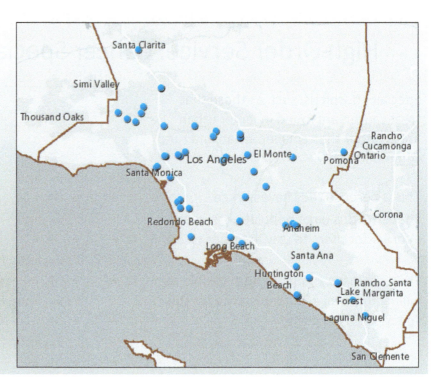

Figure 8.11. High-order service in large urban areas. Data source: ArcGIS Business Analyst.

Low-Order Service in Small Urban Areas

Small urban areas only have sufficient population to support services with low thresholds.

Thus, convenience stores are found throughout these areas.

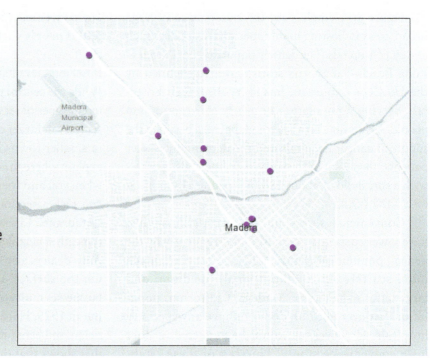

Figure 8.12. Low-order service in small urban areas. Data source: ArcGIS Business Analyst.

A common way of explaining how this works is to consider two ice cream vendors on a beach (figure 8.13). If each vendor locates at the far end of the beach, they split the customer base evenly (A). Half of the beachgoers will be closest to one vendor, and half will be closest to the other. But what happens when one vendor shifts toward the center to capture a larger segment of the beach? (B) This action forces the other vendor to move as well, so as to regain market share (C). Ultimately, each of the two vendors will move until they reach a new equilibrium that gives them each one-half of the market and where no move will allow for a larger share. The only location that allows for this is with both vendors adjacent to each other in the middle of the beach (D). While it is not the best location for customers on the far ends of the beach, it is the best location for the two vendors, who want to maximize their market share.

Hotelling's line of thought is similar to the bid-rent model introduced in chapter 5. Recall that in that model, the highest rents are found at the center of the city. This is because the center is the point of highest accessibility, and thus the highest demand by businesses, in the city. When a business is located in the center of a city, customers can access it from 360 degrees. Any location other than the center will be farther from customers on the far side of the center. Just as in Hotelling's linear location model, businesses will cluster as a natural result of competition.

Figure 8.14 illustrates clusters of wholesale flower, wholesale toy, and wholesale jewelry stores in downtown Los Angeles. Dozens of similar businesses locate within blocks of each other, forcing intense competition. Yet, a clustered location benefits them in that buyers come from all over the Los Angeles metropolitan area, knowing that there will always be an ample supply of the items they are looking for (figure 8.15). If an individual store locates in a less central location, it risks losing the volume of customers that are attracted to the cluster by good accessibility and variety of options.

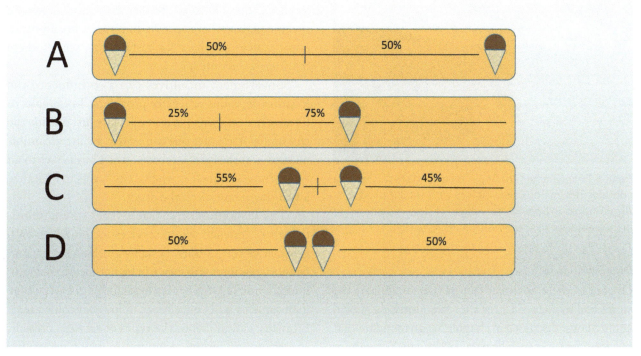

Figure 8.13. Hotelling's location model. Image by author.

Service Clusters

Services often cluster in the same location to maximize accessibility from surrounding areas.

The benefits of accessibility outweigh the cost of competition between sellers.

In downtown Los Angeles, there are clear clusters of **flower**, **toy**, and **jewelry** sellers.

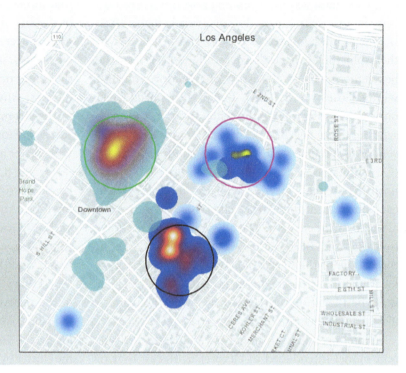

Figure 8.14. Service clusters. Data source: ArcGIS Business Analyst.

Figure 8.15. Downtown Los Angeles flower district. This area contains dozens of flower sellers with similar products. Agglomeration benefits outweigh the downside of being in the same area as competitors. Photo by author.

Other location factors

Of course, there are other factors that complicate the simplified models of Christaller and Hotelling. Local zoning laws, discussed in chapter 5, can mandate that services cluster in specific parts of the city. You cannot convert a house to a liquor store or fast-food restaurant in the middle of a residential neighborhood, even if you feel it is the best location. Rather, most cities designate specific places where specific businesses are allowed. Some zones can allow for restaurants, while others may allow only offices.

Government incentives can also influence the location decisions of service providers. Tax breaks to locate car dealerships in a smaller city just outside the city limits of a larger one can lead to a cluster. Similar incentives at the local level can attract businesses to an enterprise zone, such as when a city wants to revitalize a run-down commercial district.

The demographic characteristics of communities also influence where services locate. High-end, full-service restaurants; designer clothing stores; and financial planners will locate in higher-income communities, while fast-food, discount clothing, and payday loan services will concentrate in lower-income areas. Ethnicity can influence the types of services present, so Latino immigrant communities may have medicinal herb shops and *quinceanera* dress stores that are less common in other communities. Lifestyle, as represented by Esri's Tapestry Segmentation, has a significant influence as well. For instance, the Urban Chic segment adjacent to Vanderbilt University in Nashville, Tennessee, will likely attract organic food establishments, art galleries, yoga studios, and outdoor recreation suppliers (figure 8.16).

Population movement can also influence the location of services. Pedestrian and vehicle traffic flows vary substantially from place to place. A business in a busy downtown district can benefit from thousands of people walking by each day. The same is true for a business along a major highway thoroughfare that benefits from vehicle traffic. Population flows by time of day can also be significant. Many business districts have large daytime populations (figure 8.17) but few people after work hours. Services such as lunch-oriented restaurants and package carrier services will likely be attracted to these places. On the other hand, some districts attract a nighttime crowd. An area with movie theaters can attract restaurants and bars that benefit from the evening moviegoing crowd.

Go to ArcGIS Online to complete exercise 8.1: "Central place theory: Where can I buy a Ferrari?" and exercise 8.2: "Consumer services site selection: Where should I put my store?"

Services and clusters in the United States

In many cities of the United States, the service economy is a key basic industry. Services such as finance, entertainment, tourism, technology, and others cluster in certain cities, bringing money into local economies, directly employing many workers and supporting employment for many more through the multiplier effect. Because of the importance of services in driving local economies, many cities use subsidies, marketing

Tapestry Segmentation: Services for "Urban Chic"

Lifestyle characteristics, as identified by segmentation analysis, can influence where services locate.

MARKET PROFILE (Consumer preferences are estimated from data by GfK MRI)

- Shop at Trader Joe's, Costco, or Whole Foods.
- Eat organic foods, drink imported wine, and truly appreciate a good cup of coffee.
- Travel extensively (domestically and internationally).
- Prefer to drive luxury imports and shop at upscale establishments.
- Embrace city life by visiting museums, art galleries, and movie theaters for a night out.
- Avid book readers of both digital and audio formats.
- Financially shrewd residents that maintain a healthy portfolio of stocks, bonds, and real estate.
- In their downtime, enjoy activities such as skiing, yoga, hiking, and tennis.

Figure 8.16. Tapestry segmentation: Services for urban chic. Explore this map at https://arcg.is/eeuKj. Data sources: Nashville-Davidson Metro Government, Esri, HERE, Garmin, Intermap, USGS, NGA, EPA, USDA, NPS, US Census Bureau, Infogroup.

Population Flows and Service Location

Many service businesses base their location on population flows.

Some places have high pedestrian or vehicle traffic during the **day**, while other places have higher flows at **night**.

Business to business services and lunch establishments will be located in places like downtown Oklahoma City, where the daytime population is larger.

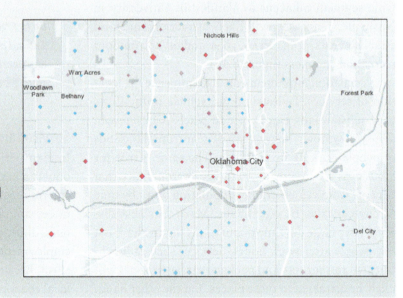

Figure 8.17. Population flows and service location. Explore this map at https://arcg.is/0HTTTG. Data sources: Esri, HERE, Garmin, NGA, USGS, NPS.

campaigns, urban redevelopment, and other policies to attract companies and workers.

Business and financial operations occupations

Some cities have been successful at creating agglomerations of high-skilled and well-paying services. New York and Chicago are world-renowned centers of finance, with large numbers of jobs in various types of banking and investment services (figure 8.18). These two areas stand out in figure 8.19, which shows cities with at least 10,000 workers in business and financial operations. This category also includes a wide range of occupations such as human resources, management, accounting, and marketing. Large urban areas such as Los Angeles also stand out, as do the largest cities in many states. Jobs in this sector tend to require higher levels of education and pay above-average wages. In 2016, the median wages in business and financial operations were 40 percent higher than the overall US median wage. These well-paying industries fuel

economic activity in other sectors of the local economy and attract workers from other places, both near and far. This increases the economic strength of large urban areas and continues to drive rural-to-urban migration. As large cities attract high-skill, high-pay jobs, the differing levels of opportunity between rural and urban America are only exacerbated.

Several cities in Texas also have a substantial number of jobs in the business and financial operations sector. While corporate location decisions involve many factors, government incentives can help play a role. In 2014, Toyota Motors decided to relocate its corporate headquarters from Southern California to the Dallas, Texas, area. The state of Texas offered $40 million in incentives, equivalent to about $10,000 per employee, to attract the company. Similarly, it offered financial incentives to attract Chevron jobs to Houston and Apple jobs to Austin. Because of the high wages paid to workers in business and financial operations, many state and local governments argue that incentives

Figure 8.18. Trading floor of the Chicago Mercantile Exchange, Chicago, Illinois. Chicago has a significant cluster of high-paying finance jobs. Photo by Joseph Sohm. Stock photo ID: 258345956.

Business and Financial Operations Employment

Finance, human resources, management, accounting, marketing, and similar jobs cluster in large urban areas.

These jobs require higher levels of education and pay higher wages than many other sectors.

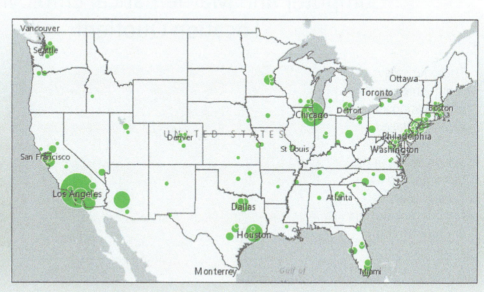

Figure 8.19. Business and financial operations employment. Explore this map at https://arcg.is/1nm14T. Data source: US Census.

are a worthwhile investment to attract corporate jobs. Over the long term, governments expect to benefit from increased tax revenue and economic development through the multiplier effect.

Computer and mathematical occupations

The role that services play in local economies can also be viewed in terms of the location quotient, first introduced in chapter 4. Figure 8.20 shows places where computer and mathematical occupations are overrepresented as compared to the US average. Naturally, Silicon Valley in California stands out. But there are also many other urban areas with substantial tech sectors. The Seattle area is home to Microsoft and Amazon. The Washington, DC, area has numerous defense- and government-related technology companies. Austin, Texas, is a growing tech hub, and Huntsville, Alabama, has many facilities for NASA's space program (figure 8.21). As with the business and finance sector, these occupations require high levels of education and pay good wages. As of 2016, wages were fully 70 percent higher than the US median, making tech hubs some of the most affluent and thriving places in the United States.

Arts, design, entertainment, sports, and media occupations

Another service sector that provides above-average wages and can be an important source of economic development is the arts, design, entertainment, sports, and media sector. Again, larger urban areas have the most jobs in this sector, which include actors, writers and editors, professional athletes, musicians, and broadcasters. Wages are 25 percent above the US median and require high levels of education or skills. Of course, Los Angeles stands out in figure 8.22, with its large film and entertainment sectors. New York has a media cluster, a thriving arts scene, its famous

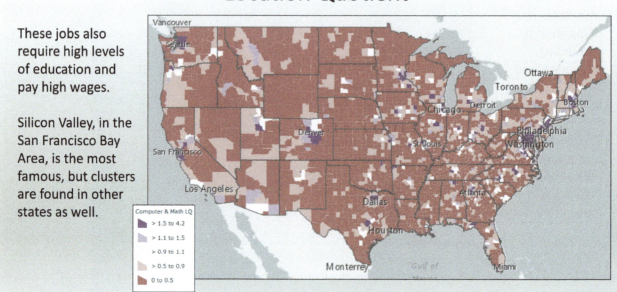

Figure 8.20. Computer and mathematical employment. Location quotient. Explore this map at https://arcg.is/1nm14T.
Data source: US Census.

Broadway theater district, major publishing houses, and much more (figure 8.23). Smaller cities also stand out, such as Nashville, Tennessee, with its influential music scene.

Figure 8.21. Davidson Center for Space Exploration in Huntsville, Alabama. A cluster of computer and math employment linked to NASA is located here. Photo by Rob Hainer. Stock photo ID: 105385199.

Looking at the film industry alone, nearly 200,000 people are directly employed in California, and close to 100,000 are employed in New York. Maybe surprisingly, Georgia ranks third, with over 25,000 industry workers. Television programs such as *The Walking Dead* and *The Vampire Diaries* have been filmed there, as have major motion picture such as the *Hunger Games* series. Georgia has actively pursued the film industry, promoting its tax incentives for production, a business-friendly regulatory environment, and free location scouting. Its effort has paid off, so that an industry agglomeration in cities such as Atlanta now offers benefits to further film production. Because of the agglomeration effect, there are now experienced film crews; studios; and firms that specialize in editing, animation, lighting, wardrobe, and all other tasks essential for film production.

Arts, Design, Entertainment, Sports, and Media Occupations Employment

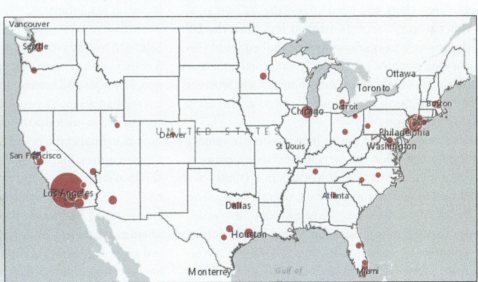

Los Angeles, with its large film and entertainment industry, dominates this sector.

New York is also important, with a media cluster, a thriving arts scene, the famous Broadway Theater district, and major publishing houses.

Smaller cities also stand out, such as Nashville, Tennessee, with its influential music scene.

Figure 8.22. Arts, design, entertainment, sports, and media occupations employment. Explore this map at https://arcg. is/1nm14T. Data source: US Census.

Figure 8.23. Theaters in Times Square, New York City. Photo by Jiawangkun. Stock photo ID: 534020845. Shutterstock.com.

Food preparation and serving related occupations

While many services jobs are high-skill and high-wage positions, driving economic development and the formation of middle-class jobs, other service occupations are just the opposite. In the case of food preparation and service jobs, skill levels tend to be lower, as are median wages. Overall, workers in this sector earn 44 percent less than the US median wage. Even chefs and head cooks, the highest paid in this category, earn 26 percent less than the US median. These examples show the challenges of service sector-oriented economies. Increasingly, there is a polarization of incomes, with some services workers earning substantial wages while others earn barely enough to live on.

Figure 8.24 shows the location quotient by county for food preparation and serving-related occupations. When a high proportion of workers in a county are employed in restaurants and bars, overall wages tend to be low. In some cases, these counties have economies focused on tourism. Las Vegas stands out, as do counties to its east that cater to Grand Canyon tourism. Likewise, Mariposa County in California has a disproportionate number of restaurant employees that serve tourists to Yosemite National Park. Much of South Florida as well as coastal communities along the Atlantic coast also have tourism-oriented economies with a large proportion of restaurant employees.

But in other cases, restaurant employment may be among the only options for work not because of tourism but because the local economy is stagnant and few other employment options exist. In these counties, the only options for work may be the local diner or fast-food restaurant.

Sales and related occupations

Another service sector that employs a large number of people is sales and related occupations. This category has median wages 11 percent below the national average but with a great deal of variation. Higher-end sales, such as in financial services and manufactured goods, often requires higher levels of education and pays above-average wages. But this category also includes retail positions, employing hundreds of thousands of lower-skilled and lower-paid workers in the US. In urban areas, there is likely a mix of both high- and low-end sales employment. Higher-educated workers sell items such as pharmaceuticals, software, machinery, and other sophisticated products, while lower-educated workers work in retail establishments.

Food Preparation and Serving Employment: Location Quotient

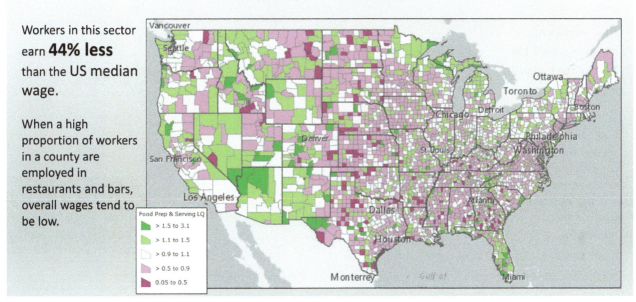

Workers in this sector earn **44% less** than the US median wage.

When a high proportion of workers in a county are employed in restaurants and bars, overall wages tend to be low.

Food Prep & Serving LQ
- > 1.5 to 3.1
- > 1.1 to 1.5
- > 0.9 to 1.1
- > 0.5 to 0.9
- 0.05 to 0.5

Figure 8.24. Food preparation and serving: Location quotient. Explore this map at https://arcg.is/1nm14T. Data source: US Census.

Problems arise in smaller rural places. As with restaurant work, in some places one of the few employment options is the local grocery store or discount retailer. Figure 8.25 shows some overlap with restaurant employment, such as in South Florida and along the Atlantic coast, where souvenir shops complement restaurants in serving tourists.

Creative class clusters

When discussing the spatial distribution of services, geographers typically look at the decisions made by firms. So far, we have seen how site and situation characteristics, such as the presence of ports and highway infrastructure, wage and skill levels, and proximity to resources and consumers, can influence where companies locate. We have also looked at the role of the government, including zoning laws and tax incentives. Such factors guide where companies invest, ultimately determining the quantity and quality of jobs

and economic growth in different places. Some places have more high-paying jobs in business, finance, and the arts, while others have more low-paying jobs in restaurants and retail sales.

Geographer Richard Florida, in contrast, flips this line of thought, focusing on how people choose places to live and work and how firms often follow. His work focuses on a group of people with the skills and education that drive the most important economic and cultural innovations in the United States. By his argument, it is the places where these people choose to live that have become the most economically vibrant centers of activity in the country, with high-paying work, innovative companies, and a high quality of life.

Florida refers to this group of people as the *creative class*. These are people who create new products and ideas and find solutions to pressing problems in science, government, business, the arts, and many other areas. The creative class includes occupations such

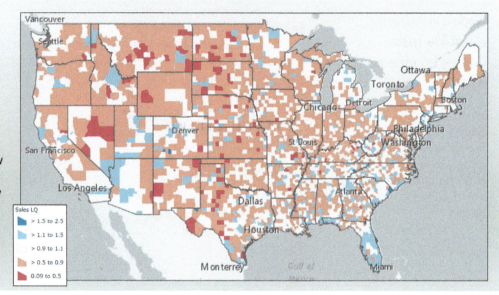

Sales and Related Employment: Location Quotient

This category has median wages **11% below** the national average.

Problems arise in smaller rural places, where one of the few employment options is in the local grocery store or discount retailer.

Sales LQ
- > 1.5 to 2.5
- > 1.1 to 1.5
- > 0.9 to 1.1
- > 0.5 to 0.9
- 0.09 to 0.5

Figure 8.25. Sales and related employment: Location quotient. Explore this map at https://arcg.is/1nm14T. Data source: US Census.

as scientists, engineers, university professors, artists, actors, designers, writers, legal and health-care professionals, financiers, and managers. These workers make up a specialized subset of the service sector. They are the most educated or skilled, and their work adds great value to the places where they cluster.

While most traditional location theory focuses on people following their economic interests, moving to places with the best job opportunities, Florida argues that the creative class is attracted to places for both economic and lifestyle reasons. Specifically, they are attracted to places with the 3Ts: technology, talent, and tolerance. Technology can be found in and around universities and where some large technology firms have located. Many of these places also have talent, or a large proportion of people with at least a bachelor's degree. The third characteristic, tolerance—openness; inclusiveness; and diversity of races, ethnicities, and lifestyles—exists to a greater degree in some cities than in others.

Places must have all three of these characteristics to attract the creative class. Some places have one or two of these characteristics, but the places that truly thrive from the creative class have all three. For example, Baltimore, St. Louis, and Pittsburgh all have technology, with important universities, yet have been unable to attract the creative class due to a lack of cultural openness. Miami and New Orleans have tolerance of different lifestyles, yet lack technology.

Examples of cities with all three of these characteristics include major urban areas such as Washington, DC, Raleigh-Durham, Boston, Austin, Seattle, and the San Francisco Bay Area. Smaller cities such as East Lansing, Michigan, and Madison, Wisconsin, can also meet the criteria. As creative class workers cluster in these places, more creative and technological companies are created and attracted, leading to powerful economic growth.

Cities with technology, talent, and tolerance tend to have microlevel urban characteristics that attract the creative class. Specifically, the creative class is drawn to places with a more organic street-level culture. Unlike the culture in formal urban cultural districts, such as those with a symphony hall or museums, street-level culture thrives in multiuse urban neighborhoods. Venues are small, with coffee shops, bars, and restaurants; small art galleries; and bookstores. People walk, intermingling with other residents of the city as they go about their daily lives in the neighborhood (figures 8.26 and 8.27). These types of communities reflect the urban revitalization, and in some cases gentrification, of cities that was discussed in chapter 5.

As the creative class clusters in attractive and economically thriving cities, there is now increasing concern about inequality. Innovation by the creative class has led to economic growth and correspondingly high wages for its members in places like San Francisco, Seattle, and New York, among other cities. Housing costs have skyrocketed, forcing those who are not members of the creative class to seek housing elsewhere. Some people see thriving creative-class cities as becoming playgrounds for the wealthy, with exciting street life and ample economic opportunity that only a segment of the population can enjoy. In fact, there appears to be little overlap between where the creative

Figure 8.27. Coffee shop in Portland, Oregon. The creative class also chooses to live in places with active neighborhoods full of coffee shops, restaurants, and bars. Photo by Joshua Rainey Photography. Stock photo ID: 391904392. Shutterstock.com.

class lives and where the working class live. Each is increasingly living in cities and/or neighborhoods completely segregated by occupation and income.

Go to ArcGIS Online to complete exercise 8.3: "Service employment growth: What should I study and where can I find a job?"

Service employment: Growth, decline, and wages

As we have seen thus far, the service sector dominates developed countries such as the United States, but within this sector is a wide range in terms of wages. Restaurant and retail sales workers face low wages and limited opportunities for advancement, while those in business and finance, computer and math, and arts and media occupations enjoy higher wages and ample room for professional development. The subsegment of professionals that are in the creative class enjoy even more opportunity for good wages, professional growth, and exciting lives in culturally diverse urban areas.

Figure 8.26. People at music festival, Austin, Texas. The creative class is attracted to places with lively arts and entertainment scenes. Photo by PiercarloAbate. Stock photo ID: 612826124. Shutterstock.com.

Given the difference within the service sector, what are the trends moving forward? What types of careers should a young person be preparing for? The US Bureau of Labor Statistics provides projections through the year 2024. The trend toward service occupations will continue, with 81 percent of Americans working in that sector by 2024. Within the service sector, the largest single-industry sector will be health care and social assistance, followed by professional and business services, then state and local government. Topping out the top five sectors will be retail trade and leisure and hospitality.

Looking at specific occupations, several stand out as having especially high growth rates. Of the fastest-growing occupations, some offer above-median wages, while others are much lower paying. The single fastest-growing occupation is projected to be for wind turbine service technicians, which pays wages substantially above the median (figure 8.28). Following that are occupational therapy assistants and physical therapy assistants, who also earn above-average wages. However, the next two fastest-growing occupations, physical therapist aides and home health aides, earn wages well below the median. Given the nature of this book, it is also worth noting that one of the

Figure 8.28. Wind turbine workers. Wind turbine service technicians are expected to see the fastest rate of employment growth in the United States by 2024. Photo by Patrizio Martorana. Stock photo ID: 374209276. Shutterstock.com.

fastest-growing occupations includes cartographers and photogrammetrists, with wages well above the median for all occupations.

While occupation growth rates are useful for showing expanding areas of employment, probably more important for American workers are the occupations with the most job growth in raw numbers (table 8.3). Looking at these numbers, employment prospects for the US workforce appear difficult. The largest number of new jobs by 2024 will be for personal care aides. In 2016 wages, these workers earned just 57 percent of the median for all workers. Next in line are registered nurses, who earn a healthy 139 percent of the median. Moving down the list, however, the employment picture looks grim. Home health aides, food preparation workers, and retail salespersons will see substantial growth, but wages are just 50 to 75 percent of the US median. Of the fifteen occupations with the most projected growth by 2024, eleven offer below-median wages. Only registered nurses, general and operations managers, accountants and auditors, and software developers have above-average wages and are expected to be among the jobs with the most growth.

This data indicates that future employment is heading in the direction of greater income polarization. Middle-wage manufacturing jobs will employ a smaller proportion of the US population. Most new jobs are projected to be in lower-income service sectors of the economy. Yet, as has been shown in this chapter, the geographic patterns of different service occupations will not be evenly distributed. At the urban and regional levels, places that see growth in higher-paying occupations, such as those that the creative class are attracted to, and others that can attract employers in the business services, technology, and scientific fields, will thrive. Cities and regions that cannot attract such employers are likely to see stagnant economies tied to lower-skill and low-wage services, such as home health aides caring for an elderly population, food servers, and retail employees. Within cities, economic segregation will continue to increase, with the educated and high-wage segment of the population enjoying vibrant neighborhoods, while less-educated and lower-paid

Occupation	Thousands of new jobs 2014-2024	Percent of Median Wage, 2016	Typical education needed for entry
Total, all occupations	9,788.90	100%	—
Personal care aides	458.1	57%	No formal educational credential
Registered nurses	439.3	139%	Bachelor's degree
Home health aides	348.4	60%	No formal educational credential
Combined food preparation and serving workers, including fast food	343.5	50%	No formal educational credential
Retail salespersons	314.2	75%	No formal educational credential
Nursing assistants	262	60%	Postsecondary nondegree award
Customer service representatives	252.9	77%	High school diploma or equivalent
Cooks, restaurant	158.9	53%	No formal educational credential
General and operations managers	151.1	152%	Bachelor's degree
Construction laborers	147.4	83%	No formal educational credential
Accountants and auditors	142.4	139%	Bachelor's degree
Medical assistants	138.9	69%	Postsecondary nondegree award
Janitors and cleaners, except maids and housekeeping cleaners	136.3	63%	No formal educational credential
Software developers, applications	135.3	213%	Bachelor's degree
Laborers and freight, stock, and material movers, hand	125.1	68%	No formal educational credential

Table 8.3. Occupations with the most job growth, 2014–24. Data source: US Bureau of Labor Statistics.

residents struggle to pay rent and face long commutes from lower-cost peripheral communities.

Challenges of automation

The United States produces vast amounts of agricultural and manufactured goods, yet the proportion of US workers employed in agriculture and manufacturing has been on a downward trend as machines replace human workers. As output and employment have shifted to the service sector, could the same trend follow there? Could large numbers of service workers soon find themselves replaced by machines?

Recent estimates are that 5 to 9 percent of jobs in the US and in the Paris-based Organisation for Economic Co-operation and Development (OECD), a group of mostly rich countries, could be automated in coming decades. Some of the service jobs that could soon be replaced are in warehouses, where robots are now able to find, pick, and package products for the rapidly growing online shopping sector (figure 8.29).

Other jobs in retail and restaurants could be replaced with scanners, touch screen orders, and automatic payment from cell phones. In transportation, rapid progress is being made in developing autonomous vehicles, which threatens the jobs of taxi drivers, truck drivers, and anyone else who drives for a living. Work is even being done to fully automate massive ocean-going cargo ships. But it is not only lower- or middle-skilled service jobs that are threatened by automation. Computers can now accurately identify tumors in x-rays, natural language algorithms can scan legal documents, and artificial intelligence can identify patterns in data.

In recent years, artificial intelligence (AI) has made great progress. In 1997, IBM's Deep Blue beat a reigning world chess champion. Then, in 2011, IBM's Watson AI system beat human contestants in the TV trivia game *Jeopardy*. By 2016, Google's AI beat a human player at Go, a boardgame invented in ancient China that is exponentially more complex than chess. In 2017, Facebook had to shut down an AI negotiating

Figure 8.29. Robotic arm for packing. Automation will make some services job, such as warehouse picking and packing, obsolete in the near future. Photo by wellphoto. Stock photo ID: 139813588. Shutterstock.com.

simulation when the two machines started to form their own language for pursuing their negotiations. As AI continues to advance, there will be pressure to use it increasingly for service tasks currently performed by humans.

In the best-case scenario, automation in the service sector will enhance productivity, allowing people to be more effective and create more output for each hour worked. In theory, higher productivity translates to higher wages for workers. In hospitals, algorithms may help diagnose illnesses, and automated gurneys may be used to move patients from room to room. In aircraft maintenance, drones may automatically scan aircraft exteriors for wear and tear, and robots may move spare parts and install new ones. In these cases, doctors, nurses, and aircraft mechanics can be more productive as automation completes some of the routine tasks, allowing human workers to focus on more challenging and complicated ones.

The pace of automation depends on several factors. First, the technology must exist. Autonomous driving technology is advancing quickly and should be available in the consumer market in a few years. Second, the technology must be cost effective to deploy.

If a technology exists but is too expensive for most firms to purchase, then it is unlikely to be used. The cost of a self-driving truck may remain high for the near future, limiting the number of truckers who are displaced. This also relates to labor markets. If an ample supply of workers is available at a reasonable cost, there may be less incentive to invest in machinery and technology. However, in cases where labor is scarce or expensive, the only option for a firm may be to switch to automation.

Challenges of inequality

The service sector faces the same challenges as agriculture and manufacturing: globalization and automation. The global outsourcing industry will continue to shift jobs to lower-cost locations if they can be done remotely without sacrificing quality. At the same time, advances in robotics and AI will replace or reshape a wide range of occupations. The risk is that these trends will further exacerbate income inequality, as those with more skills and education are able to work more productively with coworkers in other places or with robotic assistants, while those without the right skills or education become largely redundant.

Geographers and others propose myriad solutions to the growing income gap. All of them have their merits, but as with solutions to all complex issues, opinions vary greatly in the details.

Economic growth

One solution that nearly everyone agrees on is economic growth. When economies grow, the demand for workers increases. Employers hire more workers, and competition between firms drives up wages. Of course, economic growth is easier said than done, and how to promote it is beyond the scope of this book. Smart people offer widely different prescriptions for government taxation and spending, levels of regulation and deregulation, ideal levels for federal interest rates, and more. From a spatial perspective, different countries and different states have tried various mixes of these factors, yet empirically, no single combination works in all situations.

Minimum wage

Another possible, yet more controversial, solution relates to the minimum wage. Proponents argue that raising the minimum wage lifts incomes of those at the bottom of the workforce, reducing inequality and poverty. As these workers earn more, their spending spurs additional job growth through the multiplier effect. While this seems like a straightforward argument, it has many opponents. Those who oppose increasing the minimum wage say that it has the unintended consequence of reducing employment. Thus, while some workers will benefit from a higher wage, others will be worse off, since they will not have a job at all. Employers will find ways to run their business with fewer workers if labor is too expensive relative to output. Positions or hours will be cut, and the transition to automation may be accelerated.

As with economic growth, the empirical evidence on minimum wages is mixed. Some research has shown that increases in the minimum wage do not reduce employment, while others show a small decline in the number of minimum-wage jobs or a reduction in the hours that minimum wage earners work each week. Recent research in California has supported the argument in favor of raising the minimum wage, while in Seattle, Washington, some research has shown that increases have reduced the number of minimum wage jobs in that city. Most likely, the impacts depend on a variety of factors, such as the strength of the local economy and the amount of the wage increase (figure 8.30).

Negative income tax

Another solution, supported by a wide range of policymakers and politicians, is the idea of a negative income tax. With a traditional progressive income tax, a worker pays a larger tax percentage to the government as income rises, but with a negative income tax, people who earn little receive money from the government. In essence, this type of system tops off the wages of lower-income earners. In the United States, this is done via the *earned income tax credit* (EITC).

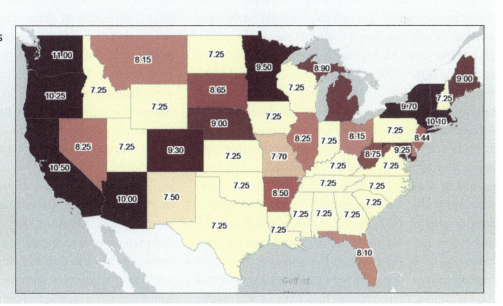

State Minimum Wage, 2017

There is still debate as to whether minimum wages help or hurt the poor.

Proponents say that higher wages reduce poverty and inequality.

Opponents say that employers hire fewer people when wages are artificially raised, leaving more with no job at all.

Figure 8.30. State minimum wage, 2017. Explore this map at https://arcg.is/0uafCb. Data source: Bureau of Labor Statistics.

This credit goes to people who work yet earn below a specified annual income. In 2017, a single person with no children could receive a modest $510, but for a family with three or more children, the EITC increased to $6,318. It has bipartisan political support, since it is effective in reducing poverty (a core goal of the left), but it is also tied to work (a core value of the right). From a geographic standpoint, it is efficient in reaching lower-income workers in a wide range of places. As discussed in previous chapters, antipoverty programs have traditionally been concentrated in denser urban areas, where clusters of poverty are easy to reach. However, this often leaves the rural and suburban poor unserved by social service programs. The benefit of the EITC is that it reaches people regardless of where they live, as it is distributed via checks issued by the Internal Revenue Service.

Universal basic income

Economic growth, minimum wage increases, and negative income taxes all rest on the assumption that people work. But what if growing automation means that an increasing segment of the population becomes permanently unemployed? An idea that is being considered recently is the universal basic income, whereby all residents of a country—rich and poor, employed and unemployed—would receive a fixed monthly minimum income, of, for instance, $1,000. Some on the political left like this idea because it would reduce poverty and inequality while some on the political right like it because it could largely replace what they see as heavily bureaucratic and inefficient welfare agencies. A basic income would allow people to pursue school and risk starting businesses without having to worry about how to pay the rent and buy food. It has been pushed by some of Silicon Valley's technology leaders, who worry that automation will increasingly replace an ever-greater share of the workforce. If that were to happen without some type of universal income, inequality could skyrocket, as the owners of firms reap a greater share of wealth with the help of robots and AI. Of course, there are also many critics who argue that a universal basic income would simply allow people to not work, becoming a burden on those who do. At least in the near future, this idea is unlikely to be implemented in any significant way, but with technological progress over the next fifty to 100 years, it may become more paramount.

Service sector unions

In the mid-1950s, 35 percent of the US workforce was in a union. For some, this was the golden age of US workers, when middle-class employment, often in manufacturing, was available to large numbers of people. But by 2016, only 10.7 percent of the workforce belonged to a union. Unionization has fallen in all industries, but those in low-wage service jobs are especially unlikely to be members of a union. If we return to the largest-growing service occupations, we can see that jobs in health-care support, such as personal care and home health aides, have lower-than-average levels of unionization, at 6.9 percent. Food preparation is even lower, with only 3.9 percent unionized. Union representation falls even more when looking at sales and related occupations, with a mere 3.1 percent unionized.

Those who support more unionization of the service sector point out that wages for union workers are higher than for their nonunion counterparts. In health-care support occupations, for example, union workers earn nearly 16 percent more, while in food preparation and serving, union workers earn close to 27 percent more. The challenge is organizing workers who are scattered geographically in different locations with different employers. Organizing a large factory with thousands of workers in the early twentieth century was much easier than trying to organize a scattered workforce.

Some states in the US are more supportive of unions than others, impacting levels of unionization (figure 8.31). Much of the South has traditionally opposed unions, seeing them as limiting the flexibility needed by firms to hire, fire, and set wages and benefits according to market conditions. By this argument, union rules can reduce competitiveness and ultimately limit economic growth.

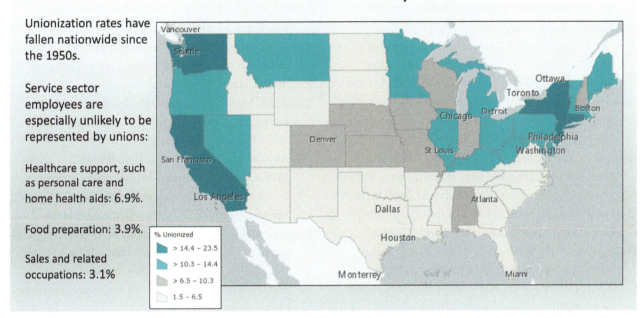

Figure 8.31. Unionization rates, 2016. Data source: Bureau of Labor Statistics.

Ultimately, new, creative solutions will be needed to address the increasing polarization of income seen in the US and much of the world. As pointed out in this section, there are many ideas but limited consensus on how to best tackle this pressing problem.

Go to ArcGIS Online to complete exercise 8.4: "Minimum wage and unemployment."

References

Arntz, M., T. Gregory, and U. Zierahn. 2016, "The Risk of Automation for Jobs in OECD Countries: A Comparative Analysis." *OECD Social, Employment and Migration Working Papers*, no. 189. Paris: OECD Publishing. http://dx.doi.org/10.1787/5jlz9h56dvq7-en.

Bajaj, V. 2011. "A New Capital of Call Centers." *New York Times*, November 25, 2011. http://www.nytimes.com/2011/11/26/business/philippines-overtakes-india-as-hub-of-call-centers.html.

Baskin, B. 2017. "Next Leap for Robots: Picking Out and Boxing Your Online Order." *Wall Street Journal*, July 25, 2017. https://www.wsj.com/articles/next-leap-for-robots-picking-out-and-boxing-your-online-order-1500807601?mg=prod%2Faccounts-wsj.

Buss, D. 2014. "It's Not About Incentives: Toyota's Texas Move Is a Corporate-Culture Gambit." *Forbes*, April 29, 2014. https://www.forbes.com/sites/dalebuss/2014/04/29/its-not-about-incentives-toyotas-texas-move-is-a-corporate-culture-gambit/#7e1fdcc736df.

Csomós, G. 2012. "Data Set 26 of the Globalization and World Cities (GaWC) Research Network." http://www.lboro.ac.uk/gawc.

The Economist. 2017. "Why Africa's Development Model Puzzles Economists." *The Economist*,

August 17, 2017. https://www.economist
.com/news/finance-and-economics/21726697-
structural-transformation-its-economies-not-
following-precedents-why.

Florida, Richard. 2003. "Cities and the Creative
Class." *Cities and Community* 2, no. 1. doi:
https://doi.org/10.1111/1540-6040.00034.

———. 2006. *The Rise of the Creative Class and How
It's Transforming Work, Leisure, Community and
Everyday Life*. New York: Basic Books.

Goel, V., and P. Mozur. 2017. "Infosys, an Indian
Outsourcing Company, Says It Will Create 10,000
US Jobs." *New York Times,* May 2, 2017. https://
www.nytimes.com/2017/05/02/business/infosys-
hire-10000-american-workers.html.

Harnett, S. 2016. "Outsourced: In a Twist, Some
San Francisco IT Jobs Are Moving to India."
NPR, December 27, 2016. https://www.npr.org/
sections/alltechconsidered/2016/12/27/
507098713/outsourced-in-a-twist-some-san-
francisco-tech-jobs-are-moving-to-india.

Holmes, N., and A. Berube. 2016. "The Earned
Income Tax Credit and Community Economic
Stability." *Brookings*, November 20, 2016. https://
www.brookings.edu/articles/the-earned-income-
tax-credit-and-community-economic-stability.

Hotelling, H. 1929. "Stability in Competition."
The Economic Journal 39, no. 153: 41–57. doi:
https://doi.org/10.2307/2224214.

Kitroeff, N. 2017. "What Will a Higher Minimum
Wage Do? Two New Studies Have Different
Ideas." *Los Angeles Times*, January 11, 2017.
http://www.latimes.com/business/la-fi-impact-
minimum-wage-20170110-story.html.

Lohr, S. 2017. "Hot Spot for Tech Outsourcing:
The United States." *New York Times,* July 30,
2017. https://www.nytimes.com/2017/07/30/
technology/hot-spot-for-tech-outsourcing-the-
united-states.html?_r=0.

Manyika, James, Michael Chui, Mehdi Miremadi,
Jacques Bughin, Katy George, Paul Willmott,
and Martin Dewhurst. 2017. *Harnessing
Automation for a Future That Works.*
McKinsey & Company. https://www.mckinsey.
com/global-themes/digital-disruption/
harnessing-automation-for-a-future-that-works.

Paris, C. 2017. "Norway Takes Lead in Race to Build
Autonomous Cargo Ships." *Wall Street Journal*,
October 12, 2017. https://www.wsj.com/articles/
norway-takes-lead-in-race-to-build-autonomous-
cargo-ships-1500721202.

US Bureau of Labor Statistics. 2017. "Industries at
a Glance: NAICS Code Index." https://www.bls
.gov/iag/tgs/iag_index_naics.htm.

———. "Employment Projections and Occupational
Outlook Handbook." Press release. https://www
.bls.gov/news.release/ecopro.toc.htm.

———. "Union Membership." Press release. https://
www.bls.gov/news.release/union2.toc.htm.

Witsil, F. 2014. "Call Center Jobs Increase as
More Return from Overseas." *USA Today*,
August 4, 2014. https://www.usatoday
.com/story/money/business/2014/08/04/
call-center-jobs-overseas/13560107.

Chapter 9
Development

Famine, poverty, illiteracy, illness, early death. These are staples of the news we see each day. In many parts of the world, people struggling to survive or improve their lives are held back by poverty and lack of opportunity. But while these are very real and very serious issues, it is informative to first see how far we have come in advancing the human condition. Despite news stories that can leave one feeling down, in many ways, life on earth is better than ever.

Figure 9.1 illustrates some of the ways in which life has improved globally in recent decades. Extreme poverty, or the percentage of the world population living on less than $1.90 per day in constant 2011 *purchasing power parity* (PPP) dollars, has fallen from nearly 42 percent in 1981 to less than 11 percent by 2013. As extreme poverty has fallen, improvements in infrastructure and health have improved. In 1990, 76 percent of people had access to an improved water source, but by 2015, this number had increased to nearly 91 percent. This means that many more people have safe drinking water from sources such as indoor plumbing, public taps, and protected wells. Clean water, among other improvements, has resulted in better health. This can be seen in rising life expectancy. In 1960, women lived an average of about 54 years and men lived an average of 50. By 2015, life expectancy had increased significantly, with women living to nearly 74 and men to almost 70. Additional improvements in

well-being are seen in literacy rates, which rose from 76 percent in 1990 to 85 percent in 2010.

The list of improvements goes on and on. Infant and child mortality are down, maternal mortality is down, more people have access to electricity, child labor is down, school enrollment is up, and immunization rates have increased. So, while it is important to continue working toward a better future, the gains of the past cannot be ignored. These are truly impressive improvements in the human condition.

The good news is that nearly all measurements of development are moving in positive directions. But the bad news is that too many people still suffer from poverty, illness, and a general lack of opportunity. A baby is twenty-three times more likely to die in Cambodia than in Canada, and a woman giving birth is over six times more likely to die. The GDP per capita in the US is over three times that in Mexico. Even between developing countries there are great variations. A young Brazilian woman is nearly four times as likely as a young woman in the Central African Republic to be literate (figure 9.2). A person in Iraq is over ten times more likely than someone in Liberia to have electricity.

This spatial variation in level of development is intertwined with much of the material from previous chapters of this book. Development affects birth and death rates, migration flows, rates of urbanization, and economic activity in agriculture, manufacturing,

Improving Human Development: Selected Measures

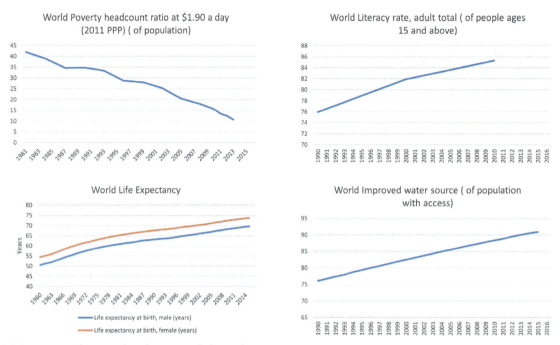

Figure 9.1. Improving human development: Selected measures. Data source: World Bank.

Figure 9.2. Rising levels of education. An indigenous school classroom in Jaqueira Village in the city of Porto Seguro, Brazil. Photo by Joa Souza. Stock photo ID: 552399136. Shutterstock.com.

and services. Understanding the spatial patterns of development and the forces that drive them can lead to further improvements in health, economics, and opportunity, improving the quality of life for people around the world.

Defining development

Geographers study these spatial patterns to understand why some places develop while others lag. Development geography and development studies are large fields, but the definition of development has no single, clear definition. The terms *more developed country* (MDC) and *less developed country* (LDC) are often used, but in which category countries fit is very subjective. Furthermore, the concept of development has changed over time, once being viewed simply as GDP per capita but at other times being seen in terms of income distribution, or the meeting of basic needs, such as food and shelter.

The United Nations focuses the definition on *human development*. Human development includes traditional economic development but moves far beyond it. The goal of human development is to create an environment where people and groups can develop their full potential and lead productive and creative lives based on choice. Economic opportunity is important, so places need adequate jobs and incomes that provide for a decent material quality of life. But health and education are also important. People must be physically and mentally healthy and have access to knowledge and skills to pursue opportunities. Choice is also key. If participation in society is restricted by gender, race, ethnicity, or any other characteristic, then material wealth, health, and education can be meaningless. Education does not confer choice if women or a minority ethnic group are not allowed to play a role in government or the workforce.

By studying multiple development variables, it is possible to get a good feel for the places where people have the economic opportunity, physical health, and education that allows members of society, both male and female, to pursue the life trajectory of their choice.

Spatial distribution at a global scale: Measuring and mapping development

Since 1990, the United Nations has been calculating the *human development index* based on a combination of data on life expectancy, income, and level of education. In a sense, these can be seen as surrogate measures of health, wealth, and opportunity.

The human development index is derived from measurements of four variables. As described in chapter 2, life expectancy at birth represents the years a newborn child is expected to live. This is strongly influenced by the infant mortality rate but can also be influenced by epidemics or conflict as well as medical care and nutrition of older age groups. Expected years of schooling is the number of years a child entering school should expect to complete given current enrollment rates. This variable gives a forward-looking indication as to how educated children are likely to be. Mean years of schooling represents the average years of education of those age 25 and older. This is more of a backward-looking indicator, showing the level of education that current adults received in the past. *Gross national income* (GNI) per capita is a measurement of all the income generated by a country, both domestically and from overseas, such as by companies that invest abroad, divided by the total population. In this case, GNI per capita is in PPP, which needs to be discussed in more detail before moving on.

Gross national income per capita in purchasing power parity

PPP is a concept used to compare prices across countries, essentially to account for differences in the cost of living. To do this, a sample basket of goods and services is used to determine the overall difference in prices. For instance, imagine that you buy a pound of apples, a pair of jeans, and a cell phone in the United States. You also get a haircut and pay your electrical bill. Now imagine that you buy the exact same goods and services in Mexico. Would you pay the same amount in dollars in the US as in Mexico? It is unlikely. Some goods and services will be cheaper in Mexico, while others may be higher. Overall, however, prices are likely to be lower in Mexico. It is for this reason that some Americans move to Mexico to stretch their retirement incomes. An American retiree living in the US may be middle class, but that same retiree may be upper-middle class when living in Mexico, since the cost of living is lower. Therefore, when comparing incomes across countries, it is important to adjust values based on PPP. Saying that a person in the US making $53,000 per year (roughly the US GNI per capita) has the same standard of living as a person in Mexico making $53,000 per year would be incorrect. An income of $53,000 in Mexico will buy much more than in the US (figure 9.3).

When viewing GNI per capita values in PPP, think of it this way: If you made $53,000 per year in the US, what would your standard of living be? Now look at the

Figure 9.3. Street market in San Andres Tuxtla, Mexico. Many goods are cheaper in places such as Mexico. Consequently, income comparisons are often adjusted for purchasing **power parity.** Photo by Gerardo C. Lerner. Stock photo ID: 327438326. Shutterstock.com.

GNI per capita in PPP in France. It is about $38,000. If you were still in the US, what would your standard of living be with $38,000? A typical person in France has the standard of living of an American living on $38,000. In terms of income, the French are poorer than the Americans.

Figure 9.4 shows wide spatial variation in GNI per capita. In the top quintile of countries, per capita GNI is over $28,000. These countries are largely clustered in Western Europe and North America, but the Pacific nations of Japan, Australia, and New Zealand also stand out, as do several countries in the Middle East. Much of Asia and Latin America fall into midrange quintiles, while at the lowest end of the income spectrum is a clear cluster in sub-Saharan Africa.

Looking in more detail, the degree of inequality between countries becomes even more apparent. To get a better grasp of these differences, it is useful to start with a familiar place such as the United States. There, the GNI per capita is just over $53,000 per year. At the high end, Qatar has an astonishing value of nearly $130,000. And remember, these values are in PPP, so an average Qatari would live like an American with $130,000. Not bad! In Singapore, the GNI per capita is over $78,000, and in Norway it is nearly $68,000. Most Western Europeans have a somewhat lower GNI per capita than residents of the United States, however, falling largely in the $30,000 to $40,000 range.

Much of Eastern Europe, Russia, and parts of the Middle East and Latin America fall into the next-highest quintile. Mexico's GNI per capita of just over $16,000 helps explain why many migrate north to the United States. The same is true for Eastern Europeans, many of whom migrate to higher-income Western Europe: Poland's GNI per capita is just $24,000 compared to Germany's at $44,000.

The relative wealth of countries such as Mexico, Chile, Argentina, and Uruguay also pulls migrants from nearby places. For a Guatemalan, where GNI per capita is only $7,000, Mexico may offer a chance to earn more money. Likewise, many Bolivians, where GNI per capita is just over $6,000, see countries to

Gross National Income Per Capita PPP

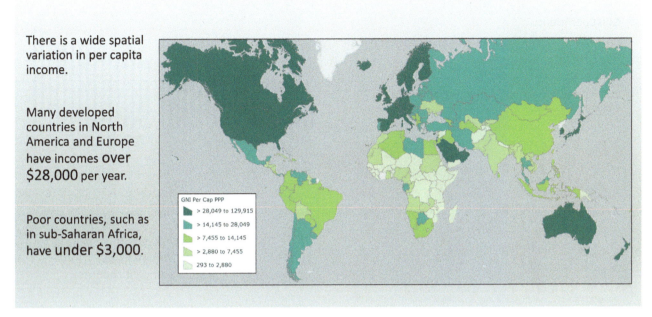

There is a wide spatial variation in per capita income.

Many developed countries in North America and Europe have incomes **over $28,000** per year.

Poor countries, such as in sub-Saharan Africa, have **under $3,000.**

GNI Per Cap PPP
- > 28,049 to 129,915
- > 14,145 to 28,049
- > 7,455 to 14,145
- > 2,880 to 7,455
- 293 to 2,880

Figure 9.4. Gross national income per capita PPP. Explore this map at https://arcg.is/G8CLq. Data source: United Nations, Human Development Report 2016.

its south as places of opportunity. In Central Asia, Uzbeks, Tajiks, and others are pulled to Russia's relative wealth.

At the bottom of the scale is sub-Saharan Africa. Here is the largest concentration of low-income countries, where low-productivity agriculture, basic manufacturing, and small-scale services offer meager wages to large segments of the population. It is here that many people live in extreme poverty. Recall that the World Bank and United Nations define extreme poverty as earning less than $1.90 per day. Taken on an annual basis, that comes to $693. In countries such as Burundi, Liberia, the Democratic Republic of Congo, the Central African Republic, and Somalia, GNI per capita falls lower, meaning the average person in these countries lives in extreme poverty. To put this level of poverty in perspective, recall that these values are in PPP, so the Central African Republic's GNI per capita of $587 per year would be like living in the United States on that amount of money. Just imagine how you would survive with only $587 to cover all your needs for an entire year.

Income distribution and the Gini coefficient

Thus, there are vast spatial differences in levels of wealth around the globe. However, GNI per capita merely shows average values. Few people actually have incomes right at the average. As a student, you probably earn less than the US GNI per capita of $53,000, while Mark Zuckerberg of Facebook earns a lot more. For this reason, an analysis of development must consider income distribution within countries. One common measure of income distribution is the *Gini coefficient*, which measures income distribution from a scale of zero to 100. Zero represents perfect equality, where everyone has exactly the same income, while 100 represents perfect inequality, where one person receives all income.

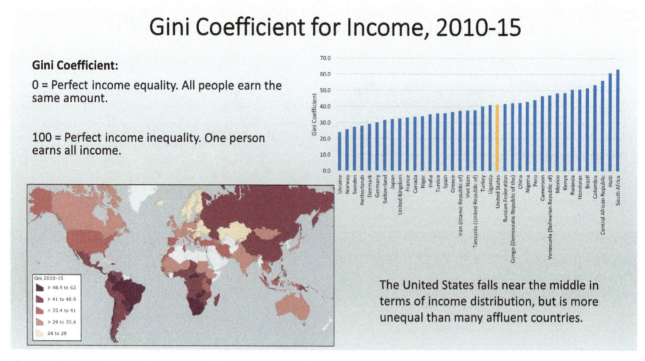

Figure 9.5. Gini coefficient for income, 2010–15. Explore this map at https://arcg.is/G8CLq. Data source: United Nations, Human Development Report 2016.

A cluster of countries with low Gini coefficients is found in Europe (figure 9.5). Among the most well-known egalitarian countries are those of Scandinavia—Norway, Sweden, and Denmark. These are relatively affluent places, with large middle classes and few people who are extremely rich or extremely poor. Here, residents pay among the highest effective tax rates in the world, but much of the money goes toward subsidizing public transportation, child day care, health care, education (including free university), and myriad antipoverty programs. Eastern Europe stands out as well. This region is not as affluent as Scandinavia, but extreme wealth and extreme poverty are limited enough that overall equality levels are high.

Among the affluent countries of Europe and North America, the United States ranks as more unequal. For historical and philosophical reasons, the United States has followed a more individualistic development model, whereby success can be rewarded with great

wealth, but programs to help those in need are more limited than in other affluent countries.

The most unequal countries lie in Latin America and sub-Saharan Africa. In countries such as South Africa, Brazil, Colombia, and the Central African Republic, the divide between rich and poor is vast. In these places, GNI per capita values paint a misleading picture, since few people have incomes near the average. Instead, incomes tend to fall at the extremes. South Africa and Brazil have GNI per capita values of $12,000 and $14,000 respectively. These are reasonable levels of wealth, but few people earn those incomes. Rather, a small group of elites in politics and business earn incomes in the millions, leaving much of the remaining population with incomes in the low thousands. For instance, in Brazil, the rich go to work by helicopter or in luxury (and sometimes armor-plated) automobiles, while the poor have brutal commutes by bus through snarling traffic. The rich live in luxury apartments with live-in servants, while the poor live in dangerous squatter settlements, some of which even the police will not enter (figure 9.6).

Figure 9.6. Inequality in Brazil. View of a squatter settlement with luxury apartments in the background, Rio de Janeiro. Photo by Andre Luiz Moreira. Stock photo ID: 611832305. Shutterstock.com.

So, per capita measures of economic development are useful, but how wealth is shared among the population can be equally important. Economic growth is essential, but high levels of inequality can negate its development impact on the general population.

Life expectancy at birth

A country's wealth and how it is distributed can translate into overall health of the population, reflected in *life expectancy at birth*. Life expectancy at birth is discussed in more detail in chapter 2, but as with per capita income, there is great spatial variation. Generally, wealthier places have higher life expectancies, while poorer places have lower ones. In most European countries, life expectancy is around 80 years or more, but in sub-Saharan Africa, it is under 60 years (figure 9.7). Latin America, falling between wealthy Europe and poor sub-Saharan Africa in terms of income, has average life expectancies of around 75 years.

Life Expectancy at Birth

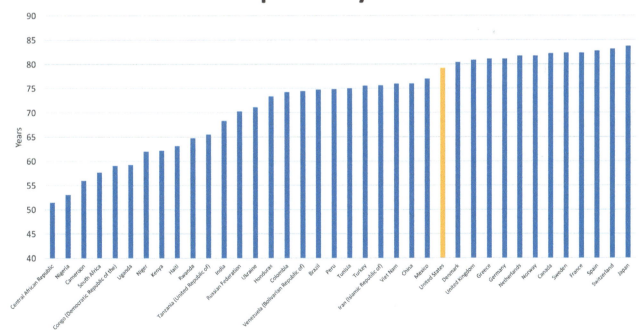

Figure 9.7. Life expectancy at birth. Selected countries. Data source: United Nations, Human Development Report 2016.

Variation in life expectancy is closely tied to the health of populations. Globally, it increased by four years between 2000 and 2015, in part due to declining deaths from HIV/AIDS and other communicable diseases, malaria, improved maternal and neonatal care, and better nutrition. Most positively, gains in life expectancy were greatest in regions where people lived shorter lives. In sub-Saharan Africa, the increase was 8.8 years, while in South Asia, it was 5.5 and in Latin America, 3.8 years. These trends are obviously positive, but much work remains to pull all societies up to the levels found in the most-developed countries.

Years of schooling

In a sense, measures of education reflect opportunity. With education, people have more options in where to work and how to pursue their goals. When education is limited, so too is choice. People are forced to work in whatever low-skill opportunity that presents itself and are more likely to remain in that line of work with limited chances for advancement or professional growth.

Mean years of schooling reflects the average number of years that those over 25 years of age have completed. After age 25, most people have finished their education, so this measure illustrates how educated the adult population is and gives a good indication of a society's level of innovation and productivity. When mean years of education are low, it is unlikely that a country will benefit from economic and technological changes in agriculture, manufacturing, and services. Rather, poorly educated workers will likely toil in low-wage and low-productivity jobs rather than in mechanized and knowledge-intensive occupations.

Again, there are vast differences in the mean years of schooling by region. In the advanced countries of Europe and North America, for instance, the adult population has over twelve years of education. This means that, on average, people have some postsecondary schooling. But moving to Latin America, it

is evident that adults have only about eight years of education, just over middle-school levels. Lack of sufficient education goes far in explaining why many places in Latin America struggle with low wages and low productivity. Limited literacy and numeracy means that workers simply do not have the skills to work in higher-wage technology and information-intensive jobs. Looking at sub-Saharan Africa, years of schooling for women averages about 4.5 years, while men do only slightly better at 6.3 years. With education hovering around elementary-school levels, economic activity will inevitably be limited.

Expected years of schooling paints a better, more optimistic picture of the future (figure 9.8). This measure looks at the expected number of years of schooling that entering students are likely to receive during their lifetime. In many cases, access to education has improved substantially in recent years, meaning that children will have much greater educational opportunity than their parents had. So, while adults in sub-Saharan Africa have only four to six years of schooling on average, a child in that region today is expected to have nine to ten. In Latin America, children are expected to have about six more years of schooling than previous generations, reaching around fourteen years of education. In the most-developed countries, children are expected to see gains as well, attaining over sixteen years of education.

Human development index

By combining measures of GNI per capita, life expectancy at birth, and levels of education, countries of the world can be classified by their level of *human development*. Figure 9.9 illustrates spatial patterns of development that vary substantially by region. Those that rank the highest are clustered in Europe and North America but also include parts of South America, Saudi Arabia, and the Pacific nations of Japan, Australia, and New Zealand. The least developed are clustered in sub-Saharan Africa, along with the war-torn states of Afghanistan, Syria, and Yemen.

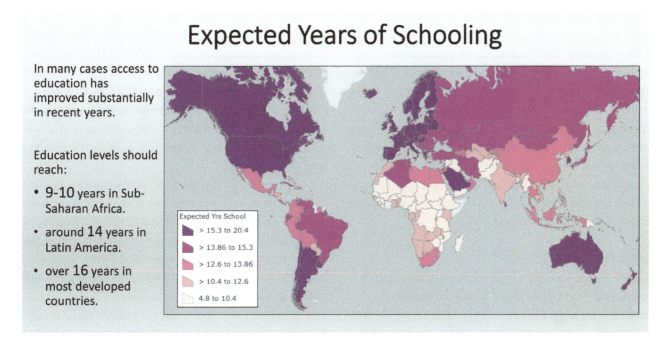

Figure 9.8. Expected years of schooling. Explore this map at https://arcg.is/G8CLq. Data source: United Nations, Human Development Report 2016.

Human Development Index, 2016

Highest ranked countries compared to the lowest:

- 25 years greater life expectancy.

- 40 times more income in purchasing power parity.

- Twice as many years of schooling.

HDI rank	Highest Human Development
1	Norway
2	Australia
2	Switzerland
4	Germany
5	Denmark
5	Singapore
7	Netherlands
8	Ireland
9	Iceland
10	Canada
10	United States
	Lowest Human Development
179	Eritrea
179	Sierra Leone
181	Mozambique
181	South Sudan
183	Guinea
184	Burundi
185	Burkina Faso
186	Chad
187	Niger
188	Central African Republic

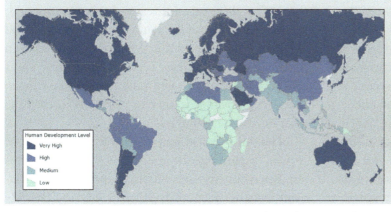

Figure 9.9. Human development index, 2016. Explore this map at https://arcg.is/G8CLq. Data source: United Nations, Human Development Report 2016.

The gap in development between those countries at the top of the list and those at the bottom is immense. Life expectancy for the most-developed countries averages around twenty-five years more than in the least. Income per capita in PPP averages over forty times greater. Looking at the future children face in terms of education, those in the most-developed countries are expected to attain an average of twice as many years of schooling. Much work is left to be done to raise the level of development for people in countries at the bottom of the index and bring them closer to those that are most well-off.

Gender inequality

Development measures at a national scale can often mask important differences within countries. As discussed previously, income distributions can be skewed so that the rich earn much more than the poor. Also, subgroups within a population can see different levels of well-being, be it by race, ethnicity, sexual orientation, religion, gender, or any number of other categories. In the case of gender, women often have lower levels of human development, especially in terms of income and education. Women face myriad cultural, political, and economic barriers that limit access to paid work outside the home, educational opportunities, and political power.

The *gender inequality index*, a subsection of the human development index, attempts to measure aspects of women's development in terms of health, empowerment, and work. Places where women have greater equality have low rates of maternal mortality (deaths related to pregnancy), low rates of adolescent births, a higher proportion of women in government positions, more women with at least some secondary education, and greater participation of women in the labor force. When childbirth is safe and done by choice, women are more likely to be able to pursue their goals. Death from childbirth is an obvious barrier to equality, but having children during adolescence also limits a woman's ability to pursue education and career opportunities. Women's empowerment comes from

both education and political power. When women have more education, their life choices are much broader than when education levels are low. Likewise, when women have political power, there is a greater chance that gender-specific issues are addressed within a country. Finally, when women participate in the paid labor market, they have greater autonomy in their lives and are better positioned to care for themselves and their children.

Figure 9.10 shows that the greatest gender inequality runs from sub-Saharan Africa through parts of the Middle East and into South and Southeast Asia. Part of this inequality is because much of those regions are poor, and in poor countries, maternal mortality and adolescent birth rates are higher and education levels lower than in more affluent places. For instance, the maternal mortality rate in sub-Saharan Africa is a shocking thirty-nine times higher than in the most-developed countries and even three times greater than in South Asia, the region with the second-highest rate. The same holds true for adolescent births. The rate in sub-Saharan Africa is six times greater than in the most-developed countries and 1.6 times higher than in Latin America. Based on sub-Saharan Africa's low levels of economic development and high proportions of rural residents, poor health outcomes and high teen fertility rates will tend to be more common.

But high gender inequality is also due to more limited female roles in government and in the workforce. This difference is greatest in the Arab world, where conservative cultural mores stress greater social distance between men and women. In sub-Saharan Africa, interestingly, these types of gender differences are much lower, with some indicators putting women in a better position relative to men than in even the most-developed countries.

The Arab world has the lowest proportion of women in parliaments, at just over 15 percent. This number contrasts with nearly 26 percent of parliament seats in the most-developed countries. Possibly surprisingly, women in Latin America hold over

Gender Inequality Index

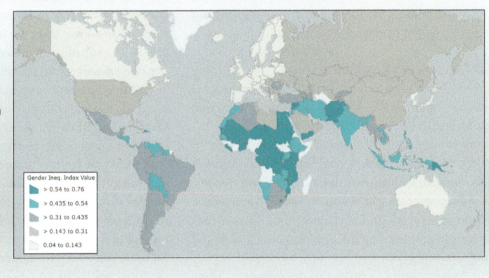

This index is based on maternal mortality rates, adolescent birth rates, proportion of women in parliament, women with at least some secondary education, and participation of women in the labor force.

The greatest inequality stretches from sub-Saharan Africa through parts of the Middle East and into South and Southeast Asia.

Figure 9.10. Gender inequality index. Explore this map at https://arcg.is/G8CLq. Data source: United Nations, Human Development Report 2016.

28 percent of the seats, higher than the average for the most-developed countries, while in sub-Saharan Africa, they hold a relatively healthy 23 percent.

The Arab world also has the smallest share of women in the workforce, at 22 percent. In the most-developed countries, the share is about 53 percent, while in sub-Saharan Africa, a much larger 65 percent of women participate in the workforce. Not only do lower proportions of women work outside of the home than men in all regions, those who do so earn less. Once again, Arab women come out poorly. In their part of the world, women earn the least relative to men, at 23 percent in GNI per capita. In the most-developed countries, the female GNI per capita is about 58 percent that of men's. However, the developing regions of both East Asia and sub-Saharan Africa have rates higher than that, where female GNI per capita is over 60 percent that of males.

The relationship between gender and education paints an interesting picture. In all regions and at all levels of development, women are less likely than men to have at least some secondary education and fewer overall years of schooling. However, the picture changes when looking at the expected years of education young children are likely to receive. For these children, girls should end up with more years of schooling than boys in all regions except those at the lowest levels of development. By region, girls in East Asia, South Asia, and Latin America will all have more education than boys. However, girls in the Arab states and sub-Saharan Africa will continue to have less schooling than their male counterparts.

Gender inequality will continue to be of concern in human development. Many countries are attempting to increase female empowerment via greater political power and education. In some places, female participation in politics is increasing through quotas, whereby some parliamentary seats are reserved for women, or where a specified proportion of candidates must be female. At the same time, as female education surpasses that of men in many parts of the world, opportunities for work may improve. However, continuing structural and cultural barriers that put more responsibility for child and home care on women will make gains in politics and the paid workforce difficult.

Sustainable development

If initial attempts to measure development focused on economic measures and later attempts focused on human measures, the idea of *sustainable development* adds a third dimension: the environment. In 2015, the United Nations established the Sustainable Development Goals. These goals are based on the view that development includes three essential components: economic, social, and environmental. To this end, seventeen areas were identified for measurement and improvement (table 9.1).

As with other measures of development, these goals begin with economic concerns, such as poverty,

economic growth, and inequality. With over 800 million people living in extreme poverty, the majority in Southern Asia and sub-Saharan Africa, and with tens of millions of young men and women without work, economic growth is an essential first step in development. However, it must be inclusive. Between 1990 and 2010, inequality increased globally, as the wealthy benefited more from economic growth than the poor.

With sufficient economic growth, countries can put resources into the development of infrastructure. Over half a billion people still lack access to treated drinking water, and nearly five times that number lack toilets or latrines (figure 9.11). Without water and sanitary infrastructure, nearly 1,000 children die daily due to diarrheal diseases. A lack of access to electricity infrastructure means than many people, especially women, must spend hours daily searching for firewood, which emits lung-damaging smoke when used for cooking and heating. Road and communications infrastructures are also important. When people

United Nations Sustainable Development Goals

No poverty	Zero hunger	Good health and well-being
Quality education	Gender equality	Clean water and sanitation
Affordable and clean energy	Decent work and economic growth	Industry, innovation, and infrastructure
Reduced inequality	Sustainable cities and communities	Responsible consumption and production
Climate action	Life below water	Life on land
Peace, justice, and strong institutions	Partnerships for the goals	

Table 9.1. UN Sustainable Development Goals. Data source: United Nations.

have access to these foundational services, they can get goods to market, find the best price for their produce, access more information and education, and generally improve their lives.

Development of economies and infrastructures feeds improvements in health and education. With

Figure 9.11. Need for clean water and sanitation. Women draw water from a well in Nepal. Untreated water is a serious source of illness in much of the developing world. Photo by Aleksandar Todorovic. Stock photo ID: 155539829. Shutterstock.com.

one in nine people still undernourished and over 100 million children still lacking basic literacy skills, development programs must still focus on these areas. And as discussed previously, programs often need to explicitly target women, who typically lag behind men in education and opportunity.

What makes the sustainable development approach different from other schemes is its focus on making improvements long lasting so that generations can benefit well into the future. This means that climate change must be addressed, along with conserving resources on the land and in the sea. Rising sea levels, extreme weather, and changing weather patterns will disproportionately impact the poor, who lack resources to mitigate the impacts. Without financial resources, they will struggle to relocate, purchase irrigation systems, and build more resilient homes to withstand environmental change. Furthermore, since three billion

people depend on the oceans for their primary source of protein, and since 80 percent of human diet comes from plants, development must focus on sustainable maintenance of oceans and land.

Sustainability is also tied to the political arena in terms of peace, justice, and good institutions. Over $1 trillion are lost annually to corruption, tax evasion, bribery and theft. This is a huge sum of money that could be used to reduce poverty if only countries move toward better governance. Reducing war and conflict is also a necessary condition for developing in a sustainable manner, since improvements in health, wealth, and opportunity are nearly impossible without peace. Lastly, fair and efficient judicial systems are needed if development is to be inclusive. When the law supports only those with wealth and power, those without have no recourse to fight for their rights, be it to protect their land from unlawful expropriation or to enforce protections for women, children, and other disadvantaged groups.

Go to ArcGIS Online to complete exercise 9.1: "The human development report: Patterns of human well-being."

Spatial distribution at a national scale: Development within the United States

The United States scores well on most measures of development. It ranks in tenth place out of 188 countries in the human development index, with relatively high levels of income, life expectancy, and schooling. However, within the United States, as within all countries, there is great variation in levels of development. Zooming in to a state level of analysis, or even larger scales, differences in development become more apparent; the wealth, health, and opportunity of people are influenced by where they live.

Figure 9.12 shows median earnings per household, the proportion of people age 25 or older with at least

Development Indicators in the United States

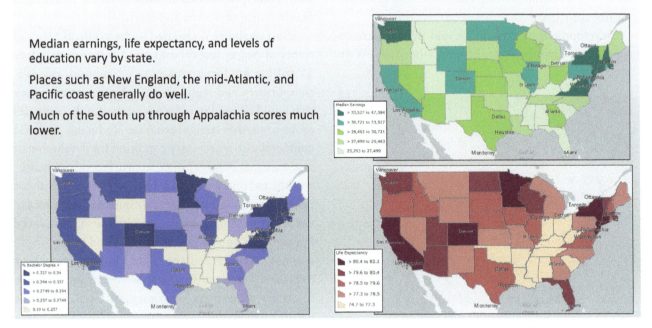

Figure 9.12. Development indicators in the United States. Explore this map at https://arcg.is/nDeWH. Data sources: Centers for Disease Control and Prevention, US Census Bureau.

a bachelor's degree, and life expectancy at birth by state. The New England and Mid-Atlantic regions score well in all three areas, reflecting generally high levels of human development from Vermont down through the Washington, DC region. The Pacific Coast also does relatively well, with combined levels of education, earnings, and life expectancy that reflect decent levels of development. But looking at much of the South up through Appalachia, the picture is very different. Here, life expectancy, earnings, and education levels are substantially lower.

The states that rank high all have major urban areas that drive their economies and overall well-being. Cities such as Los Angeles, New York, San Francisco, Boston, and Washington, DC, are globally connected places where major corporations and other high-paying employers concentrate. Education levels are high, and both skilled and unskilled immigrants flow in to their thriving and dynamic economies. Global and domestic

investment flows to these places as well, feeding a cycle of growth.

But outside of globally connected urban areas, levels of development can be much lower. Where education levels are low, so too are earnings. Few cities in the South and Appalachia have the human capital or levels of education that attract global corporate headquarters and their corresponding high wages. The same is true for rural places and smaller urban areas even in states with high development. California's Central Valley, just a couple of hours drive from Silicon Valley or Los Angeles, has a population of low-income agricultural workers with low levels of education.

Levels of education and earnings are closely related to health and life expectancy. In the most thriving states, life expectancy is on par with that in the affluent countries of Western Europe, such as Sweden, France, and Spain. But in the least-thriving states, it is closer to that of much less-developed places. For instance, life expectancy in Mississippi is similar to that in Iraq, Thailand, and El Salvador. Life expectancies in other pockets around the country are similarly at levels of much poorer countries. Poor health is often related to income and access to health care, but those who live in areas of the US that are underdeveloped also suffer disproportionately from risky health behaviors, such as smoking, poor diet, physical inactivity, and drinking to excess.

Theories of development: Why do spatial differences exist?

From Adam Smith in the 1700s and Karl Marx in the 1800s to more recent policy debates at the United Nations and in national capitols, there has been a wide range of views on the causes of *development* and *underdevelopment*. For some, underdevelopment is due to exploitation by those in power, where a small elite squeezes the wealth out of people and places for their own benefit. Others see underdevelopment as due to a lack of sufficient private investment and trade, so that poverty is found where people and places are not well integrated into broader national and international economic systems. Measuring and mapping levels of development is relatively easy, but explaining why it happens is much more contested.

Exploitation-based theories

In the 1950s, many development specialists viewed poverty and underdevelopment in Africa, Asia, and Latin America as being the result of exploitation stemming originally from the colonial era but continuing on well into the twentieth century. *Dependency theory* argued that the colonial powers of Europe, and later the United States, established trade relationships to extract the wealth from colonized territories. Colonies were used to produce raw materials and food, such as minerals and agricultural goods, which were sent to the colonial powers to feed growing urban populations and as inputs to their expanding manufacturing

sectors. Thus, Great Britain imported tea, tobacco, sugar, coffee, and cotton from its colonies around the world. The wages paid to workers in colonial territories was minimal, meaning that these goods were obtained relatively cheaply and that workers would remain poor. At the same time, colonial powers used their territories to expand markets for manufactured goods. For example, cotton that was imported cheaply could be turned into manufactured clothing that was then sent back to colonial territories for sale.

In essence, cheap imported raw materials were extracted from the colonies, and then higher-value manufactured goods were sold back to them, enriching the colonial powers. Back in the colonial territories, the vast majority of the population remained poor, while a small elite that worked closely with the colonial powers became rich and perpetuated the systems of uneven spatial development.

Dependency theory was expanded and refined into the *world systems theory*. This theory uses the dependency framework, whereby trade relationships shape the "winners" and "losers" in global development. This framework is conceptualized as countries that fall into the *core*, *semi-periphery*, and *periphery*. The core countries are those with great economic, political, and military influence. They contain the most sophisticated industry, financial services, and information-intensive sectors. Core countries include the United States and most Western European countries as well as places such as Australia, Japan, and New Zealand. The periphery includes countries that largely export raw materials (figure 9.13). These tend to be the poorest countries, where wages and skill levels are low. For instance, in much of Africa, agricultural and mineral exports still dominate. Sierra Leone earns fully 61 percent of GDP from agricultural exports, especially cocoa for feeding the chocolate cravings of the developed world, while Botswana exports diamonds and Zambia exports copper. The semi-periphery falls in-between and is where lower-skill and lower-wage manufacturing takes place alongside the production of raw materials.

The concept of core and periphery is also seen at more local scales in the form of *dualism*, where there

Commodity Dependency

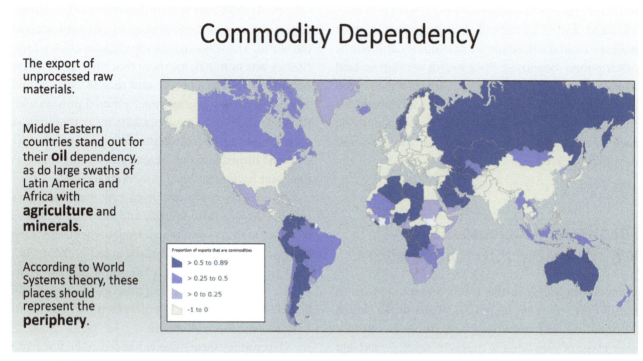

The export of unprocessed raw materials.

Middle Eastern countries stand out for their **oil** dependency, as do large swaths of Latin America and Africa with **agriculture** and **minerals**.

According to World Systems theory, these places should represent the **periphery**.

Proportion of exports that are commodities

- > 0.5 to 0.89
- > 0.25 to 0.5
- > 0 to 0.25
- -1 to 0

Figure 9.13. Commodity dependency. Explore this map at https://arcg.is/1WWfej. Data source: United Nations.

are two distinct levels of development within a country. In this case, dualism can reflect exploitative relationships between urban areas and rural hinterlands. Raw materials from rural farms and mines are shipped to core urban areas for manufacturing, while finished goods are then sold back to rural residents. This pattern has been seen in thousands of cities around the world, from Chicago in the United States to Santiago in Chile to Guangdong in China. In each case, wealth is extracted from the countryside (the periphery) and accumulates in the city (the core).

Policy responses to exploitation

With a diagnosis of exploitative trade relationships behind underdevelopment, a set of policy solutions to counter this exploitation became popular in many less-developed countries. One that was widely adapted was *import substitution industrialization* (ISI). As its name implies, this policy sought to substitute imports with locally made industrial goods. The logic was that it was a losing strategy to export raw materials

and import higher value-added manufactured goods. Why import expensive goods when you can make them domestically yourself? India, in the 1950s, for instance, told General Motors India and Ford Motor Company India that they could no longer import car parts for mere assembly in India. If they wanted to continue operating in India, both companies would have to manufacture automobiles, not just assemble them with foreign-made parts. Both companies decided not to manufacture in India and left the country. This gave a huge boost to locally produced Indian cars, such as the Hindustan Ambassador (figure 9.14).

In addition to manufacturing licensing requirements, ISI policies used tariffs and quotas to limit foreign-produced goods. Import tariffs are essentially taxes placed on imported goods, while quotas limit the total quantity of a good that can be imported each year. Brazil, during the 1950s and 1960s, imposed an average tariff on manufactured goods of over 100 percent, so many goods that were imported were twice as expensive as they would be without a tariff. Obviously,

Figure 9.14. An Indian-made Ambassador automobile. As part of India's import substitution industrialization policies, the country made its own cars while restricting imports of foreign made cars. Photo by Anandoart. Stock photo ID: 665948647. Shutterstock.com.

this tariff priced imported goods beyond the reach of most Brazilians, forcing them to purchase locally manufactured goods instead.

Even in the mid-2010s, some countries continued using ISI-inspired protectionist trade policies. Brazil and Argentina during this time had numerous import restrictions and taxes, resulting in high prices on foreign goods. For instance, in 2014, an iPhone that cost $815 in the US sold for $1,196 in Brazil. The price of an iPhone was even higher in Argentina where its sale was banned. There, a law required that cell phones be produced locally. While some manufacturers agreed to open factories in Argentina, Apple refused, making it available only on the black market. As a result, in many cases, it was cheaper to fly from Argentina to Miami to buy an iPhone than to buy one in Argentina.

Market-based theories
Modernization theory

Others view market forces as the cause of spatially uneven development. By this line of thought, it is countries and places that promote private property ownership, entrepreneurship, trade, and technology that thrive, while those that limit them remain underdeveloped.

One of the most influential development theorists in the twentieth century was W. W. Rostow, who wrote *The Stages of Economic Growth: A Non-Communist Manifesto* in 1960. His work drew on the successes of the Marshall Plan in rebuilding Europe after World War II and became highly influential in President John F. Kennedy's foreign policy and development of the Peace Corps. Rostow's work contributed significantly to *modernization theory*, claiming that all countries follow a process of stages from traditional to modern.

In the Rostow model, societies can be divided into five stages of development. In the first stage, the *traditional society*, places are prescientific, meaning that advances in technology are haphazard and limited in their ability to substantially improve productivity. Social mobility to higher or lower status is limited, and social organization is based on family and clan. Political power is held regionally by the landowning class. Examples of traditional societies were ancient Chinese dynasties and Middle Eastern civilizations as well as medieval Europe.

The second stage is the *precondition for take-off*. In this stage, political change results in the formation of centralized nation states that replace landed regional power structures. Modern science evolves and is applied to new production techniques in agriculture and industry. Importantly, private entrepreneurs and the government become willing to invest profits and take risks in the pursuit of modernization, such as new transportation and communications networks. Yet progress is slow, as wide sections of the economy continue with low-productivity production and old social structures and values. Western Europe entered this stage in the late seventeenth and early eighteenth centuries, as powerful states began to expand trade and seek resources in far-flung colonies.

Third comes the *take-off* stage. In this stage, both technological innovation and political priorities lead to greater diffusion of economic progress throughout a country. Technology, such as railroads and other industrial machinery expand, but so too does the power of a political class that expressly promotes modernization. Investment as a proportion of income increases, fueling advances in agricultural and industrial productivity. Great Britain was the first to go through this this stage in the late 1800s, followed by other European countries and Japan through the early 1900s. By the 1950s, both China and India were in this stage of development as well.

Stage four is the *drive to maturity*. By this time, investment rates have increased substantially, leading to agricultural and industrial output that grows faster than population. Countries in this stage are tightly tied to the international economy, with sizable levels of imports and exports. Technological innovation becomes more complex, shifting production from coal, iron, and heavy industry to machine tools, chemicals, and electrical equipment. Typically, this stage comes sixty years or so after the beginning of the take-off stage.

Finally, the last stage is that of the *age of high mass consumption*. In this stage, consumer goods and services are commonplace. People have incomes that allow them to purchase many items beyond basic food, shelter, and clothing. Most people live in cities and work in offices or skilled factory jobs. Social welfare spending increases, as governments attempt to soften the impacts of poverty, malnutrition, inadequate housing, and illness. By the 1950s, the United States, Western Europe, and Japan had entered this last stage of development.

Policy response to modernization theory

In Rostow's view, investment and technology propelled societies to modernization through the effort of both private entrepreneurs and the government. This approach was more top-down than later market-based explanations of development in that national and international organizations helped focus and promote capital investment in modernization programs.

From a policy standpoint, this meant that governments of the most-developed countries could accelerate modernization in lower-stage places through technology transfer and technical assistance. Development would result from greater diffusion of technology

and know-how. Given the geopolitical situation of the 1960s, much development policy that arose from Rostow's modernization theory was oriented toward assistance that would help contain the expansion of communism. National governments in the developed world, such as the United States and many countries of Western Europe, increased foreign aid. At the same time, global organizations such as the World Bank and International Monetary Fund, which were formed at the end of World War II to help foster global economic development, helped finance and promote projects.

Many projects in the 1960s focused on technology transfer and technical assistance in the form of large infrastructure projects such as power plants, telecommunications, and transportation (figure 9.15). For instance, the United States helped finance and build electrical distribution systems in rural Vietnam, while the World Bank financed railway construction in Tanzania. All over the developing world—from Latin America to Africa, Asia, and the Middle East—cooperation between national governments, international organizations, and private business expanded electrical systems, built highways and railroads, financed the construction of factories, expanded water and sewer

Figure 9.15. Guri hydroelectric power, Venezuela. The World Bank helped finance numerous power generation projects as part of a push to modernize developing countries. Photo by Paolo Costa. Stock photo ID: 297629225. Shutterstock.com.

systems, developed rural irrigation, and much more. Again, the idea was that by transferring technology of this sort and training local workers to manage and operate it, developing countries would more quickly move through the stages to modernization.

Neoliberalism

Proponents of the *neoliberal model* argue that development is best achieved through market forces, with as little government intervention as possible. This philosophy grew from Adam Smith's famous eighteenth-century idea of the "invisible hand," whereby individuals working for their own benefit inadvertently contribute to the common good of society. Unlike modernization theory, which saw benefits in private-public cooperation, and very unlike the statist ideas of dependency and world systems theory, neoliberalism sees the greatest development in places that unleash the power of the private sector. Individual entrepreneurs contribute to societal development in a political framework whereby government's role focuses on protecting private property rights, free markets, and free trade.

This development theory gained strength in the 1980s and was promoted strongly by Ronald Reagan in the United States and Margaret Thatcher in the United Kingdom. The World Bank and International Monetary Fund also began using this framework in promoting development around the world. By the end of the 1980s, neoliberal development theory became known as the *Washington Consensus*, a set of ideas agreed upon by most major multinational agencies and many national governments.

Policy response to neoliberalism

Neoliberal development theory, as implemented via the Washington Consensus, involved a range of policies meant to guide development. Countries seeking to receive development assistance, be it from national governments such as the United States or from multinational organizations such as the World Bank and International Monetary Fund, had to follow neoliberal prescriptions as a condition of aid.

Fiscal discipline was required so that government budget deficits were kept low. This could be accomplished, in part, by reorienting public expenditure away from inefficient spending, such as gasoline subsidies that go disproportionately to higher-income car owners, and toward more efficient spending in health, education, and infrastructure. Wasteful government spending and inefficient production could also be achieved through privatization, whereby government assets are sold to private investors. In many cases, telecommunications companies, water and power companies, and many other companies were owned and operated by national governments. However, these companies were often run inefficiently, with managers and workers receiving their positions because of political connections rather than merit and state subsidies covering ongoing financial losses.

Figure 9.16 shows how influential privatization became through the 1990s. Prior to 1990, no World Bank project explicitly involved privatization, but that quickly changed, with dozens being funded through the 1990s and into the early 2000s. Most of the early privatization projects took place in the former Soviet Bloc countries that were just shedding their communist systems. These projects included privatization of electrical utilities, banks, agribusinesses, and more in places such as Albania, Kazakhstan, Serbia, and Romania. Other countries throughout Africa, Asia, and Latin America received assistance as well, under the neoliberal view that private companies operate more efficiently than the state, facilitating more rapid growth and development.

Property rights were to be protected, as well, so that governments could not unjustly confiscate land, firms, or other assets held by individuals. With strong property rights, individuals can invest without fear of losing hard-earned wealth, contributing to development of the country. Furthermore, deregulation was prescribed. Government rules and regulations were to be kept to a minimum to allow greater freedom for firms to invest, build, hire and fire workers, and perform other economic functions. Many aid programs

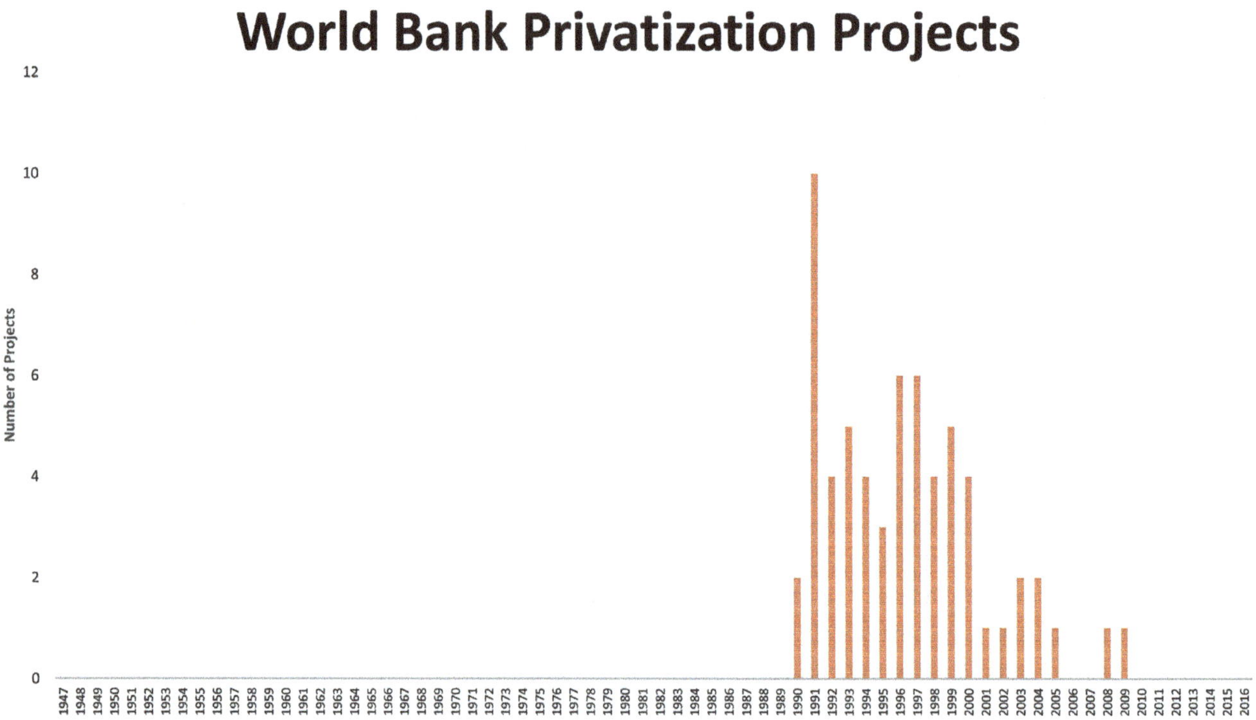

World Bank Privatization Projects

Figure 9.16. World Bank privatization projects. Neoliberal development theory gained traction in the 1990s when the World Bank and other institutions began promoting privatization of state-owned assets and agencies. Data source: World Bank.

thus focused on assisting governments with legal and regulatory reforms to codify these changes.

Trade liberalization was encouraged so that goods and services could be imported and exported freely, without heavy import tariffs, quotas, or other restrictions. Comparative advantage was encouraged so that each country could focus on what it produces most efficiently. As part of trade liberalization, foreign direct investment was promoted. This allowed multinational companies to invest around the world, seeking the best locations for production of agricultural goods, manufactured goods, and services.

As more countries adopted free trade policies by reducing tariffs, quotas, and other trade barriers, global trade increased substantially. The promotion of global trade was coordinated first by the *General Agreement on Tariffs and Trade* (GATT), which was founded in 1947, and then by the World Trade Organization

(WTO), which replaced GATT in 1995. GATT, in 1947, consisted of just twenty-three countries, but by 2016, 164 countries were members of the WTO. As more countries joined these organizations, trade expanded (figure 9.17). Bumps and troughs along the way correspond to new countries joining the organization, as well as fluctuations in the global economy, such as the Great Recession of 2008.

In the United States, trade has been promoted in several important ways. First, it is a member of the WTO. This means that the United States, along with all the member countries, has a common set of rules for trade and dispute resolution. Members cannot put up trade barriers that disadvantage some countries, so trade flows more freely among all.

Going beyond WTO requirements, the US also has bilateral and multilateral free trade agreements. These agreements further lower trade barriers

World Trade (% of Global GDP)

Figure 9.17. World Trade (percent of global GDP). Data source: World Bank.

between participating countries, bringing down tariffs and other policies that limit trade. The first US free trade agreement was with Israel in 1985. It was followed by the North American Free Trade Agreement (NAFTA) in 1989, which promotes trade between Canada, the United States, and Mexico. Since then, free trade agreements have been made with a total of twenty countries (figure 9.18). Most are in Latin America, where free trade agreements run from Mexico through Central American and along the Pacific coast of South America. The remaining partners are in North Africa and the Middle East as well as the Asia-Pacific region with South Korea, Singapore, and Australia.

Go to ArcGIS Online to complete exercise 9.2: "The invisible hand: How do market forces impact development?"

Density, distance, division: A geographic framework for understanding development

Most development approaches today are more aligned with market-based theories than with those that are exploitation based, although pure neoliberalism has lost some of its allure. While some politicians, economists, and development specialists see markets in a negative light and push for protectionist policies that limit imports and restrict markets, the vast majority understand that, historically, the best route to prosperity has been through open economies and trade.

The World Bank developed a framework for understanding development that is explicitly geographic, focusing on the role played by *density*, *distance*, and *division*. Within this framework, development can

US Free Trade Agreements

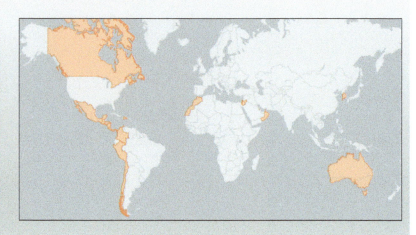

Australia	Israel
Bahrain	Jordan
Canada	Korea
Chile	Mexico
Colombia	Morocco
Costa Rica	Nicaragua
Dominican Republic	Oman
El Salvador	Panama
Guatemala	Peru
Honduras	Singapore

Figure 9.18. US free trade agreements. Data source: Office of the United States Trade Representative.

be understood at three scales of analysis. At the local scale, dense cities function as engines of development. At a country scale, the connectedness between leading urban areas and rural hinterlands drives the spatial distribution of development within countries. At an international scale, divisions, or the lack of them, in terms of trade influence the degree of development between countries.

Density

As has been discussed in previous chapters, urban agglomerations offer great benefits to firms and workers. Productivity increases with density, as large groups of people and companies interact in complex webs of economic activity. Knowledge and innovation are passed more readily between people. Companies have access to necessary goods and services used in production. Infrastructure costs, such as for water, electricity, and roads, are shared among a large population.

The resulting productivity gains from this interaction create much greater economic output per square mile than in less-dense areas. For this reason, urban areas are typically wealthier and more developed than lower-density places. Figure 9.19 illustrates how China's largest prefectures, those with over thirteen million residents in 2016, had the highest disposable incomes. Places such as Shenzhen, Shanghai, and Beijing consist of massive urban areas with dense networks of people and companies, which function as leading areas in the economy. In contrast, most low-density prefectures in the interior of the country have substantially lower disposable incomes.

At the aggregate level, countries with higher levels of urbanization also have higher GDP per capita (figure 9.20). When a greater share of a country's population lives in denser cities, more people benefit from agglomeration, working in higher-productivity industries and sharing knowledge and innovation. This is especially significant in the early stages of urbanization, with the most rapid gains in income accrued as the first 50 percent or so of the population becomes urban. This represents a period when large

Density and Development in China

China's largest prefectures, those with over 13 million residents in 2016, had the highest disposable incomes.

These massive urban areas have dense networks of people and companies, which drive economic growth.

Low-density prefectures in the interior of the country have substantially lower disposable incomes.

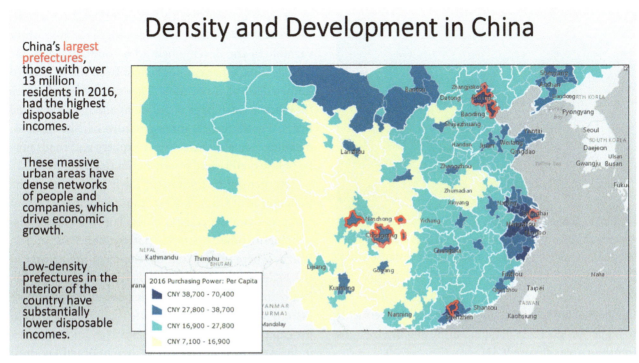

Figure 9.19. Density and development in China. Data sources: Esri, HERE, Garmin, NGA, USGS | Copyright: © Michael Bauer Research GmbH 2016 based on © National Bureau of Statistics of China.

GDP Per Capita (PPP) by Percentage Urban

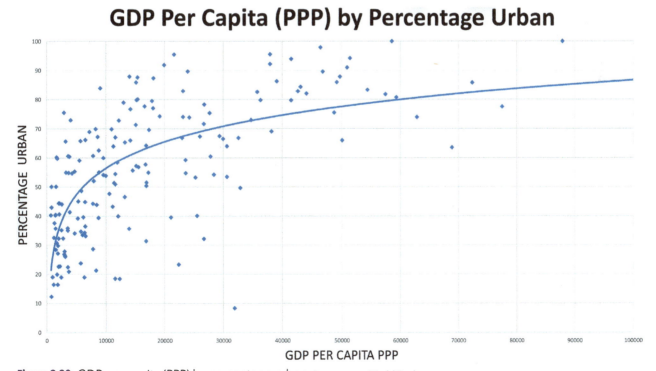

Figure 9.20. GDP per capita (PPP) by percentage urban. Data source: World Bank.

Figure 9.21. Shanghai, China. Dense urban areas are more productive, innovative, and wealthy than less dense areas. Photo by DB_Crazyhorse. Stock photo ID: 555822043. Shutterstock.com.

numbers of low-productivity agricultural workers shift to higher-productivity manufacturing and services jobs in urban areas (figure 9.21).

Some development specialists once attempted to restrict urban growth by establishing policies that resisted the market forces of agglomeration. The belief was that rapid urbanization led to too many deglomeration effects, such as shortages of housing and infrastructure. But urbanization is now seen as having more benefits than costs and so should be promoted, albeit with careful planning. This means that while market forces should be the primary drivers of development, the government should play a role as well. This role includes providing law and order via strong police, court, and regulatory institutions. It also includes developing urban infrastructure, such as roads, water, and sewer systems. Housing policies to upgrade slums are implemented as well, along with flexible zoning laws that guide development without stifling it. By harnessing the forces of growth conferred by the private sector, but with guidance of the state in areas of market failure, cities can form as powerful nodes of development.

Distance

Cities function as leading areas of development. What this implies is that places farther from cities often lag in terms of income, health, education, and other indicators of development. Thus, distance is an important variable in the development of a country. Places far from leading cities cannot benefit from the sharing of knowledge and innovation. They cannot benefit from the efficiencies of large populations and interconnected firms. Furthermore, infrastructure can be limited, meaning that electricity may be spotty and roads may be of low quality and poorly maintained (figure 9.22).

Returning to figure 9.19, it is evident that the prefectures in China's West, those which are farther from its dynamic coastal cities, lag in terms of disposable income. A similar pattern can be found in Brazil, where wealth and development are greater in the Federal District and southern cities such as Sao Paulo and Rio de Janeiro, while poverty is in the more distant northeast and Amazonia (figure 9.23). In fact, per capita disposable income is nearly 4.5 times greater in the wealthy Federal District than in Brazil's poorest state of Maranhao.

Figure 9.22. Guilin, China. Places with lower levels of density are poorer and less developed than those with high population densities. Photo by Feiyuezhangjie. Stock photo ID: 116434333. Shutterstock.com.

Distance and Development in Brazil

Places far from leading cities are less likely to benefit from core-area knowledge and innovation, the efficiencies of large populations and interconnected firms, and modern infrastructure.

In Brazil, wealth and development are greater in the Federal District and southern cities such as Sao Paulo and Rio de Janeiro, while poverty is in the more distant northeast and Amazonia.

2016 Purchasing Power: Per Capita

- BRL 43,000 - 1,029,100
- BRL 25,000 - 43,000
- BRL 18,000 - 25,000
- BRL 12,000 - 18,000
- BRL 0 - 12,000

Figure 9.23. Distance and development in Brazil. Data sources: Esri, HERE, Garmin, NGA, USGS | Copyright: © Michael Bauer Research GmbH 2016 based on © Instituto Brasileiro de Geografia e Estatística (IBGE).

Overcoming distance and its barrier to development requires work in several areas. First, people should be allowed, and even encouraged, to move to dynamic and growing cities. Rural-to-urban migration should not be restricted, since people become more productive and have better access to services when they live in urban areas. This also relieves pressure on labor markets in rural areas, as fewer workers remain there to compete for limited work. Second, there must be investment in infrastructure that connects places. Interstate highways, railroads, and public transportation help link places together. With high-quality transportation, the costs of moving goods and people diminish, reducing the relative distance between places. Communications infrastructure, such as cell phone connections and internet networks, are important for transmitting information. When farmers can find the best market rate for their crops rather than relying on the word of a single trader from the city, they are more likely to get a good price for their produce. Third, programs that target lagging areas can help boost education, skills, and health of workers. These programs can improve rural productivity and further integrate distant economies into the network of larger urban areas.

Division

Dense cities function as nodes of development at a local scale, while diminishing the friction of distance facilitates development at a country scale. At the international scale, divisions between countries influence levels of development. Generally, the free flow of goods and capital, people, and ideas around the world has stimulated economic growth and development. When countries are divided from each other through restrictions in these areas, they tend to lag behind others that are more integrated into the global economy.

Reducing divisions in terms of goods and capital means opening economies to free trade and investment. As discussed previously, free trade allows for countries to focus on their comparative advantage, producing goods and services in areas where they are most efficient. In turn, they import goods and services they are less efficient at making. This efficiency boosts productivity and growth, providing resources for improved education, health, and material standards of living. Countries with more open borders grow faster than those with closed ones.

The World Bank's Doing Business project ranks countries by ease of trade, measured by the time and cost to import and export goods. Figure 9.24 shows how countries rank on this variable and illustrates that, generally, open countries are more developed and closed countries are less developed. The countries of Europe and North America are the most open to trade, while much of Africa and the Middle East as well as parts of Latin America remain more divided

from the global economy. In the United States, for instance, it takes less than eight hours and under $300 to import or export a shipment of goods in New York. In contrast, importing and exporting a shipment in Lagos, Nigeria, costs over $1,000 and between 250 and 450 hours for compliance. This difference in time and cost has a dramatic impact on moving goods, increasing Nigeria's costs for both exports sold to other countries and imports to be consumed by the Nigerian people.

The economic case for removing the divisions between people is clear as well. As discussed in chapter 3, when there are free flows of migration, people move in pursuit of opportunity. In aggregate, this helps receiving countries by providing additional workers. Be it a Middle Eastern country that relies on foreign labor to run oil fields and construct buildings or a European country that needs young workers to offset declining birth rates, many places benefit from immigration.

Trading across Borders Rank

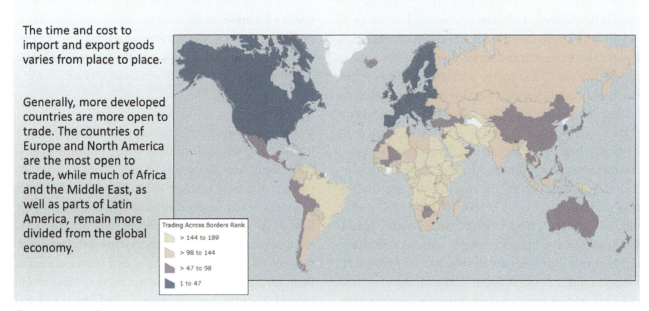

The time and cost to import and export goods varies from place to place.

Generally, more developed countries are more open to trade. The countries of Europe and North America are the most open to trade, while much of Africa and the Middle East, as well as parts of Latin America, remain more divided from the global economy.

Trading Across Borders Rank

> 144 to 189

> 98 to 144

> 47 to 98

1 to 47

Figure 9.24. Trading across borders rank. Data source: World Bank, Doing Business.

Sending countries, which tend to be those that are less developed, also benefit from migration. Migrants send remittances back to family, and they acquire new skills and knowledge while living and working abroad. In many cases, these skills include habits of good governance. Given that much migration is circular rather than permanent, when workers return to their native country, their skills and habits can help transform local business and institutions. In fact, it is believed that liberalizing migration would do more to help developing countries than any other form of aid.

But while the overall economic benefits of free migration are clear, it has numerous critics. Economic benefits for receiving countries are diffuse, while costs are concentrated. Immigration may help boost the GDP for a country, but it is individual neighborhoods and workers that have to deal with an influx of newcomers that compete in specific job categories and may bring different languages and cultural habits. For this reason, immigration policy remains highly contested in countries around the world.

Finally, a free flow of ideas between countries is important in promoting development. When knowledge passes easily from place to place, it facilitates the development of new ideas and new technologies. A free press is one key component of the free flow of ideas. When information of all sorts, even information that is critical of those in power, can be shared, it increases public accountability, challenging corrupt political and business elites that stifle competition for their own gain.

According to Reporters without Borders, in 2017, European countries were predominately good or fairly good with press freedom (figure 9.25). However, in no other region did the majority of countries score in the top two most-free categories. In the Asia-Pacific, Middle East and North Africa, and Eastern Europe and Central Asia regions, over half of the countries

Press Freedom: Percentage of Countries by Region and Category

Region (% of countries)	Good	Fairly Good	Problematic	Bad	Very Bad
Africa	0	15	40	33	13
Americas	7	29	39	21	4
Asia-Pacific	3	13	28	44	13
Eastern Europe and Central Asia	0	0	38	38	23
European Union and Balkans	33	35	28	5	0
Middle East-North Africa	0	0	21	42	37

Figure 9.25. Press freedom: Percentage of countries by region and category. Data source: Reporters without Borders. © 2016 Reporters without Borders.

ranked bad or very bad. These regions include China, where the state controls nearly all media, many Western social media platforms are banned or highly restricted, and journalists are frequently jailed. It also includes Saudi Arabia, where there is no independent media at all and social media platforms are heavily monitored for posts critical of powerful political and religious leaders. Punishment can include public flogging for writers who publish critical views. Russia also falls into this group, where independent media have been increasingly brought under control of the government and where numerous bloggers and others have been jailed and, in some cases, assassinated for their writing.

Ultimately, development is the result of economic forces guided by geography and government policy. Entrepreneurs and market forces can lead to efficient, wealth-producing organizations that finance a high quality of life, but the state needs to provide a fair and stable legal framework in which to operate. The state must also play a role in areas of market failure, ensuring that services for the common good benefit all, not just those with the immediate means to pay for them. Market forces will drive dense urban development, but adequate housing, transportation, and municipal infrastructure must keep pace. Firms in distant regions will contribute to national development, but they need good transportation and communications connections with leading urban areas. Workers and employers thrive when divisions between countries are reduced. Government restrictions on the free flow of information should be close to zero, while trade and immigration should be encouraged for their positive economic outcomes, keeping in mind those groups who may be negatively impacted.

World development has made great strides in terms of wealth, health, and opportunity, yet much more is needed. Millions have been lifted out of poverty and are living longer, healthier lives, but many millions remain poor and sick. Oppression and lack of opportunity have improved in many countries around the world, yet women and minority groups still face discrimination and limited life choices in too many places. Constant effort needs to be made to further improve the lives of an ever-greater number of people on our planet.

Go to ArcGIS Online to complete exercise 9.3: "Density: The rural-urban divide in the United States," exercise 9.4: "Distance: How does proximity to urban areas impact development?," and exercise 9.5: "Division: Globalization or protectionism? Trade and development."

References

Abreu, M. de P. 2004. *The Political Economy of High Protection in Brazil Before 1987.* Buenos Aires: IDB-INTAL.

Aires, B. 2016. "Why Black Market iPhones Are a Hot Commodity in Argentina." *CNNMoney,* September 14, 2016. http://money.cnn.com/2016/09/14/technology/argentina-iphone-black-market/index.html.

Burd-Sharps, S., and K. Lewis. 2015. *Geographies of Opportunity. Measure of America.* Social Science Research Council. http://www.measureofamerica.org/congressional-districts-2015.

Kathuria, S. 1987. "Commercial Vehicles Industry in India: A Case History, 1928–1987." *Economic and Political Weekly* 22, no. 42/43: 1809–23. http://www.jstor.org/stable/4377638.

Peterson, A. P. 2011. "Academic Conceptions of a United States Peace Corps." *History of Education* 40, no. 2: 229–40. doi: https://doi.org/10.1080/0046760X.2010.526966.

Reporters without Borders. 2016. "2016 World Press Freedom Index." https://rsf.org/en/news/2016-world-press-freedom-index-leaders-paranoid-about-journalists.

Rostow, W. W. 1999. *The Stages of Economic Growth: A Non-communist Manifesto.* Cambridge, MA: Cambridge University Press.

United Nations Development Programme. 2017. "Human Development Reports." http://hdr.undp .org/en.

United Nations. 2017. "Sustainable Development Goals." http://www.un.org/sustainabledevelopment/ sustainable-development-goals/#prettyPhoto.

Williamson, J. 2004. "The Washington Consensus as Policy Prescription." https://piie.com/ publications/papers/williamson0204.pdf.

World Bank. 2009. *World Development Report 2009: Reshaping Economic Geography.* Washington, DC: World Bank.

———. 2017. "Doing Business—Measuring Business Regulations." https://www.loc.gov/item/ lcwa00095490/.

Chapter 10

Cultural geography–folk and popular culture, language, religion

One of the most significant forces that shapes human society is culture. It affects how people think and act, how they transform landscapes, and how they relate to other social groups. Linguistic, religious, and cultural norms shape peoples' identities, regions, and sense of place, giving each locale its unique character. It is largely because of diverse cultural landscapes that we love to travel: sights, sounds, and smells vary from place to place because of cultural differences.

Culture is often divided into two broad categories: *material* and *nonmaterial*. Material culture involves things we can sense, be it hearing music, tasting food, or seeing architecture. Nonmaterial culture involves the things in our heads, such as religious beliefs and the way language structures our views of the world.

Human culture, both material and nonmaterial, undergoes constant change. As humans migrate and communicate between places, new cultural traits arise from the constant borrowing and stealing of ideas. This exchange affects musical styles, architectural design and building techniques, culinary habits, religious views, ways of speaking, and much more. Although modern transportation and communication have shrunk the world in a sense, allowing cultural traits to diffuse rapidly from one place to another, places still retain their uniqueness. Indigenous culture and traditions remain a powerful force, resulting in local adaptations of global trends.

Culture regions

In chapter 1, you were introduced to the concept of regions and how they can be identified as formal, functional, or perceptual. In the context of culture, formal and perceptual techniques are commonly used to identify regions. Formal regions are often differentiated by mapping some commonality, such as ethnicity, language, religion, or other aspects of culture. At the same time, perceptual regions can be identified by surveying what name people use for where they live and how it contrasts with the names used for surrounding areas. Thus, regions represent places with common cultural habits, beliefs, ways of life, and cultural landscapes.

We use *culture regions* as shorthand to describe places, visualize them in our mental maps, and simplify the world for purposes of analysis. Globally, for instance, geographers and others often write about the Middle East on the basis of its geographic location. This region often includes countries around the Persian Gulf and the Arabian Peninsula as well as those forming a crescent around the eastern Mediterranean Sea from Turkey to Egypt. Other writers focus on the Arab world using ethnicity and language to identify a region. This focus overlaps with much of the Middle East but is distinct. It includes countries such as Morocco at the far western edge of North Africa but excludes Israel, Turkey, and Iran. Still others may use religion as a unifying regional criterion by examining the Islamic world. The Islamic world is much broader and includes the Muslim nations stretching from North Africa through the Middle East to part of Central, South, and Southeast Asia.

Culture regions are also identified at a more local scale. Within Europe, we can talk about Scandinavia, the Iberian Peninsula, the Mediterranean, the British Isles, Eastern Europe, the Balkans, and more. Each of these European regions stands apart from the others because of characteristics such as language (Slavic, Germanic, Romance), religion (Catholic, Protestant, Eastern Orthodox), history, and more. Based on these differences, we can see cultural landscapes that vary in terms of architectural styles, social differences in terms of family and social life, and political difference in terms of government and institutions.

Culture regions in the United States

As can be imagined, a large country such as the United States can also be divided into distinct culture regions (figure 10.1). For instance, the geographer Wilber Zelinsky, in 1980, published an influential article on the fourteen vernacular regions of the United States. These regions, such as Dixie, Pacific, and Midland, were identified through an analysis of regional words used in businesses, schools, churches, and other organizations. Zelinsky's paper did not detail how each region's culture was unique, but other authors have analyzed regional differences by subregions similar to his. Based on these analyses, differences become apparent in the way people speak, their religious beliefs, foods, architecture, social behavior, and much more. For instance, we are all familiar with regional accents, be it a Texas twang, a southern drawl, a New York accent. We also know that foods vary substantially, with fried okra in the South, spicy Tex-Mex in Texas, and clam chowder in New England.

What we may be less familiar with is how the historical evolution of each region impacts aspects of culture. For instance, much of New England and the Northwest was originally settled by Calvinists, who

Figure 10.1. Culture regions of the United States. Image by author.

valued work for the common good and a greater denial of self. This cultural trait has survived for hundreds of years, so that the political culture still gives stronger support to government social programs than in many other regions of the United States. This lies in direct contrast to the South and its northern border in Appalachia. There, historical settlement patterns led to populations that favor individual liberty and that resist federal intrusion from Washington, DC. In Appalachia, this came from immigrants from the borderlands between Scotland and England and between Northern Ireland and Ireland. These conflict-ridden regions in the British Isles helped create a warrior-ethic culture that resisted those in power, characteristics that are still common to this day. In the South, early settlement was modeled after the slave societies of the Caribbean Basin. It was based on power for a small elite and disenfranchisement for most of the rest of the population. As the elite did not need government assistance, and the disenfranchised were excluded from political decision making, a culture that resists taxes and other government interference remains a strong force.

Another interesting regional difference in the United States relates to violence. Since the colonial era, violence has been higher in the Appalachian and Southern regions than in regions such as New England where Puritan society put priority on societal order. The state dealt harshly with violent acts, creating an institutional and cultural framework that kept it under control. In contrast, Appalachia's warrior ethic led to a culture of clan and family loyalty, with a high value placed on exacting vengeance when honor was violated. The South, too, had a relatively high tolerance for violence. There, violence was sanctioned culturally, and individuals used it to a greater degree than the state. It was also tied to honor, but it had a hierarchical component as well. Superiors used violence against inferiors, husbands used it against wives, and parents used it against children.

Regional cultural patterns of violence reach from the colonial past to the present. In 2014, the states with the highest homicide rates were in the South: Louisiana, Mississippi, Alabama, and Arkansas. In fact, the homicide rates in Louisiana and Mississippi were nearly seven times higher than in Massachusetts. Some research has even shown that homicide rates correlate more strongly with cultural region than with poverty, urban density, or other material variables.

Tapestry segmentation

Broad culture regions still have a strong influence on how people think and act in the United States, but with changes in communication and transportation, regional differences have become somewhat more blurred. Northerners move to the South, Southerners move to the West, and Westerners move to the Northeast. As people move, they bring their regional cultures with them but also absorb elements of the region to which they move. People across the country can also see the same television shows, web pages, and social media, regardless of the region in which they live. Thus, in many ways, cultural differences can now vary more by urban or rural location, income and education, age, ethnicity, or other demographic and socioeconomic factors. We still form clusters with other like-minded people, but often at more local scales.

Tapestry Segmentation, first mentioned in chapter 4, identifies places with common characteristics along social, economic, and demographic lines. It divides neighborhoods in the United States into fourteen broad Life Mode groups, which are further divided into sixty-seven more detailed tapestry segments. The logic behind segmentation analysis is that people with similar tastes and behaviors cluster in specific places. Our political and social beliefs; the food, clothing, and other items we buy; and the music we listen to tend to be like that of our neighbors. Where we grow up influences us, and when we move, we tend to choose places where others think similarly.

At a county level, clear Life Mode clusters are seen, but they are not always confined to traditional culture regions (figure 10.2). The Rural Bypasses segment is largely clustered in the South, but the Soccer Mom segment is found in suburban communities throughout

Culture Regions: Segmentation Analysis

Culture regions can be seen at a local scale, such as by county.

Rural Bypass clusters are found largely in the South. Income and education is lower. Traditional values and religion are strong. Hunting and fishing is common.

Soccer Mom clusters are found throughout the country. College education and higher incomes are more common in these suburban communities. Jogging, cycling, and golf are popular pastimes.

Figure 10.2. Culture regions: Segmentation analysis. Explore this map at https://arcg.is/1ai4SG. Data sources: Esri, HERE, Garmin, NGA, USGS, US Census Bureau, Infogroup.

the country. In the largely Southern Rural Bypasses counties, people live in rural and small-town places where incomes are relatively low, as are levels of education. Values are traditional, and religion is a central part of people's lives. Outdoor activity such as hunting and fishing is popular, and people prefer trucks over sedans. The South contains a substantial number of Soccer Mom communities as well, but these are also found scattered throughout the country. Here, people live in higher-income suburban communities. Those with college educations are more prevalent than in surrounding rural areas, and technology is heavily integrated into their lives. Outdoor activities consist of those common to suburban communities, such as jogging, bicycling, golfing, and boating as well as target shooting.

Thus, cultural regions can be seen at different scales, from the broad region to smaller neighborhoods. Regardless of the scale, people with common beliefs and behaviors often cluster with other like-minded people, creating cultural mosaics on the landscape.

Folk and popular culture

Geographers often divide culture into two broad categories: *folk* and *popular*. Folk culture originates in more isolated and homogenous societies. Music, food, building styles, and other components of material and nonmaterial culture develop organically in folk cultures. There is no author, designer, or inventor; rather, culture evolves from the shared life experiences of the group and the natural environment in which they live. There is no copyright or patent on folk artifacts, but instead traditions pass orally from person to person and generation to generation as common knowledge.

Popular culture, on the other hand, develops in larger, more heterogeneous societies. Typically, the person or group that creates the cultural artifact, be it a song or a building, is known. Copyrights and patents are common, allowing creators to gain fame and wealth from their creations. Popular culture, as its name implies, diffuses widely and is not tied to a specific place or environment.

Music: From Appalachia folk to New York rap

Folk music, as with all folk culture, is music passed on through oral tradition, with origins in rural communities. It is not written but rather is learned via exposure to others in a social group. Consequently, folk songs can change over time, as people add and subtract material to suit changing circumstances. In the United States, the Appalachia region was a source of much research on folk music. Early-twentieth-century researchers sought to collect and catalog folk songs from this region because of the relative isolation of many communities in remote mountain and valley settlements that had limited exposure to urban cultural influences. The Appalachian culture consisted largely of folk traditions from the English and Scots of the British Isles who had migrated and settled there in eighteenth and nineteenth centuries.

As Western society urbanized, folk traditions weakened. Mass communication and mobility meant that people increasingly shared culture, including musical styles. Radio allowed music to be heard across wide areas; thus, folk music began to transform into popular music. Traditional folk instruments, rhythms, and lyrics were incorporated into music aimed at wider audiences. Gaining wider audience appeal often involved adding political and social movement themes to songs, supporting the folk, or common people, and their struggles in modern society.

The origin and diffusion of rap music illustrates how folk traditions blend into new forms of popular music. African folk customs diffused to the Americas with the forced migration of the slave trade. Many of these traditions, including those of *griots*, or traditional West African storytellers/singers, remained integral parts of African American life. Along with other musical traditions, it evolved into blues, jazz, and R&B. By the 1970s, these influences, along with a Jamaican tradition of speaking over beats, began to take form as rap in isolated African American urban neighborhoods. This blending of musical styles is known as *syncretism*, the process of combining cultural traditions into a new form. Thus, rap music is based on syncretism of African, Caribbean, and African American musical traditions.

Rap followed a classic pattern of hierarchical diffusion. It originated in New York City, and through the 1980s, this was its clear cultural hearth, led by groups such as The Sugarhill Gang, Grandmaster Flash, and others. By the 1990s, Los Angeles had become an important source of new rap talent, second only to New York, where Ice T and Easy-E's debut albums were seeing huge sales. Following New York and Los Angeles, new rap artists gained traction in other large and medium-sized urban areas, such as New Orleans, Houston, Oakland, Atlanta, Chicago, and Detroit. By the 2000s, artists were gaining popularity in a number of cities outside of the coastal cities and the handful of larger cities where rap first had success.

From its origins in segregated neighborhoods of New York City, rap is now a global phenomenon. In many ways, it has become a *lingua franca*, or common language, for youth around the world. Rap competitions can now be found in Shanghai, China; Paris, France; Moscow, Russia; Cape Town, South Africa; and major urban areas on every continent (figure 10.3).

Although it is considered popular music enjoyed by people across the country and around the world, rap has strong associations with specific places. Sense of place is an important component of the genre, so that where the rapper is from advances or hinders his/her street cred and, ultimately, fame and fortune. Typically, coming from a poor inner-city neighborhood grants authenticity, something that Eminem, from inner-city Detroit, had but Vanilla Ice, from a more suburban location, did not. Rivalries, often promoted by record labels, have often been tied to geography, pitting neighborhoods, cities, and regions against one another. These place-based rivalries resulted in homicide for

Figure 10.3. A Russian rapper performs in Moscow. Rap evolved from folk traditions to a global popular culture phenomenon. Photo by: Hurricanehank. Stock photo ID: 681052426. Shutterstock.com.

some, such as Tupac and Notorious BIG in the 1990s East Coast–West Coast feud.

Food: Beef and potatoes to Subway

As with music, culinary traditions can be seen in terms of folk and popular culture. Folk foods are closely tied to local physical environments, whereby people's diets are based on the foods and animals that thrive in their locale. With the advent of agriculture, farm crops and animals quickly diffused to new places. Through trade and migration, they spread from origin points across continents and, ultimately, around the world. Thus, foods considered folk in nature do not necessarily originate in the place in which they are consumed.

Food has a strong association with place and often forms an important part of people's regional identities. For example, collard greens originated in the eastern Mediterranean region but thrived in the Southern climates of the United States. Dishes with collards are now promoted in restaurants as "traditional cooking" and "soul food," reflecting their long history in that region. In the Great Plains region of the United States, beef, potatoes, and bread constitute folk meals, although none of these items originated in North America; wheat and cattle originated in the Old World, and potatoes are from Andean Latin America. In many ways, foods of the Great Plains are seen as "American food," although

in reality, they represent diets of a region stretching from Montana south and eastward through Kansas. Even dominance of the all-American apple pie has a limited spatial distribution. Moving to the southern plains of Oklahoma and Texas, one crosses the "Apple pie line," beyond which desserts begin to include peach cobbler and Mexican-influenced flan (a custard with caramel sauce) and sopapillas (fried dough with sugar and cinnamon).

In Tijuana, Mexico, a large border city just south of San Diego, California, some restaurants now promote folk foods that represent "Mexicanness." In a region that is often dominated culturally and economically by its large northern neighbor, traditional Mexican dishes serve as a means of reasserting cultural boundaries. Traditional foods from rural haciendas consumed by the Spanish-indigenous mestizo population are promoted as an alternative to European influences among the elite and Americanized versions of Mexican food. In these restaurants, one does not find tacos, chimichangas, and enchiladas. Rather, maguey worms and *champolines* (grasshoppers), squash blossoms, nopal (cactus), and other traditional ingredients are used to prepare dishes tied to folk customs (figure 10.4).

Figure 10.4. Folk foods in Tijuana, Mexico. Here, mezcal red worms (also known as *chinicuil*, maguey worms, and *gusano rojo*) are served with traditional Mexican tlacoyos, a stuffed corn dough. Photo by VVDVVD. Stock photo ID: 639126400. Shutterstock.com.

In contrast to folk foods are popular culture foods—those marketed and consumed over wide areas by diverse populations. Soda such as Coca-Cola and Pepsi, fast-food restaurant chains such as McDonald's and Subway, global food conglomerates such as Nestlé and General Mills, all produce for mass markets. Companies in the mass-market food sector operate in countries around the world, selling products that are often developed, manufactured, and marketed far from the place in which they are consumed.

Subway, in 2017, was the largest restaurant chain in the world, with over 44,000 stores in 111 countries (figure 10.5). McDonald's is also found in over 100 countries, and Nestlé sells its products in nearly 200. Coca-Cola is sold in over 200 countries, and it claims that its logo is recognized by fully 94 percent of the world's population!

Some argue that these types of companies contribute to *placelessness*. If one can travel the world and continue to drink Coca-Cola and eat at McDonald's,

so the argument goes, then why bother traveling? Others argue that global companies that ignore local cultures are doomed to fail. Just because a business model and product work in one culture does not mean it will work in another. Therefore, companies must use a strategy of *glocalization*. This somewhat awkward portmanteau combines globalization with localization. The idea behind it is that global companies must consider local cultural differences if they are to succeed. Despite the concerns of critics, local culture is still a powerful force that drives what people consume. Just like Southerners, residents of the Great Plains, and Tijuanenses have different culinary traditions, so too do people in all regions.

Successful global companies have embraced glocalization. Nestlé Kit-Kats, for instance, come in cherry blossom, green tea, and strawberry cheesecake flavors in Japan (figure 10.6). Pringles potato chips in Asia include soft-shell crab flavor, and Oreos in Argentina come in dulce de leche (a type of caramel). One of

Popular Culture Food: Subway Sandwich

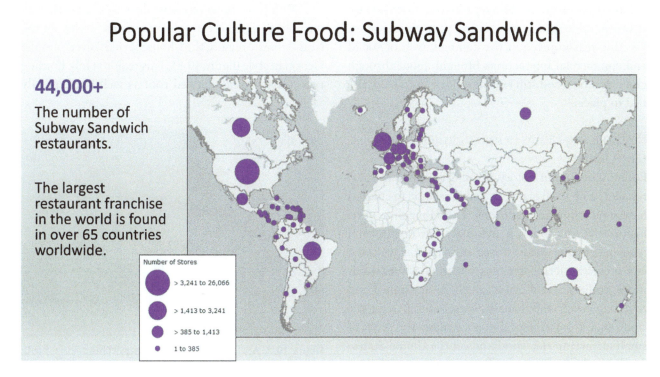

44,000+
The number of Subway Sandwich restaurants.

The largest restaurant franchise in the world is found in over 65 countries worldwide.

Number of Stores
- \> 3,241 to 26,066
- \> 1,413 to 3,241
- \> 385 to 1,413
- 1 to 385

Figure 10.5. Number of Subway sandwich restaurants. Explore this map at https://arcg.is/0KKvGj. Data source: Subway.com.

Figure 10.6. Glocalization of Kit-Kat candy in Japan. Flavors come in cherry blossom, green tea, and strawberry cheesecake. Photo by Gnoparus. Stock photo ID: 366302453. Shutterstock.com.

the most successful companies to use glocalization is McDonald's. Its traditional menu of burgers and fries is substantially modified around the world, so that you can order a *pulut hitam* (black rice pudding) pie in Korea, vanilla McNuggets in Taiwan, and an Eggcellent Silly Double Beef Burger (with an egg on top) in Hong Kong. Likewise, beef and pork are off the menu in India in deference to its large Hindu and Muslim populations.

Cultural foodways still exert a powerful influence in human culture, and culinary folk traditions remain strong and integral parts of many people's identities. For this reason, even in the complex web of social and commercial interactions brought about through globalization, food still retains unique characteristics tied to place.

Housing: Log cabins to the suburban ranch house

For most of human history, housing was built by local residents with local materials and in relation to cultural habits and climatic conditions. The Mongolian yurt, historically and today, is made of animal skins and woolen felt from locally raised animals. It is windowless, protecting its inhabitants from brutal winter winds in Central Asia, and can be easily disassembled and moved in synch with nomadic lifestyles. Likewise, the traditional frontier log cabin once suited the needs of small farmers in the woodlands of the United States.

But folk housing typically changes as spatial interaction increases and communities become less isolated over time. In the United States, from the time of early colonization up until about 1760, housing was largely folk in nature, where Old World forms and techniques dominated construction and were tied to specific ethnic immigrant groups. One common type of folk house during this time was the I-house (figure 10.7). This type of house, which was symbolic of economic attainment at the time, consisted of at least two side-by-side rooms in two stories, with gables on the side (triangular shaped rooflines).

With time, however, new ideas worked their way into old designs, creating new regional styles. In the Mid-Atlantic region, the I-house style evolved into the Georgian style (figure 10.8). This regional folk housing type usually had a gabled roof as well but typically

Figure 10.7. An I-house made of logs. This type of folk housing is among the most widely distributed in the United States. Photo by David Ross. Stock photo ID: 1569030. Shutterstock.

Figure 10.8. Georgian style house in Newport, Rhode Island. Typical of this type of housing is a façade with five openings for windows and doors on each floor. Photo by Jiawangkun. Stock photo ID: 702450331. Shutterstock.

Figure 10.9. Suburban ranch house. This form of popular housing is found in regions across the United States. Photo by rSnapshotPhotos. Stock photo ID: 172812272. Shutterstock.

included four rooms per floor, separated by a central hall. It is often distinguished by two window openings per floor on the end of the house and five (including the door) on the front. This style, as with the I-house, represented economic success during the eighteenth and nineteenth centuries.

As the United States became further developed and spatially integrated, first with the construction of the intercontinental railways, then the national highway system, and ultimately the interstate freeway system, regional influences further eroded, and what can be referred to as *popular housing* styles gained prominence.

The epitome of popular housing in the United States can be seen in the ubiquitous suburban ranch house, which became the most popular form of American housing by the 1950s (figure 10.9). In contrast to folk housing, the ranch house represented the modern industrial economy. It was made almost entirely of mass-produced rather than locally produced materials. Construction was done at an industrial scale, with hundreds and even thousands of nearly identical homes built in subdivisions at the same time. Its design was the same regardless of the region in which it was built, consisting of a single-story, rectangular form with stark, minimalist interiors.

The lot it was built on was flattened and scraped clean of natural vegetation, which was replaced with green lawns and shrubbery. Some allowance was made for exterior variation, such as Spanish, Tutor, or colonial style, but the overall shape and function varied very little.

The ranch house also reflected the modern automobile age. It was built outside of the city center in new suburban locations where modern life centered on the car. Architecturally, this was represented by a large, visually dominant garage on the front of the house. Never before had a mode of transportation played such a prominent role in residential architecture.

As with the I-house and the Georgian house of previous centuries, the ranch house represented economic achievement. The difference is that by the mid-twentieth century, a large segment of the US population could now attain this level of material success. It allowed for ownership of a single-family home in a relatively natural setting, a valued tradition in American life. The ranch house in an automobile-oriented suburban neighborhood represented the individualism and freedom of American culture.

Go to ArcGIS Online to complete exercise 10.1: "Popular music trends: Which songs top charts around the world?" and exercise 10.2: "Food and drink: How healthy is your town?"

Language

Language is arguably the most important component of human culture. It is how people interact, for instance, sharing elements of culture such as musical styles, ways of producing and preparing food, and techniques for building homes. Those who speak a common language can more easily conduct commerce, worship together, organize politically, and form families and communities. Ties based on this interaction mean that language often serves as a key component of people's identities, guiding whom they consider to be or not to be part of their group. Without a common language, interaction is inhibited, making commerce and social interaction more difficult, often fomenting cultural divisions between people.

Most cultures have language-origin myths that attempt to explain why people from different places often speak in different tongues. The Judeo-Christian Bible tells the story of the Tower of Babel in ancient Mesopotamia, where God imposed multiple unintelligible languages upon those building the tower and scattered them across the land (figure 10.10). The aborigines of Australia have a different myth. They say that long ago there was an ill-tempered old woman who would scatter the fires that warmed those who were sleeping. When she died, the people celebrated, with many coming from afar to express their joy. They began to eat her flesh, and as each group consumed her, they began speaking in distinct languages. Myriad other societies have their own language-origin stories, but their commonality is that people from different places speak different languages.

Over time, linguists have developed new ideas as to why we speak different languages. It is now understood that languages evolve over time, diversifying as new words and pronunciation are added, others are dropped, and syncretic blending occurs as societies migrate and interact. In a globalized world, languages continue to change. Demographic and economic strength is making some languages more prevalent, while weakness in these areas is leading to the extinction of others.

Number of languages and speakers

Schools in the United States often offer classes in a handful of languages. Spanish, French, German, and

Figure 10.10. Language origin myths: The Tower of Babel as painted by Pieter Brueghel in 1563. In the Judeo-Christian Bible, God is said to have imposed multiple unintelligible languages upon those building the tower, and scattered them across the land. Image by Jorisvo. Royalty-free stock photo ID: 88154914. Shutterstock.com.

Italian have been some common staples, with Chinese and Arabic added more recently. Occasionally, a less-common language will be taught, such as Korean, Vietnamese, or Farsi, but for the most part, students can choose from a half dozen or fewer.

Given the limited number of foreign languages offered in schools, it may be a surprise to know that there are about 7,100 living languages in the world today. But with that said, only a handful are spoken as first languages by the vast majority of the world's population. Over 40 percent of people speak one of just eight languages: Chinese, Spanish, English, Arabic, Hindi, Bengali, Portuguese, and Russian. Adding in another eighty-two languages incorporates over 80 percent of all people. That leaves roughly 20 percent of the world's population speaking the remaining 7,000 or so languages. What this tells us is that a few dominant languages are used by many people over frequently large areas, and thousands of smaller languages are spoken by very few people in small, isolated areas.

Of the eight largest languages, spatial distributions vary. Figure 10.11 shows the countries where each is either officially recognized by the government or is the dominant language spoken by the population. Chinese, the single largest language in terms of native speakers, is spatially concentrated in China. The same can be said for Bengali and Hindi, which cluster in and around densely populated India and Bangladesh, and Russian, which is found in Russia and a handful of adjacent states. Arabic is more broadly distributed, stretching across North Africa into East Africa and the Middle East. The three European-origin languages, Spanish, English, and Portuguese, have wider distributions, the result of European colonial influence that began in the fifteenth century. English is the most widespread, dominating or being officially recognized in all the

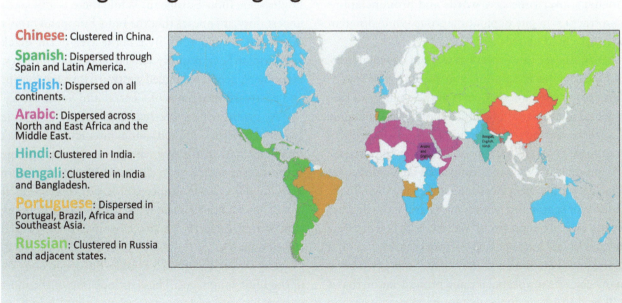

Eight Largest Languages: Official or Dominant

Chinese: Clustered in China.

Spanish: Dispersed through Spain and Latin America.

English: Dispersed on all continents.

Arabic: Dispersed across North and East Africa and the Middle East.

Hindi: Clustered in India.

Bengali: Clustered in India and Bangladesh.

Portuguese: Dispersed in Portugal, Brazil, Africa and Southeast Asia.

Russian: Clustered in Russia and adjacent states.

Figure 10.11. Eight largest languages: Official or dominant. Explore this map at https://arcg.is/aaa50. Data source: CIA, The World Factbook.

populated continents. Spanish, too, is relatively widespread, although predominately in Latin America. Portuguese is found in several regions, although in many fewer countries: Portugal in Europe, Brazil in Latin America, Angola and Mozambique in Africa, and East Timor in Southeast Asia.

Language families and their spatial distributions

Despite ancient language-origin myths, we now know that language diversity is analogous to diversity in plant and animal life. Just as life on earth originated with single-cell organisms that evolved over hundreds of millions of years into the plant and animal species we see today, languages go through their own sort of evolution. Linguists generally believe that human language evolved from an initial human proto-language many thousands of years ago. As the human population grew and spread over larger parts of the planet, this initial language went through a process of divergence. This occurs when human populations migrate and separate, losing all or most contact with groups in other locations. Although they once spoke a similar language, over time new words and pronunciations gain traction in one population group while different ones take hold in another.

You can see this process at work in our own lives. Just think about slang words you use with your friends that are not used in other parts of the country or that older generations do not understand. Now imagine that there was no interaction with people outside of your local geographic area—no travel, no phones, no TV, no internet. As your group adds new words over time, people in other places are adding their own. In the absence of spatial interaction, more and more distinct words and pronunciations will be added separately in each isolated group. Fast-forward several hundred years, and there is a good chance that there will be so many changes in vocabulary and pronunciation that it will be difficult or impossible for people in one group to communicate with people from another.

Evolutionary changes of this sort have led to the formation of *language families*, groups of languages

that evolve from a common linguistic ancestor. Families further evolve into subfamilies and finally individual languages. There are over 140 language families, but just six of them constitute languages spoken by 85 percent of the world's population (figure 10.12). The family with the most speakers is Indo-European, which includes six of the top eight most commonly spoken languages. Bengali, English, Hindi, Portuguese, Russian, and Spanish are all part of this family, gaining numerical strength from large population clusters in South Asia and European colonial expansion. The other language families that include the top eight spoken languages are Sino-Tibetan, with Chinese, and Afro-Asiatic, which includes Arabic. The Niger-Congo and Austronesian language families are less dominant in terms of speakers, but they represent the most linguistically diverse parts of the world, each consisting of over a thousand separate languages. The smallest of the large language families is Trans-New Guinea, clustered on the island of New Guinea.

The Indo-European language family

The largest language family—and that which includes English—is Indo-European. While there is still some uncertainty, it appears that the Indo-European family developed around 4000 BCE in the area of the Black and Caspian Seas, possibly around modern-day Turkey. From there, people expanded westward through Europe and eastward into parts of Central and South Asia.

As groups of people scattered across this wide swath of territory, limited spatial interaction between them resulted in divergence into smaller language subfamilies, which then diverged into many more individual languages (figure 10.13). For this reason, it is no coincidence that the word for mother has many similarities in Dutch *(moeder)*, Spanish *(madre)*, and Russian *(mat)*.

The far eastern branch of Indo-European consists of the Indic languages. These include two of the most commonly spoken languages, Bengali and Hindi, in South Asia. A bit closer to the family's point of origin is the Iranian branch, which includes Persian and Kurdish, among others, while Eastern Europe and

Language Point of Origin: Six Largest Language Families

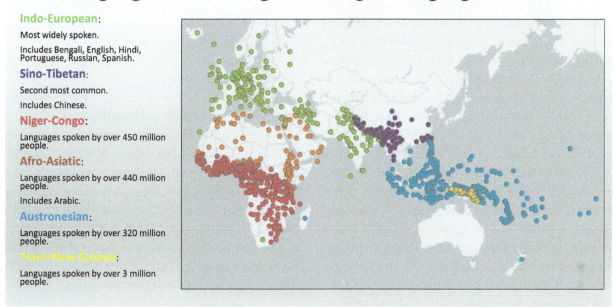

Indo-European:

Most widely spoken.

Includes Bengali, English, Hindi, Portuguese, Russian, Spanish.

Sino-Tibetan:

Second most common.

Includes Chinese.

Niger-Congo:

Languages spoken by over 450 million people.

Afro-Asiatic:

Languages spoken by over 440 million people.

Includes Arabic.

Austronesian:

Languages spoken by over 320 million people.

Trans-New Guinea:

Languages spoken by over 3 million people.

Figure 10.12. Point of origin for languages: Six largest language families. Explore this map at https://arcg.is/1nSvyr. Data source: Dryer, Matthew S. & Haspelmath, Martin (eds.) 2013. Simons, Gary F. and Charles D. Fennig (eds.). 2017.

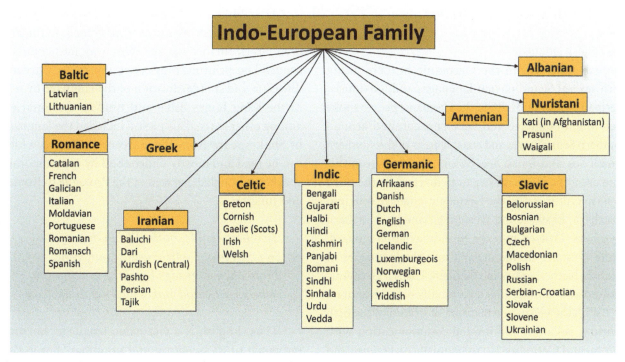

Figure 10.13. Indo-European language family. Sample of languages. Data source: Dryer, Matthew S. & Haspelmath, Martin (eds.) 2013.

Russia contain the Slavic languages, such as Russian, Polish, and Czech.

Among the most geographically widespread languages are some from the *Romance* and *Germanic* branches. The Romance languages evolved from Latin, the language of the Roman Empire, and therefore stretch largely along the northern Mediterranean Sea, where tongues such as Italian, French, Spanish, and Portuguese are spoken. The exceptions to this geographic pattern are Romania and Moldova, which retained their Latin roots in the eastern portion of Europe. Each of these languages evolved from Latin after the fall of the Roman Empire, which left far-flung former Roman provinces relatively isolated as Europe entered the Middle Ages. Given that Romance languages are commonly spoken in the United States and are often taught in high schools and colleges, many students may notice commonalities between these languages. For instance, the verb "to sing" in both Spanish and Portuguese is *cantar*, in French is *chanter*, in Romanian is *canta*, and in Italian is *cantare*. Many other words, as well as grammatical structures, are similar, so that speakers of one Romance language can relatively quickly learn another.

The English language

English falls within the Germanic branch, along with several Scandinavian languages and, obviously, German. Around 750 BCE, the Germanic people spoke a common language and were concentrated around southern Scandinavia and coastal areas of the southern North Sea and Baltic Sea. With migration from this core area, the language began to diverge. By the third to sixth centuries CE, the Germanic language split into Northern, Eastern, and Western Germanic. Northern Germanic evolved into the Scandinavian languages, while Western Germanic split into the languages of English, Netherlandic, and German. Eastern Germanic evolved into Gothic, which eventually became extinct.

With the Germanic invasion of Britain in the fifth century, isolation and limited communication with continental Germanic led to linguistic divergence and the evolution of Old English. The well-known poem "Beowulf" was written in this tongue, but although considered an early form of English, it is unintelligible to the modern speaker:

Hie dygel lond
warigeað, wulfhleoþu, windige næssas, frecne
*   fengelad,*
ðær fyrgenstream

From there, English continued to evolve. In 1066, the Normans, people from the north of modern-day France, invaded and occupied Britain. Especially among the aristocracy, the language of the Norman invaders took hold. Their Latin-based language merged with Old English, adding new words and pronunciations. This evolved into Middle English, the language of Chaucer in *The Canterbury Tales*. While difficult for modern English speakers to understand, Middle English is at least partially recognizable:

A cook they hadde with hem for the nones
To boille the chiknes with the marybones,
And poudre-marchant tart and galyngale.

After 1400, further evolution led to an early form of Modern English, as words from French, Latin, and even Greek further worked their way into common usage. The printing press contributed to a greater diffusion and standardization of the language by way of the King James Bible and English dictionaries, among other texts. This period included the language of Shakespeare from the sixteenth century, which can be read and mostly understood by contemporary speakers of English, as seen in the Prologue to Romeo and Juliette:

Two households, both alike in dignity,
In fair Verona, where we lay our scene,
From ancient grudge break to new mutiny,
Where civil blood makes civil hands unclean.

Thus, English is a Germanic language, borrowing much of its structure and grammar, but it also incorporates a substantial Latin influence. As with all languages, it continues to evolve. New words are added as technology changes—bicycle, camera, astronaut—while others are added as people migrate and communicate around the world—ninja (Japan), bagel (Yiddish), wok (Cantonese), latte (Italian).

Language variation
Dialects

English, like all languages, constantly evolves, but the spatial pattern of change varies from place to place. New words, grammar, and pronunciations gain traction in one place and lose dominance in another. Based on the concept of distance decay, a change can gain strength in the core area in which it evolved, but its use becomes increasingly less common as one moves away. This process of language evolution results in the formation of dialects, which are variations in a single language that remain mutually intelligible. They can be found at different scales, so that, for example, globally, an Australian dialect differs from an American one. Then, within a country, there can be further regional divisions, say between the US South and West. Finally, there can be local dialects, such as Boston, New York City, or Long Island.

The formation of dialects is the same as that of languages. In fact, with enough time and enough spatial isolation, dialects can eventually become unintelligible to outsiders, eventually making them distinct languages. Generally, variations in dialects are found in places settled prior to mass transportation and communication. When it was difficult to interact with people the next town over, dialects formed at a much more local scale. For this reason, dialects in Great Britain can vary from city to city, and the East Coast of the United States, being settled earlier, has a greater diversity of dialects than the West Coast.

In the United States, researchers have identified various dialect districts, including Northern New England, New York City, Southern, Midland, and Western, among others. Words and pronunciations vary within these regions, so, for example, much of the South and the East Coast pronounces caramel with three syllables ("car-ra-mel"), while much of the Midwest and North Central part of the country uses two syllables ("car-ml"). Similarly, along much of the West Coast, the thing you drink water out of at school is a "drinking fountain," while the majority of the East calls it a "water fountain." Then there is the unusual "bubbler" used nearly exclusively in Wisconsin.

Figure 10.14 shows the commonly used "soda, pop, or Coke" example of regional dialects. This represents the generic word used for carbonated soft drinks. For example, in the South, many people ask what type of Coke a restaurant serves, even if they want something other than a cola.

In many Latino immigrant communities in the United States, and to some degree in border cities within Mexico, the *Spanglish* dialect is commonly used. While there is still some debate as to whether it is truly a dialect or some other language hybrid, it represents a means of communication used by millions of Latinos to varying degrees. Spanglish involves several different linguistic processes in its use. English loanwords are commonly interspersed with a Spanish sentence, so that a worker may "*tomar un break*" (take a break) at work or "*dejar el coche en el* driveway" (leave the car in the driveway). English verbs can also be converted to Spanish verbs, so to go to yard sales becomes "yardear," and to park a car becomes "parquear." Another common feature of Spanish is code-switching, whereby speakers randomly switch between languages in the same conversation. "When we went to the beach I saw my friend" becomes "*Cuando fuimos a la playa* I saw my friend." As with all dialects, the continued use of Spanglish will depend on the degree of spatial interaction between its users and others. If immigration from Latin America continues to slow, in all likelihood Spanglish will fade away, with traditional American English dialects superseding its use. On the other hand, if steady streams of new immigrants continue to come from Latin America, Spanglish is likely to remain in use, serving as a common dialect for those with varying levels of English and Spanish competency and as a common means of communication for Latinos with different Spanish dialects from throughout the Spanish-speaking world.

Regional Dialects: Soda, Pop, or Coke?

When you ask for a carbonated soft drink, which word do you use?

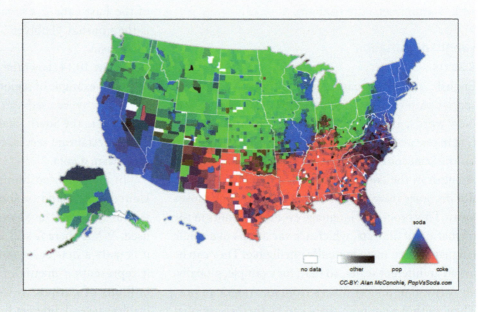

Figure 10.14. Regional dialects: Soda, pop, or Coke? Image from Alan McConchie, PopVsSoda.com.

Lingua franca and global English

Language-origin myths go back millennia, signifying that incomprehension between culture groups has been of concern for most of human history. While of little importance when human societies were small and isolated, this changed once populations grew, migration flows increased, and transportation and communication technology brought people into regular contact. One common solution to mutually unintelligible languages has been to use a lingua franca, a second language used by different communities to communicate with each other. In the past, when the Roman Empire held sway over large parts of Europe and North Africa, Latin served as a lingua franca among diverse cultures. Later, with European exploration and trade, Portuguese served this purpose in many places.

British colonial expansion and economic might planted the seeds for a new global lingua franca in the eighteenth and nineteenth centuries. With colonies and territories across the globe, the English language gained foothold in diverse locations, used by colonial administrators in capital cities and by traders in ports that linked together an expanding global economy. After World War II, the British Empire began to fade but was replaced by an ascendant United States, which through the end of the twentieth century and the beginning of the twenty-first century was dominant politically, militarily, and economically. This gave the English language further strength for business and political communication, establishing it as the dominant global lingua franca.

It is now estimated that one-fourth of the world's population can communicate to some degree in English. Throughout the non-English speaking world, it is used by companies and institutions such as the European Central Bank in Germany and the Association of Southeast Asian Nations based in Indonesia. Japan-based Nissan uses English as its official corporate language, as does German-based Siemens. Many other companies across the globe either use English as an official language or require that all employees have fluency in it. In academia, it is the default language of many conferences, and universities around the world offer entire programs taught in English. Ship and airline pilots also use English as their universal language. Eighty percent of the information on the internet is in English.

But while English has become the most influential lingua franca in the world, its global usage is often distinct from that of native speakers. What is sometimes referred to as a global English, or *Globish*, is more common than fluency in formal standard English. This is a simplified version of the language, with a smaller vocabulary and less-rigid rules on pronunciation and grammar. The English used by airline pilots is really a simplified *Airspeak*, while ship pilots use *Seaspeak*. What is known as *Special English*, which consists of just 1,500 words, is used by the Voice of America in its international broadcasts. Many who use English on a daily basis for business transactions and work-related functions know just enough to get the job done; they are unlikely to be able to debate the merits of Karl Marx and Adam Smith.

Ultimately, it is unlikely that English will become a universal world language that all people speak. Rather, it will probably serve the needs of business transactions and work-related communication in a basic, simplified form. Only a small proportion of the world's population will be truly fluent. This also leaves open the possibility that another language could gain influence over time. The position of English as a lingua franca looks relatively secure now, but Latin looked the same way during its heyday as well.

Language, identity, and assimilation

Language is often an important part of people's identity. It acts as a social marker that expresses where someone is from geographically and often socially. A common language can tie people together by facilitating communication and a sense of community and belonging. The flip side is that different languages sometimes can keep groups of people separate, inhibiting communication and a sense of community. For this reason, there are cases of government restrictions on the use of different languages within their borders

and resentment by some toward those governments for doing so. Cultural battles can occur between visions of a country that differ between those that see linguistic assimilation as essential and those who see multicultural linguistic diversity as a national strength rather than a weakness.

Examples of government suppression of language are myriad. In the 1800s, the US government prohibited the use of Native American languages in reservation and mission schools, and after World War I, many states enacted anti-German language laws. In twentieth- and twenty-first-century Turkey and Syria, the Kurdish people have seen their language banned from being used in education and politics as well as prohibited from public broadcasts. Under the dictatorship of Franco in Spain, regional languages such as Basque and Catalan were prohibited. In these and many more cases, governments were attempting to repress multiculturalism, fearing that it posed a threat to national unity. In 1992, the United Nations passed the Declaration on the Rights of Persons Belonging to National or Ethnic, Religious and Linguistic Minorities, which explicitly states that minority groups have the right to use their own language in private and in public. Linguistic discrimination and repression continue, however, as the declaration is nonbinding and used only as guidance by national governments.

The United States has never had an official, legally proscribed language. However, thirty-one states have enacted official English legislation. None of these laws prohibit the use of foreign languages in public or private. Rather, they typically state that all official government proceedings are to be conducted in English, although most government agencies continue to provide official state materials in additional languages as well (figure 10.15).

One reason that there is no official English policy at the national level in the United States is that, despite being a nation of immigrants, language assimilation has occurred naturally. In fact, assimilation has been so strong that the United States is sometimes called the graveyard of languages, a place where people come from around the world to lose their native tongues

Figure 10.15. Bilingual signage. Despite official English laws in many states, government materials are often provided in multiple languages. Photo by Leonard Zhukovsky. Stock photo ID: 164333684. Shutterstock.com.

within a relatively short time period. Historical research shows that language assimilation by the massive waves of European immigrants from the 1920s was nearly complete by the third generation (the grandchildren of immigrants), as 95 percent of third-generation immigrants were English monolingual.

More recent immigrants continue to follow the same pattern. In Southern California, a place with large, often relatively homogenous immigrant neighborhoods, both Asian and Latino language assimilation is mostly complete by the third generation. English quickly becomes the preferred language spoken at home, with those in the second generation (US-born children of immigrants) preferring it to their parent's native tongue. Monolingualism comes a bit later and tends to occur more quickly in Asian than in Latino groups. Most Asians can no longer speak their parent's languages well by the second generation, but Latinos, especially Mexicans, can speak it well into the third generation.

Language diversity: Endangered languages

As human populations migrated and settled across the globe, our languages diverged into myriad tongues. Throughout history, new languages have formed, evolving from parent languages through physical separation

and contact with other languages. This process continues today but with a significant difference: the rate of language extinction is accelerating rapidly. Modern transportation and communications link disparate communities like never before, encouraging the use of common languages with wider groups of people. While over 7,000 languages are spoken in the world today, it is estimated that roughly 3,000 are endangered, and there is a possibility that 50 to 90 percent of all languages could be lost in the next 100 years. Within North America, 312 languages were spoken when Europeans first arrived. Through genocide and assimilation, 40 percent of those languages have been lost, and only about twenty are currently being learned by children.

The risk factors for language extinction are small geographic ranges, being spoken by a small number of people, and a rapid decline in new speaker numbers (figure 10.16). Small populations living in clusters in the tropics, from Central America and the Amazon Basin through tropical Africa and into Southeast Asia, Northern Australia, and New Guinea, have large numbers of endangered languages. The mountainous region of the Himalayas also has a significant number of languages in danger of extinction. In North America, the native languages that remain are concentrated along the Pacific Coast, from California into Canada.

The loss of languages means a loss of culture and knowledge. Language and oral tradition are a common means of passing these on to future generations. The stories that are told, the histories of groups, the knowledge of local plants and animals for food and medicines, all pass from generation to generation orally in local languages. When these languages disappear, much of this information can be lost as well, as a more dominant language and culture supersedes it. As in the biological world, extinction implies the loss of a branch of our evolutionary history and information about how groups and places have interacted over time.

Language and landscapes

Language can play a role imbuing meaning to a place, contributing to its sense of place through *toponyms*.

Endangered Languages

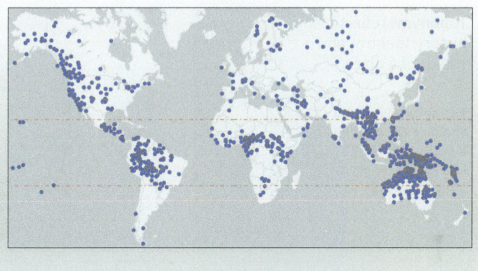

Risk Factors:

- Small area.
- Small population.
- Declining number of new speakers.

Major locations:

- Tropical zones.
- The mountainous Himalayas.
- Pacific coast of North America.

Figure 10.16. Endangered languages. Explore this map at https://arcg.is/0W1LbO. Data source: Catalogue of Endangered Languages. 2017.

Toponyms are essentially place-names, or the names that people give to features on the landscape, be it a city, mountain, river, school, street, library, or any other object commonly included on a map. Through toponyms, we can get clues as to the history of a place and the values of its residents. For instance, names such as Los Angeles and San Francisco reflect early Spanish settlement in California, while New York and New Hampshire, named after British cities, reflect settlement by people from the United Kingdom. Places with Hangman in their name can reflect a rough history in former mining boom towns, while Starvation can be found along historic migration and settlement routes. Also, places named after important historic figures illustrate who is valued in contemporary society, such as Washington, Lincoln, and other important politicians, civic leaders, artists, or generals.

But since toponyms reflect history and values, they can be controversial, often changing as new groups gain power. In Russia, Saint Petersburg was changed to Leningrad under the Soviet Union, only to be changed back to Saint Petersburg with the end of communism. In the colonized regions of Africa and Asia, many colonial names were replaced with local ones, so that Bombay became Mumbai in India and British Honduras became Belize. In Iraq, place-names went from those of the early-twentieth-century monarchy to those of the revolutionary Baath Party, then to many named after Saddam Hussein and, finally, to the removal of all signs of the fallen strongman.

Controversy is found in the United States as well, often driven by themes of race and ethnicity (figure 10.17). Over the years, many offensive names that include derogatory terms for African Americans, Jews, Native Americans, and others have been changed. Recently, controversy over Confederate place-names have grown, with many people saying it is offensive to commemorate Jefferson Davis, Robert E. Lee, and other leaders who defended or participated in slavery. Instead, many say it is more just to commemorate those who struggled against slavery and discrimination and that parks, schools, libraries, and other places should

Toponyms: Race and History

Toponyms reflect history, identity, and values of a place.

The battle over history and race can be seen in places named after **Robert E. Lee** and **Martin Luther King Jr.**

Figure 10.17. Toponyms: Race and history. Data source: US Board on Geographic Names.

reflect modern concepts of racial equality rather than historic oppression.

> *Go to ArcGIS Online and complete exercise 10.3: "Language assimilation of Asians and Latinos," and exercise 10.4: "Toponyms and sense of place."*

Religion

Along with language, religion is one of the most salient and significant elements of a culture. It forms a central part of many people's identity and can facilitate social cohesion within groups. It shapes how people view the world and afterworld and guides behavior. It can shape how people dress, what they eat, architectural styles, marriage and sex, and myriad other aspects of culture. In fact, religion is so universally tied to human culture that no society has been found that lacks some form of religious beliefs.

Geographers typically categorize religion as *universalizing* or *ethnic*. Ethnic religions are particular to a group of people, such as a tribal or ethnic group. They tend to form organically within a community and are not tied to a specific spiritual founder. Given that they are part of a specific cultural group, ethnic religions do not proselytize, or actively seeking new members from other groups. Rather, membership in an ethnic religion usually comes from being part of the ethnic group. Universalizing religions aspire to be accepted by all. They have a founder who played a key role in the origin and diffusion of the religion, and members actively proselytize to win over new adherents.

Spatial distributions and change over time
Distribution of universalizing and ethnic religions
Of the major world religions, only Buddhism, Christianity, and Islam are universalizing, whereby anyone from any background can join the faith.

As would be expected, these religions have expanded widely across the globe, dominating the religious landscape in terms of geographic coverage and number of adherents (figure 10.18 and table 10.1). Today, nearly 62 percent of the world's population belongs to one of these three universalizing religions.

Buddhism, founded roughly 2,500 years ago, is the oldest of the universalizing religions. It began in northern India, founded by Siddhartha Gautama, the Buddha. Through trade and missionary activity, Buddhism spread through Southeast and East Asia, reaching northward through Nepal and China to Mongolia, east to Japan, as well as south into India, Sri Lanka, and Indonesia.

Founded in the first century CE, Christianity began in the region of modern-day Israel with the teachings of Jesus of Nazareth. Missionaries spread Christianity through the region, but it expanded more widely when it became the official religion of the Roman Empire in the fourth century. From there it spread throughout Europe. As the religion of Europe, Christianity diffused globally through colonial conquest, as missionaries and migrants spread it through the Americas, into Africa, and around Pacific states such as Australia, New Zealand, and other smaller islands.

In the seventh century CE, Islam formed from the teachings of Muhammad in the area around Mecca and Medina in modern-day Saudi Arabia (figure 10.19). It spread across North Africa into Spain, through the Middle East, and into parts of South and Southeast Asia, such as India, Pakistan, Bangladesh, Malaysia, and Indonesia, among others.

Of the major ethnic religions, Hinduism dominates in size, with fewer adherents than Christianity and Islam but roughly twice as many as Buddhism. Hinduism developed in the Indus Valley region around modern-day Pakistan and India in the third to second centuries BCE. It is older than Buddhism and possibly

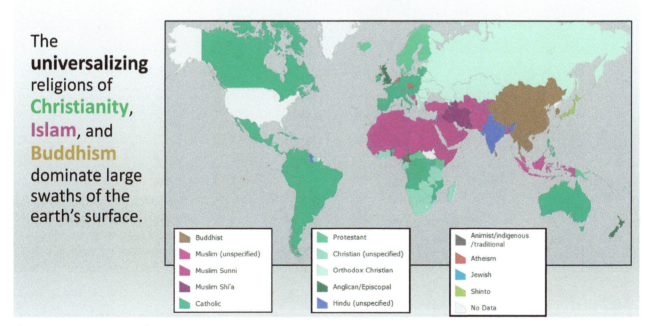

Figure 10.18. Dominant religion. Data source: The Association of Religion Data Archives.

	2010 Population	% of World Population
Christians	2,168,330,000	31.4
Muslims	1,599,700,000	23.2
Unaffiliated	1,131,150,000	16.4
Hindus	1,032,210,000	15.0
Buddhists	487,760,000	7.1
Folk (Animist) Religions	404,690,000	5.9
Other Religions	58,150,000	0.8
Jews	13,860,000	0.2

Table 10.1. Size of major religious groups, 2010. Data source: Pew Research Center, 2015.

Figure 10.19. Mecca, Saudi Arabia. All Muslims are supposed to make a pilgrimage to Mecca at least once in their lifetime. Photo by Zull Must. Stock photo ID: 590946917. Shutterstock.com.

the oldest religion on earth. It spread through South and Southeast Asia with migrants and traders and now dominates India and Nepal. Other substantial Hindu populations are found around the world in places where large numbers of South Asian migrants settled. These include places such as Suriname and Guyana in South America and Mauritius in Southern Africa.

After Hinduism, *animist* religions are the most commonly practiced of the ethnic religions and are practiced, for example, by Native Americans in the United States, aborigines in Australia, and countless other people on all continents. Animist religions view the world as inhabited by countless spiritual beings. These beings are not all-powerful deities, as in many other religions, but rather are tied to particular things and events. There are spirits in rocks, thunder, the sun, the moon, wind, trees, birds, even childbirth, and more. In addition, witches, ghosts, and monsters can be part of the animist belief system. The spirits can be benevolent or malevolent, helping or interfering with humans in their daily lives. Spiritual leaders, such as shamans, interact with the spirits, helping to bring good fortune or to pacify an angry spirit. Animist beliefs can be the dominant religion in some societies, while others blend them syncretically with other religions, be it with Christianity in Brazil or Buddhism in Vietnam and Mongolia.

Judaism is a smaller, though still significant, ethnic religion. Its origins are in the region around modern-day Israel, where Abraham, the founder of the Hebrew people, lived in the second century BCE. Various empires, from the Assyrians to the Romans and beyond, conquered the region, expelling the Jews and forcing them to emigrate to new lands. The Jewish diaspora settled throughout Europe and the Middle East and North Africa, then later immigrated to countries throughout the Americas. Today, there are nearly fourteen million Jews, of whom roughly six million live in Israel.

Future changes in religious composition

The religious composition of the world's population depends on two key variables: *population growth* and *conversion*. Religions in high-growth regions are likely to see their numbers increase, while those in low- and negative-growth regions should see their numbers decline. Likewise, religions that gain more converts than they lose will grow, while those that lose more members will shrink.

As you recall from chapter 2, the total fertility rate is a good indicator as to how fast a population is growing. Table 10.2 shows the total fertility rate by religion in recent years. Both Muslims and Christians have rates above the world average of 2.5, indicating that they are likely to constitute a greater share of followers. Buddhists and the unaffiliated (atheists, agnostics, and others) have rates well below replacement level, suggesting that their proportions will decline in coming years.

	TFR, 2010-15
Muslims	3.1
Christians	2.7
Hindus	2.4
Jews	2.3
Folk (Animist) Religions	1.8
Other Religions	1.7
Unaffiliated	1.7
Buddhists	1.6

Table 10.2. Total fertility rate by religion. Data source: Pew Research Center, 2015.

The variation in total fertility rates by religion can be explained by where each is concentrated. Islam is concentrated in North Africa, the Middle East, and parts of South and Southeast Asia. These regions have relatively high fertility rates, where the transition to urban society has not yet fully taken place and where opportunities for female education and employment is often constrained. The same holds true for many Christians. While low-fertility Europe and North America are Christian, high-fertility sub-Saharan Africa also has large Christian populations. South Asia, with its large Hindu population, also has high fertility rates, contributing to the likely growth of that religion. Then there are Buddhism and the unaffiliated. Buddhists are concentrated in East Asia, such as in China and Japan, where fertility rates are very low. Likewise, the unaffiliated, such as atheists (those who do not believe in a god) and agnostics (those who say it is impossible to know if there is a god), are found primarily in low-fertility regions such as North America, Europe, China, and Japan.

Conversion rates are difficult to predict, but based on past trends, some believe that Christians, Buddhists, and Jews will see a net loss of adherents in coming decades, while the unaffiliated, Muslims, and folk/animist religions will see a net gain.

Between population growth and conversions, the Muslim population is likely to grow the most by 2050, reaching a proportion similar to that of Christians (figure 10.20). These two religions will continue to be the most popular, constituting roughly 60 percent of the world's population between them. The proportion of unaffiliated will decline significantly, given that many in this group reside in low-growth regions such as North America, Europe, and East Asia. Buddhists and folk (animist) religions will decline proportionately as well, but to a lesser degree, while the other major religions will roughly maintain their share.

Religion in the United States

With settlement by European colonial powers, the United States formed as a Christian country, and to this day Christianity is the dominant faith. In 2014, 70.6 percent of the population identified as Christian, although that was down from 78.4 percent in 2007. Regionally, the South and Midwest are more Christian, while the West and Northeast are less so (figure 10.21). Non-Christian faiths make up 5.9 percent of the US population, with Jews being the most prominent, followed by Muslims, then Buddhists, then Hindus.

Beyond the population that identifies with a specific faith, the largest and most rapidly growing segment of the population are those who identify as unaffiliated. In 2014, 22.8 percent of the US population was unaffiliated with any specific religion. This included about 7 percent who were atheist or agnostic, but the largest group (15.8 percent) simply identified with "nothing in particular." The unaffiliated are found especially among the younger generations, indicating that organized religion may be losing its appeal with time. Nevertheless, Americans remain religious overall. Regardless of the faith people identify with (or do not), nearly 90 percent still report that they believe in God, a number much higher than in other advanced industrial countries.

Religious intolerance: Government and social restrictions on religion

Freedom to worship as one pleases is often taken for granted in the United States, where the right is protected by the First Amendment of the Constitution. However, that freedom is not universal, as various degrees of government preferences and restrictions influence which religions can or cannot be followed within their borders. At the same time, society can informally influence participation in religion. Regardless of government policies, some countries have social environments where people tolerate religious differences, while in others, people do not.

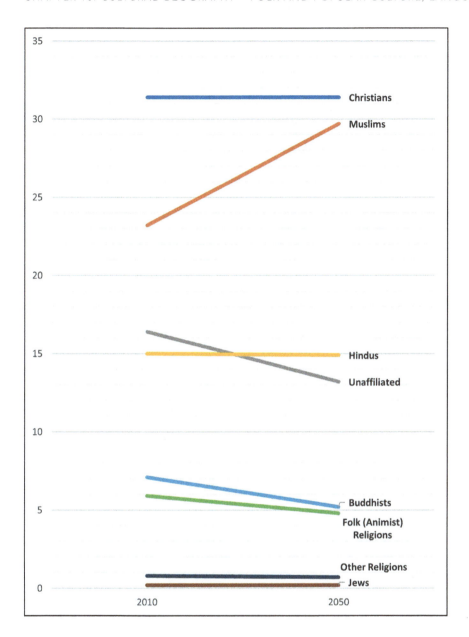

Figure 10.20. Projected change in religion: Percentage of global population. Data source: Pew Research Center, 2015.

Figure 10.22 maps countries by level of government regulation of religion in 2011. Some allow nearly unrestricted freedom of religion, while others have high levels of restrictions. Australia, for instance, protects religious freedom in its constitution. The same is true in other countries, such as Argentina and Norway. However, both Argentina and Norway give some preference to one religion, with the Catholic church receiving various types of subsidies in Argentina and

the state paying the salaries of Church of Norway employees.

Moving up one category toward less religious freedom, countries such as France and Nigeria stand out. France in recent years has enacted laws to prohibit the use of face coverings, such as the Islamic niqab, in public places and has banned conspicuous religious symbols, such as headscarves and Sikh turbans, in public schools. In Nigeria, freedom of religion is

Percentage of Adults Who Are Christian

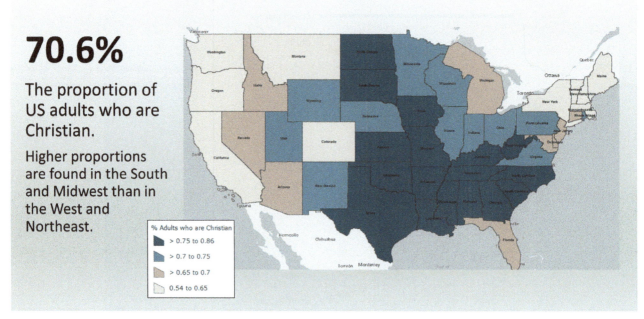

70.6%

The proportion of US adults who are Christian.

Higher proportions are found in the South and Midwest than in the West and Northeast.

Figure 10.21. Percentage of adults who are Christian, 2014. Data source: Pew Research Center, Religion and Public Life.

Government Regulation of Religion

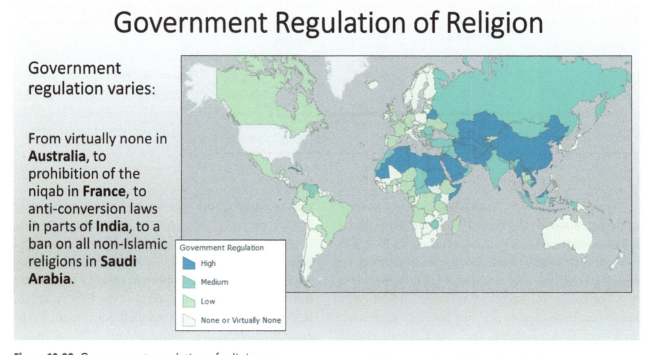

Government regulation varies:

From virtually none in **Australia**, to prohibition of the niqab in **France**, to anti-conversion laws in parts of **India**, to a ban on all non-Islamic religions in **Saudi Arabia**.

Figure 10.22. Government regulation of religion. Data source: Harris et al. 2011. Association of Religion Data Archives.

protected in the constitution, but several northern states have implemented Islamic-based Sharia law, using it to convict clerics and others of blasphemy for allegedly insulting the Islamic religion.

India and Iraq represent countries with medium levels of government restrictions. Religious freedom is protected in theory in India, but several states have anti-conversion laws that prevent people from changing religion. There are also numerous cases in which the government failed to protect people from religious persecution. For instance, "cow protection" groups of Hindus, for which the cow is sacred, have attacked Muslims and others accused of selling beef with impunity. In Iraq, Islam is the official religion, but the constitution guarantees freedom for Christians and a handful of other smaller religions. The country is dominated by the Shia branch of Islam, and the government regularly persecutes members of the Sunni Islam minority, using "religious profiling" in searches and arrests and discriminating against them in employment.

The greatest levels of government restriction on religion are concentrated in two areas: communist and former communist countries and countries of the Middle East and North Africa. The communist countries of Cuba, China, Vietnam, and North Korea, as well as several ex-Soviet republics, are actively hostile to all religion. In these places, religion is discouraged, although not necessarily banned. Religious leaders are often the targets of government surveillance, and proselytizing can be severely restricted. North Korea stands out as the worst in this group, essentially banning religious activity. Those accused of praying, singing hymns, or reading the Bible have been imprisoned, tortured, and executed. Even family members of those accused of being Christian have been imprisoned regardless of their personal beliefs. Anti-religious activity of the North Korean state may even take place beyond its borders; in 2016, a Christian pastor operating in China near the North Korean border was allegedly assassinated by North Korean agents.

Much of the Middle East and North Africa has high levels of government restriction on religion. Islam is the official religion of most states, enjoying a strong preference over other religions. It is often supported financially, with mosques and clerics receiving funds from the state. Non-Islamic religions, on the other hand, can face myriad restrictions, from outright bans to limits on construction of places of worship. Legal systems are often based on Sharia, or Islamic law, which holds that church and state are essentially one and the same.

Saudi Arabia stands out as one of the most religiously restrictive countries in the world. Its constitution is based on the Quran (Islamic holy text) and Sunna (teachings of Muhammad), and all non-Muslim religion is banned. Because it is a Sunni Islam state, in Saudi Arabia, even the Shia branch is persecuted, with discrimination against Shia Muslims in terms of education, employment, and more. It is illegal to violate Islamic law or insult Islam, and punishments include lashings, imprisonment, and in some cases, execution. Punishments have been imposed for posts on social media that promote atheism, witchcraft, sorcery, and other violations of Sharia law. Foreign workers are not exempt from strict religious laws. For instance, Lebanese Christians working in Saudi Arabia have been arrested and expelled for privately holding religious celebrations. The government of Saudi Arabia also promotes religious intolerance outside of its borders through active funding of ultraconservative and puritanical Wahhab Sunni teachings. By funding Wahhabist mosque construction and paying clerics' salaries in countries around the world, Saudi Arabia has spread its influence to places with traditions of tolerant and open Islam. In African states such as Mali and Niger, and in Southeast Asian countries such as Malaysia and Indonesia, many Muslims are shedding their traditional regional versions of Islam and replacing them with Saudi Arabia's conservative theology.

Iran stands out as another of the most religiously restrictive countries in the world. The Shia branch of Islam is its official religion, and its constitution stipulates that all laws must be based on Islamic criteria and official interpretation of sharia. While Zoroastrians,

Jews, and Christians are recognized as official religious minorities with the right to worship, their activities are severely restricted. Building permits for construction of places of worship by these religions are regularly denied, and there is discrimination in education and employment. The government frequently uses anti-Semitic rhetoric in official statements and promotes Holocaust denial. Proselytizing to convert Muslims to another religion, even those that are officially recognized, is strictly prohibited and even punishable by death. It is illegal for Muslims to renounce their faith or convert to another, and as a means of enforcement, identity checks are made at Christian worship services to find Muslim converts. Even Muslims who do not convert or renounce their faith are subject to punishment for violating the government's strict interpretation of religion. The minority of Iranian Muslims who practice Sunni Islam face raids and destruction of their places of worship as well as arbitrary arrests and physical abuse by security services. Muslims who do not fast as required during the holy month of Ramadan are subject to flogging. Even more severe is punishment for insulting Islam or the prophet Mohamed, which is punishable by death.

While government action or inaction can lead to religious intolerance, social intolerance can also come from individuals, organizations, and groups. Within Europe, intolerance has grown less from state action than from the people. Religious buildings such as Jewish synagogues and Islamic mosques have been vandalized, and assaults based on religion of the victim have increased. Organized marches by hate groups have been held, some by neo-Nazi groups rallying against Jews but also, increasingly, by those opposed to the rising number of Muslim immigrants on the continent.

Intolerance can come from all faiths. Hindu nationalists in India have led hundreds of attacks against Muslims and Christians. Cow protection groups have raped and killed Muslims accused of transporting or selling beef, while Christian priests and missionaries have been attacked and beaten by Hindu mobs (figure 10.23). Buddhists in Myanmar

Figure 10.23. Holy cows. Cow details at the Kapaleeswarar temple, Chennai, India. Cows are considered holy in the Hindu faith. Some militant Hindus attack those from other religions whom they accuse of selling beef. Photo by Katarina S. Stock photo ID: 261666224. Shutterstock.com.

have protested and encouraged violence against the Muslim Rohingya minority, preventing them from buying land in Buddhist villages and discouraging business and social interaction between the two groups. In Israel, ultra-Orthodox Jews have spat on other Jews, protested a Christian church choir at a mall, and discouraged ultra-Orthodox women from dating non-Jewish men.

Religiously motivated terrorism has killed and displaced millions of people across the globe. In countries such as Pakistan, Syria, and Iraq, Islamic State and other terrorist organizations have targeted Shia Muslims, Christians, and other religious faiths, killing them, forcing them to convert, or expelling them from homelands. Bombings at churches and mosques have been common, as have attacks on festivals, funerals, and minority neighborhoods.

References

Amano, Tatsuya, Brody Sandel, Heidi Eager, Edouard Bulteau, Jens-Christian Svenning, Bo Dalsgaard, Carsten Rahbek, Richard G. Davies, and William J. Sutherland. 2014. "Global Distribution and Drivers of Language Extinction

Risk." *Proceedings of the Royal Society B*, September 3, 2014. doi: https://doi.org/10.1098/rspb.2014.1574.

Ardila, A. 2005. "Spanglish: An Anglicized Spanish Dialect." *Hispanic Journal of Behavioral Sciences* 27, no 1: 60–81. doi: https://doi.org/10.1177/0739986304272358.

Ash, A. 2017. "Animal Spirits." *1843 Magazine*, August/September. https://www.1843magazine.com/features/animal-spirits.

Borzykowski, B. 2017. "The International Companies Using Only English." *Capital*, March 20, 2017. http://www.bbc.com/capital/story/20170317-the-international-companies-using-only-english.

Catalogue of Endangered Languages. 2017. University of Hawaii at Manoa. http://www.endangeredlanguages.com.

Chang, J. 2009. "It's a Hip-Hop World." *Foreign Policy*, October 12, 2009. http://foreignpolicy.com/2009/10/12/its-a-hip-hop-world.

Davis, E. H., and J. T. Morgan. 2005. "Collards in North Carolina." *Southeastern Geographer* 45, no. 1: 67–82. doi: https://doi.org/10.1353/sgo.2005.0004.

Dryer, Matthew S., and Martin Haspelmath (Eds.). 2013. *The World Atlas of Language Structures Online*. Leipzig, Germany: Max Planck Institute for Evolutionary Anthropology. Available online at http://wals.info.

Fischer, D. 1989. *Albion's Seed: Four British Folkways in America*. New York: Oxford University Press.

Frayer, L. 2017. "For Catalonia's Separatists, Language Is the Key to Identity." NPR, September 29, 2017. https://www.npr.org/sections/parallels/2017/09/29/554327011/for-catalonias-separatists-language-is-the-key-to-identity.

French, K. 2015. "Geography of American Rap: Rap Diffusion and Rap Centers." *GeoJournal* 82, no. 2: 259–72. doi: https://doi.org/10.1007/s10708-015-9681-z.

Glassie, H. 1986. "Eighteenth-Century Cultural Process in Delaware Valley Fold Building." In Upton, D., and J. M. Vlach (eds.), *Common Places: Readings in American Vernacular Architecture*. Athens: University of Georgia Press.

Gold, J. R., and G. Revill. 2006. "Gathering the Voices of the People? Cecil Sharp, Cultural Hybridity, and the Folk Music of Appalachia." *GeoJournal* 65, no. 1–2: 55–66. doi: https://doi.org/10.1007/s10708-006-0007-z.

Harris, J., R. Martin, S. Montminy, and R. Fink. 2011. "Cross-National Socio-Economic and Religion Data, 2011." Association of Religion Data Archives. http://www.thearda.com/Archive/Files/Descriptions/ECON11.asp.

Hickey, W. 2013. "22 Maps That Show How Americans Speak English Totally Differently From One Another." *Business Insider,* June 5, 2013. http://www.businessinsider.com/22-maps-that-show-the-deepest-linguistic-conflicts-in-america-2013-6/#the-pronunciation-of-caramel-starts-disregarding-vowels-once-you-go-west-of-the-ohio-river-1.

Hubka, T. 1995. "The American Ranch House: Traditional Design Method in Modern Popular Culture." *Traditional Dwellings and Settlements Review* 7, no. 1: 33–39.

Human Rights Watch. 1999. "Restrictions on the Use of the Kurdish Language." https://www.hrw.org/reports/1999/turkey/turkey993-08.htm.

———. 2015. "Group Denial: Repression of Kurdish Political and Cultural Rights in Syria." https://www.hrw.org/report/2009/11/26/group-denial/repression-kurdish-political-and-cultural-rights-syria.

Hookway, J. 2016. "Clairvoyants See Good Fortune in Capitalist Vietnam." *Wall Street Journal,* October 14, 2016. https://www.wsj.com/articles/clairvoyants-see-good-fortune-in-capitalist-vietnam-1476437581.

Joyce, A. 2012. "Disease Maps Pinpoint Origin of Indo-European Languages." *Scientific American,* August 23, 2012. https://www.scientificamerican

.com/article/disease-maps-pinpoint-origin-of-indo-european-languages.

Kniffen, F. 1965. "Folk Housing: Key to Diffusion." *Annals of the Association of American Geographers* 55, no. 4: 549–77. doi: https://doi.org/10.1111/j.1467-8306.1965.tb00535.x.

Liu, A. H., and A. E. Sokhey. 2014. "When and Why Do US States Make English Their Official Language?" *Washington Post,* June 18, 2014. https://www.washingtonpost.com/news/monkey-cage/wp/2014/06/18/when-and-why-do-u-s-states-make-english-their-official-language/?utm_term=.be8ef32e4298.

Meyer, H. A. E. 2000. *Manners and Customs of the Aborigines of the Encounter Bay Tribe, South Australia.* Blackwood, S. Australia: Second Edition History.

Mydans, S. 2007. "Across Cultures, English Is the Word." *New York Times*, April 9, 2007. http://www.nytimes.com/2007/04/09/world/asia/09iht-englede.1.5198685.html.

Price, M. 2003. "Middle East: Confusion on the Streets of Baghdad." *BBC News*, August 4, 2003. http://news.bbc.co.uk/2/hi/middle_east/3122729.stm.

Rumbaut, R. C. A. G., D. S. Massey, and F. D. Bean. 2006. "Linguistic Life Expectancies: Immigrant Language Retention in Southern California." *Population and Development Review* 32, no. 3: 447–60. doi: https://doi.org/10.1111/j.1728-4457.2006.00132.x.

Shortridge, B. G. 2010. "A Food Geography of The Great Plains." *Geographical Review* 93, no. 4: 507–29. doi: https://doi.org/10.1111/j.1931-0846.2003.tb00045.x.

Simons, Gary F., and Charles D. Fennig (eds.). 2017. *Ethnologue: Languages of the World, Twentieth Edition.* Dallas: SIL International. Available online at http://www.ethnologue.com.

Trofimov, Y. 2016. "Jihad Comes to Africa." *Wall Street Journal,* February 5, 2016. https://www.wsj.com/articles/jihad-comes-to-africa-1454693025.

Walker, M. A. 2013. "Border Food and Food on the Border: Meaning and Practice in Mexican Haute Cuisine." *Social & Cultural Geography* 14, no. 6: 649–67. doi: https://doi.org/10.1080/14649365.2013.800223.

Wallraff, B. 2014. "What Global Language?" *The Atlantic* (November). https://www.theatlantic.com/magazine/archive/2000/11/what-global-language/378425.

Woodard, C. 2013. "Up in Arms." *Tufts Magazine* (Fall). http://emerald.tufts.edu/alumni/magazine/fall2013/features/up-in-arms.html.

Wormald, B. 2015. *The Future of World Religions: Population Growth Projections, 2010–2050.* Pew Research Center's Religion & Public Life Project. http://www.pewforum.org/2015/04/02/religious-projections-2010-2050.

———. *US Public Becoming Less Religious.* Pew Research Center's Religion & Public Life Project. http://www.pewforum.org/2015/11/03/u-s-public-becoming-less-religious.

Zelinsky, Wilbur. 1980. "North America's Vernacular Regions." *Annals of the Association of American Geographers* 70, no. 1: 1–16. doi: https://doi.org/10.1111/j.1467-8306.1980.tb01293.x.

Chapter 11
Political geography

When we view maps, one of the most salient features we see are divisions into political units, be they countries, states, counties, cities, or neighborhoods. These are some of the most important divisions that influence our lives and how we interact with places. The country we live in influences economic opportunity and political rights. More locally, housing, entertainment, and many other elements of our lives are shaped by our cities and neighborhoods. One or more of these elements can also be an integral part of a person's identity, be it a patriotic American, a proud Texan, a die-hard Bostonian, or a through-and-through Brooklynite. Political geography is about how humans carve territory into distinct areas of control, such as these types of political units. It examines boundaries and borders and how groups exert control over space. It looks at the formation of political units, such as countries and voting districts, and how groups compete for power within and between them. It also studies the effects that political boundaries and control have on spatial interaction, such as in terms of political and economic cooperation and competition.

Political geography explores the division of space and its control at myriad scales, from the local to the global. Within cities, borders and boundaries can impact the quality of urban spaces, making some places more desirable to live in and visit and others less so. Political geography comes in many forms, from urban design to the presence of street gangs. Moving to a broader scale, political districts play a profound role in who governs the places where we live. Finally,

at smaller scales, the development of states, empires, and multinational organizations influences where power is centered and to which sets of laws people become subject. Political geography thus sheds light on myriad historical and contemporary issues, such as the formation of states and empires, underlying causes of civil wars, struggles for independence, battles over voting districts, and people's emotional ties to specific pieces of land.

Territoriality is the process of enforcing control over a geographic area. Arguably, this is one of the most basic of animal instincts, found in ants and in elephants and in just about everything in between. Of course, human culture complicates our natural instincts, but it is safe to say that territorial control for power over resources and defense is a key feature of human society. Through the control of territory, people can make use of natural resources, such as minerals and agriculture, devise policies that govern industrial and service sector employment, control space for defense against outsiders, set rules on acceptable cultural norms, and more.

Political geography at a local scale

Jane Jacobs's border vacuums
At an urban scale of analysis, borders and boundaries are often less apparent than at smaller scales. While many maps show county and city limits, other borders

exist that are less visible and widely acknowledged yet can still have profound impacts on how people relate to their cities and neighborhoods. The renowned urban writer Jane Jacobs describes development of what she calls *border vacuums*, places that suck vital urban life out of neighborhoods, leaving them stagnant and absent of vibrant street life. Control over these spaces is minimal, as limited use by people means few eyes on ground, limited informal supervision, and thus, often dangerous environments.

These types of borders are created by single massive land uses that stretch over multiple city blocks. Railroad tracks, large hospital and university complexes, civic centers, freeways, large parks, and industrial developments that inhibit the free flow of people through them divide the city, cutting off vital street life that makes cities interesting, productive, and safe (figure 11.1). Some borders of this sort inhibit the movement of people by preventing them from crossing, as with railways and highways. Others, such as a university campus, park, or civic center, have limited pedestrian activity at night. On the other hand, a municipal concert hall may be used at night but be mostly vacant by day (figure 11.2). In any case, street life is sucked away by these border vacuums.

Figure 11.2. Border vacuum 2. Large block-long developments, such as concert halls, discourage street activity except for limited performance times. This creates a minimally used border that can divide neighborhoods. Photo by Dedo Luka. Stock photo ID: 655184335. Shutterstock.

Border vacuums lie in contrast to diverse city blocks, with regular intersections that allow people to connect with other parts of the city and a variety of businesses. When people are passing through places and visiting businesses at different hours of the day, street life is more interesting and safety is enhanced through the informal control of space by myriad pedestrians (figure 11.3).

Figure 11.1. Border vacuum 1. A freeway underpass makes for an uncomfortable walk. This creates a border with the urban fabric of the city. Photo by Maria Koriakovtseva. Stock photo ID: 662661706. Shutterstock.

Figure 11.3. A vibrant city street in Chicago. Blocks with regular intersections and diverse businesses tie the city together, promoting movement and social interaction. Photo by Lissandra Melo. Stock photo ID: 129054737. Shutterstock.com.

Street gangs

Invisible borders and boundaries in urban areas are also created by street gangs that carve up and control space for economic activity, safety, and pride. For street gangs, territory is a key component of identity and opportunity. Within a gang's territory, it can control illicit activity, such as drug sales, extortion, and prostitution. It also provides safety for gang members so they can move relatively freely within their territory knowing that members defend the area against intrusion from rival gangs. Gang members also see their territory with a sense of pride. Being from a particular neighborhood infers membership in a social support network with other gang members. In fact, the term used for gang and the term used for neighborhood can be the same, as when a Latino gang member talks about both as "mi barrio."

Gang borders often lie along border vacuum areas identified by Jacobs (figure 11.4). In Los Angeles, for example, many borders follow the Los Angeles River, freeways, large industrial blocks, and other urban features that are rarely transited by the population at large. Gangs on either side of these places rarely interact, meaning that violence between them is minimal. Danger lies with invisible borders, regular city streets that serve as contested lines of gang demarcation. It is at these places, where people move freely as they navigate the city and live their lives, that violence can be most intense. Since these borders are not clearly fixed by physical features, gangs can try to push them outward and expand their territorial control. Graffiti can be used to mark gang territory, and rival gangs often write over each other's markings in contested border areas. Violence, such as shootings and other assaults, occur more frequently in these contested border areas than within secure gang territory.

Electoral geography

Political power in the United States is tied much closer to geography than to the popular will, given that the Electoral College ensures that presidents are elected on the basis of state results, not national ones. For this reason, the presidential candidate with the most total votes lost the election five times in US history, twice since the year 2000.

But geography is even more important for other political offices. Seats for the US House of Representatives, as well as state-level legislative positions, are based on political districts that divide states into groups of voters. But there are no fixed rules as to how these districts are drawn, meaning that battles over their borders can be nearly as intense (although much less violent) as border conflicts between street gangs. Interestingly, the art and science of drawing political districts can be more important than any candidate or political platform for winning seats.

The reason for this is that people of different political parties are not randomly distributed throughout cities and across states. Rather, as you have seen in this book, people tend to cluster together on the basis of common characteristics. Support of political parties varies depending on combinations of different socioeconomic factors, such as race and ethnicity, level of education, religious beliefs, income, and more. Democrats tend to cluster in certain neighborhoods, and Republicans cluster in others. Therefore, if political districts can be drawn that bring lots of Democratic neighborhoods together, then that party is likely to win more elections. The same holds true when Republican neighborhoods are combined in their own district.

Gerrymandering is the term used to describe the political partisan drawing of electoral districts. The term was coined in 1812 when Massachusetts Governor Elbridge Gerry redrew state senate districts to link together a string of communities in his favor. The shape of the district was so odd that some said it resembled a salamander, thus the term *gerrymander* (figure 11.5). Since then, political districts with unusual shapes that favor one group over another have been said to be gerrymandered.

Two of the most common means of gerrymandering are *packing* and *cracking*. Packing involves drawing lines around communities that are likely to

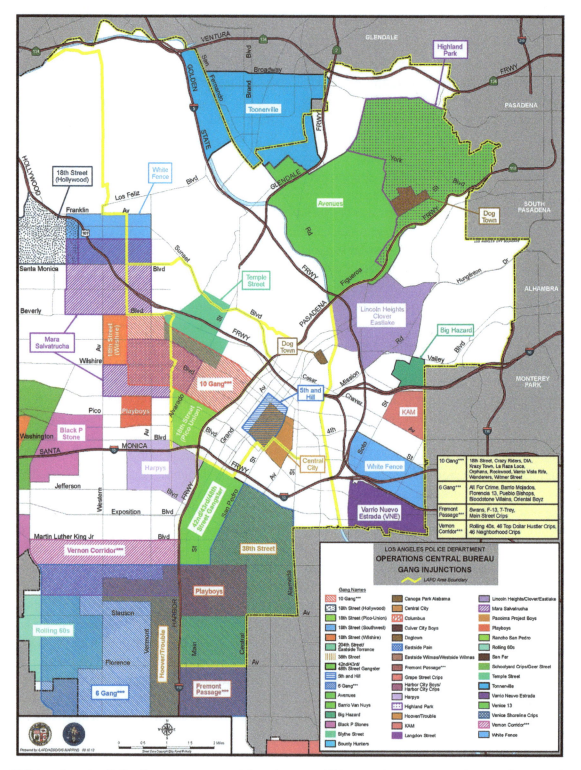

Figure 11.4. Los Angeles Police Department gang injunction map. This map shows gang territory in the central area of Los Angeles. Note how freeways can function as border vacuum areas for gangs, inhibiting interaction between them.

Data source: Los Angeles Police Department.

Figure 11.5. The original gerrymander. The 1812 state senate district drawn by Massachusetts Governor Elbridge Gerry was so oddly shaped it was said to resemble a salamander. Thus, the name *gerrymander*. Image by Elkanah Tisdale; drawing first appeared in the Boston Gazette, 1812.

vote disproportionately for one party or the other (figure 11.6). When likely Republican voters are packed into a district, then Republican candidates are likely to win there. The same holds true for Democrats. If one party dominates a state, packing in enough districts ensures that one party wins control of the majority of seats, be it in the US House of Representatives or in state legislatures. On the other hand, packing can lump together most voters of a minority party, ensuring that they win only a minority of seats and are unable to exert political power. Cracking involves dividing people with similar political views into different districts, thus diluting their voting power and ensuring that the opposing political party wins.

The US Supreme court has ruled that gerrymandering based on race is illegal. Thus, parties cannot draw boundaries to pack or crack racial minorities. However, politically based gerrymandering has not been deemed illegal, and some argue that it is a fair

weapon to use in the ultra-competitive world of politics. When a party wins enough power at the state level, some believe it should redraw boundaries to benefit it in future elections. The future of political gerrymandering was being contested in 2017, with the US Supreme Court evaluating its legality.

Given the persistence of gerrymandering in US politics, what have the results been? Often, the popular vote for one political party has little to do with the number of seats that it controls. For instance, in 2012, New York Democrats won 66 percent of the vote for the House of Representatives. However, due to House district boundaries, they won 78 percent of the seats. Thus, the Democrats' political power in the House was larger than if seats were allocated according to the popular vote. But then, in Pennsylvania, the Republicans held the advantage. There, in 2012, Democrats won 51 percent of the popular vote but held only 28 percent of House seats; over half of the

Electoral Geography: Gerrymandering

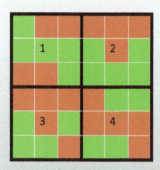
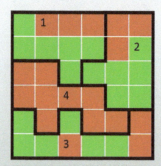
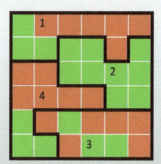

- Four competitive districts.
- Each has 4 to 5 of each party.

- Gerrymandered for the green party.
- Districts 1, 2, and 3: 6 greens to 3 oranges.

- Gerrymandered for the orange party.
- Districts 1, 3, and 4: 6 oranges to 3 greens.

Figure 11.6. Electoral geography: Gerrymandering. Image by author.

state voted Democrat, but the party controlled less than one-third of House seats.

Another impact of gerrymandering is more extreme with uncompromising political candidates. Landslide victories are more common in autocratic societies than in true democracies. This means that for a political candidate, the greatest threat comes not from the opposing political party but from a primary challenger from the same party. A centrist Democrat or Republican, one who understands the concerns of voters of all political stripes and is willing to compromise to pass legislation, can be outflanked by a candidate from the same party who promises to stick to the party line and take no political prisoners. Given that primary elections tend to draw more partisan voters, the more extreme candidate is likely to win. There is no incentive to win the votes of people from the other party when their numbers are miniscule in your district. Their votes just don't matter. And in fact, many political-minority voters in gerrymandered districts know this and do not even bother to vote, reducing political participation and

leading to greater voter cynicism that the system is rigged. The end result is representatives who promise to legislate only along party lines. Compromise is seen as surrender, radical political views are held, and legislative inaction becomes the norm.

> *Go to ArcGIS Online to complete exercise 11.1: "How gerrymandered is your congressional district?"*

States and nations: Spatial distributions at a global scale

Moving to a broader scale, political control of territory has evolved into what we now call *states*. All the earth's land, except for the internationally managed Antarctica, is now under sovereign control of a state. But this was not always the case. For most of human history, small tribes and clans controlled areas of land without fixed borders. People moved relatively freely, and many places were not under regular control by anyone. With time, however, the advantages of uniting into larger political units became clear. A more centralized political hierarchy could marshal economic and military resources for defense, conquest, and economic growth. People would pay taxes to the centralized government in exchange for safety. These revenues could then be used to finance armies, which defended territory from outside invaders and conquered and incorporated new lands. Revenues could also be used for economy-enhancing infrastructure, such as large irrigation systems for agriculture and road networks for trade.

The *city-state* was the earliest type of state, consisting of a sovereign area with an urban core and surrounding farmland. Around 4000 BCE, city states first appeared in Mesopotamia, within modern-day Iraq. The benefits of this type of political control of territory proved so useful that it appeared independently in many other places as well. Tenochtitlan, at modern-day Mexico City, formed as a city-state, while numerous Mayan city states also developed in Mesoamerica. Athens and other city-states formed around the Mediterranean.

Over time, some city-states gained strength and expanded their influence over wider areas. This led to the formation of the next level of state, the *empire*. Empires exerted political and economic control over multiethnic territories well beyond their core urban area. These, too, formed independently in different areas, including the Aztecs and Incas in the Americas, the Romans around the Mediterranean, the Persians and Ottomans in parts of Asia and the Middle East, and more.

While there is no single definition, the modern state generally possesses several distinctive features. First, it has a territory with fixed boundaries. Within that territory, a government administrative apparatus manages and controls activity. Two of the most important of these activities are the ability to raise tax revenues and a monopoly on the use of force. Through taxes, the state provides security and facilitates economic

growth. A monopoly on violence means that only the state has the right to enforce rules, including the right to detain and punish offenders. No other individual or group has the right to use violence, and those who do face sanction by the state. In the common vernacular, the term *state* is typically synonymous with *country*.

The concept of the modern state is often seen as originating in Europe after the Middle Ages. From there, it diffused worldwide, initially via relocation diffusion with European colonialism in the Americas, Africa, and Asia, and then through contagious diffusion as additional territories near colonial states consolidated. Thus, by the end of the twentieth century, the state had become the nearly universal form of political organization.

When discussing the formation of states, it is essential to differentiate the idea from a *nation*. A nation refers to a group of people with a common history, culture, religion, language, or homeland. The people of a nation have a common identity and view themselves as part of a single, distinct group. This contrasts with a state, which is a political entity that may or may not have a citizenry with these common characteristics. A nation may lie completely within a state or be split among different states. These terms are somewhat confusing given the way in which we commonly use them; the United Nations is actually an organization consisting of many states, not nations.

If a state is a political unit with a government and fixed territory, and a nation is synonymous with a unified cultural group, then a *nation-state* is a place where cultural and political boundaries are largely one and the same. It is a place where a people exercise sovereign self-government without the imposition of power by another state. In its strictest interpretation, the nation-state consists of a single homogenous people living within a single state. However, this situation is rare. States encompass diverse populations, and perfectly aligned borders that correspond with both a state and single cultural group exist in theory more than in reality. A handful of countries come close to meeting this definition; examples are North and South

Korea, where nearly all residents are Korean, and Egypt, where nearly everyone is Egyptian. In the case of pure nation-states, the people have long histories of occupying the land, speaking the same language, practicing the same religion, and following the same customs. They are united, and their national identity matches that of their state.

More commonly, nation-states are actively created. By promoting a national language, common literature, music and folklore, historic events and personalities, and more, countries build patriotism and national pride that tie people together. This is often done through the educational system, where national history is taught that stresses the commonalities of people within a country and builds solidarity. Most Spaniards, French, Americans, Mexicans, Brazilians, and others will identify themselves as members of their corresponding nation-state. It is typically their highest-level identity, even if they have subidentities tied to race, ethnicity, religion, social class, or some other group (figure 11.7).

But while nation-states with diverse populations attempt to build national unity and pride, tensions can exist when strong national identities are held among minority groups. *Multinational states* are states with

Figure 11.7. Mexico's Revolution Day. National unity is created and reinforced through festivals and the teaching of historical events. Photo by Byelikova Oksana. Stock photo ID: 523997776. Shutterstock.com.

two or more national identities. Often, but not always, these national identities can be superseded by an overarching identity with the nation-state. In the Americas, for instance, serious movements for national sovereignty among minority groups are limited, yet still exist. Indigenous groups from Chile and Argentina in the south, through Central America and Mexico, and into the United States and Canada periodically form movements for more autonomy in their ancestral lands. In Europe, Spanish identity is challenged by the Basque and Catalan people, who have distinct languages and desires for independence. Aside from Egypt, in much of the Middle East and North Africa, nation-states struggle to maintain cohesion among people who identify more with their ethnic or religious group than with the state in which they live. Nation-states are further challenged by multistate nations, where the traditional homeland of a people is split between different countries. The Kurdish people, for instance, occupy parts of Turkey, Syria, Iraq, and Iran. Likewise, the Basque people are split between Spain and France.

In some cases, a people do not belong to any state, making them a *stateless people*. Some argue that the Kurds are a stateless people because they do not have their own state of Kurdistan. However, nearly all do have citizenship in their country of residence, be it Turkey, Iran, Iraq, or Syria. The Rohingya people in Myanmar, on the other hand, are one of the largest truly stateless people (figure 11.8). They are a Muslim minority living in Buddhist-dominated Myanmar who have been denied citizenship for generations. Their lack of citizenship results in discrimination, lack of government services, and in 2017, brutal forced eviction into Bangladesh at the hands of the Myanmar military. The Haitians in the Dominican Republic are another example of a people left stateless. In 2013, the Constitutional Court of the Dominican Republic ruled that those born in the country to undocumented parents were not entitled to Dominican citizenship. What made this decision especially draconian is that it was applied retroactively to 1929, leaving thousands of people with Haitian descent instantly stateless. This ruling deprived them of the right to work, travel, own land, attend school, and receive health care. Some migrated to Haiti, lacking official documents in both countries, while others continue to risk deportation or violence from vigilante groups as they stay in the Dominican Republic.

Figure 11.8. Stateless people: The Rohingya. Members of Myanmar's Muslim Rohingya minority in Bangladesh. In 2017, thousands were violently forced from their homes in Myanmar, where they have lived for generations despite having no citizenship rights. Photo by Sk Hasan Ali. Stock photo ID: 714606577. Shutterstock.com.

State borders and boundaries

One of the essential components of the nation-state is fixed territorial borders. Because of this, state borders are the most commonly recognized boundaries in the world, subject to international law that recognizes the sovereignty of territory within them and attempts to adjudicate their location when disputes arise. Their level of permeability varies significantly; in some cases, the flow of people and goods through them is extremely fluid, while in other cases, it is nearly impossible. Thus, borders can be points of tension between restricting flows for security and facilitating flows for economic opportunity. For instance, goods and people can flow relatively easily through most of Europe as part of its integration since the end of World War II. In contrast, waits to cross the border from Mexico to the United States can take hours as US Customs agents interview and inspect all people and vehicles entering the country.

While the border between the United States and Mexico is hardened to keep out contraband and unauthorized immigrants, other borders are hardened due to animosity between neighboring states. This can be seen at border crossings between India and Pakistan as well as at the most secure border in the world, that between North and South Korea (figures 11.9 and 11.10).

Figure 11.9. The India-Pakistan border. Ongoing mistrust and rivalry between these two countries has resulted in a heavily fortified border. Photo by Pavel Chepelev. Stock photo ID: 760379995. Shutterstock.com.

Figure 11.10. The South Korea-North Korea border. This is arguably the most heavily fortified border in the world. The demilitarized zone is four kilometers wide, with US and Korean troops on the south and North Korean troops to the north. Land mines, watch towers, razor wire, and electrified fences divide the two countries. Photo by Meunierd. Stock photo ID: 135695417. Shutterstock.com.

Borders have been described in various ways. Those that existed prior to development of most of the cultural landscape are known as *antecedent boundaries*. Often, these are based on physical boundaries. Clearly, the great oceans have acted as physical boundaries that predate human settlement, but other physical features can also divide people and states. The Andes Mountains have separated the people of South America for centuries and now form the border between states such as Argentina and Chile. The Saharan Desert has also served as a boundary between the Arab states and cultures of North Africa and the people of sub-Saharan Africa.

While there are numerous examples of antecedent boundaries, most borders today can be described as *subsequent* in that they were drawn after human settlement. As nation-states replaced earlier city-states and empires, their borders were drawn with the goal of encompassing people with common national identities. Sometimes, physical boundaries served as natural places for borders, such as the Rio Grande between Mexico and the United States, and the Pyrenees Mountains between Spain and France. Other times, borders are based on cultural boundaries. The modern borders of Poland, for instance, encompass a population that is over 96 percent Polish, with other groups such as Germans and Ukrainians largely in their own nation-states on either side. Likewise, Czechia and Slovakia are divided along ethnolinguistic lines.

In addition to physical and cultural borders, some are *geometric*. Geometric borders are those drawn on a map that do not account for physical or cultural features on the ground. Rather, they form straight lines and sharp angles that cut across the landscape. One of the first geometric boundaries was drawn with the Treaty of Tordesillas in the late fifteenth century. This treaty was an agreement between Spain and Portugal that divided the New World along a line from north to south at about 46 degrees longitude. All land to the west was to be under Spanish control, and all land to the east was to be Portuguese (figure 11.11). While the

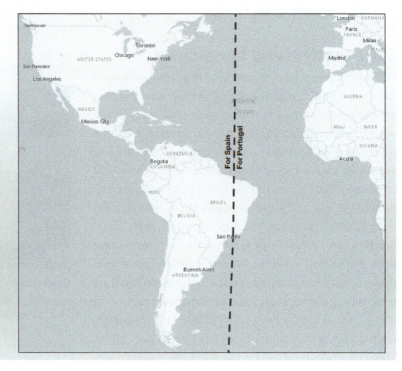

The Treaty of Tordesillas

- An early example of a **geometric boundary**.
- 15th century Spain and Portugal agreed to divide the New World near 46 degrees longitude.
- This set the stage for Brazil to be under Portuguese control.
- Spain gained control of most of the rest of Latin America.

Figure 11.11. The Treaty of Tordesillas. Data sources: Esri, HERE, Garmin, NGA, USGS.

line did not stick as a boundary, it did set the stage for Portugal to control the eastern portion of South America now known as Brazil, while Spain gained all Middle and South American lands to the west.

Another geometric border lies at forty-nine degrees latitude between the United States and Canada. This border was established in the nineteenth century in uncharted lands of the western half of North America, running from Minnesota to Washington. Conflicts between natural physical boundaries and geometric boundaries are seen in a couple of locations, where small protrusions of land are attached to Canada but lie in US territory south of the forty-ninth parallel (figure 11.12).

Superimposed borders are those drawn on the landscape by outside powers. They often follow physical features and geometric lines. In theory, they can also follow cultural boundaries, but in reality, they tend not to. Certainly, borders such as those under the Treaty of Tordesillas and along the forty-ninth parallel were superimposed in the Western Hemisphere with no consideration made for the Native American nations that existed throughout the region.

Superimposed borders dominate the regions colonized by European powers and can be clearly seen in much of the Middle East and Africa. There, land was divided among European states searching for raw materials to feed their industrial growth, with consideration of local cultural groups an afterthought at best. In Africa, formal boundaries were negotiated by Great Britain, Germany, Belgium, France, and Portugal at the 1884 Berlin Conference (figure 11.13). This conference, which took place in Berlin, Germany, clarified boundaries that were being superimposed on Africa as part of Europe's Scramble for Africa. Of course, no African leaders were invited to this conference. Boundaries split some unified African cultural groups into different colonial territories and combined some

Geometric Border: 49ᵗʰ Parallel

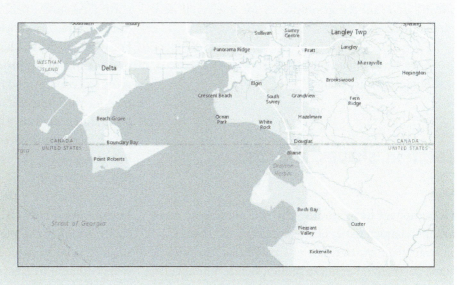

- Geometric boundaries ignore the physical and cultural landscape.

- Many US students in Point Roberts must take a 40-minute bus ride through Canada to reach their school in Blaine, WA.

Figure 11.12. Geometric border: 49th Parallel. Data sources: Esri, HERE, Garmin, NGA, USGS, NPS.

The Berlin Conference and the Political Division of Africa

- **European colonial powers carved up Africa.**
- **Even as of 2013, many ethnic groups are split between different states.**
- **While many others are combined in a single state.**
- **All too often this has resulted in conflict.**

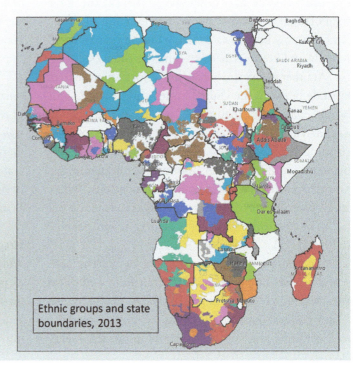

Ethnic groups and state boundaries, 2013

Figure 11.13. The Berlin Conference and the Political Division of Africa. Explore this map at https://arcg.is/j5zmW. Data source: Vogt, Manuel et al 2015 and Wucherpfennig et al, 2011.

rivals into a single territory. This disrupted African society and laid the groundwork for later civil and interstate conflict.

> *Go to ArcGIS Online to complete exercise 11.2: "Borders and boundaries."*

Location, shapes, and sizes

Size

Geographers have studied the role that basic geographic characteristics of size, location, and shape play in state viability. In terms of size, Russia, the largest state, covers over seventeen million square miles, while tiny Monaco, the smallest, covers just two square miles (table 11.1). The size of a state can potentially impact its economic viability in a couple of ways.

First, size can influence the amount and diversity of natural resources available for economic development. Larger states are more likely to have more of these. For instance, Russia has oil, natural gas, a wide range of minerals, wood products, and agricultural land to grow its economy and feed its population. The same is true for the United States, Canada, Brazil, and other large states. On the other hand, small states can have very limited natural resources. For instance, Monaco has none, while the island state of Tuvalu has only fish and coconuts.

Second, size can impact ease of self-defense. During World War II, Nazi Germany was able to easily conquer the majority of Europe, which consisted of small- and medium-sized countries. However, when it came to Russia (which at the time was the core of the even larger Soviet Union), the Nazis met their match. Russia's larger population meant that it could fight a war of attrition, sending millions to

State Size: Large to Small

Some of Largest Countries	Sq. Km.	Some of Smallest Countries	Sq. Km.
Russia	17,098,242	Monaco	2
Canada	9,984,670	Tuvalu	26
United States	9,833,517	Bermuda	54
China	9,596,960	Liechtenstein	160
Brazil	8,515,770	Singapore	697

Table 11.1. State size: Large to small. Data source: CIA Factbook.

their deaths as they held the line again Nazi offenses. At the same time, its size allowed for its industrial production facilities to be moved eastward into the Volga region, the Urals region, and Siberia, beyond the reach of German forces. This allowed the Soviets to maintain industrial production to supply weapons and equipment to its armed forces.

But while size matters sometimes, it is not always the case, as some microstates have highly successful economies. While Monaco has no natural resources, it has one of the highest per capita incomes in the world, based on banking, insurance, and tourism. The same is true for Singapore, which focuses on global trade, business, and finance, and for other small, resource poor yet affluent states such as Lichtenstein and Bermuda. In all of these cases, a lack of size and natural resources was compensated with development of human resources through education and ties to the global economy.

Location

A key locational difference lies between states that are landlocked and those that are not. Landlocked countries are those with no access to open ocean. In a global economy that relies heavily on oceangoing trade, this can present a massive disadvantage. There are forty-five of these countries in the world, and most of them are poor (figure 11.14). In fact, ten of the twenty lowest-scoring states on the 2016 Human Development Report are landlocked. While many of these are poor because they are located in sub-Saharan Africa, they still have even lower GDP per capita levels than adjacent ocean-facing neighbors.

The reason landlocked countries typically have lower levels of development is that trade costs are higher without direct ocean access. This happens in several ways. First, a landlocked country must rely on transport infrastructure of a neighboring country. Goods must be moved to port along highway or rail,

Landlocked Countries

- Landlocked countries generally have **lower** per capita **incomes** than those with sea access.

- This is especially true for landlocked developing countries.

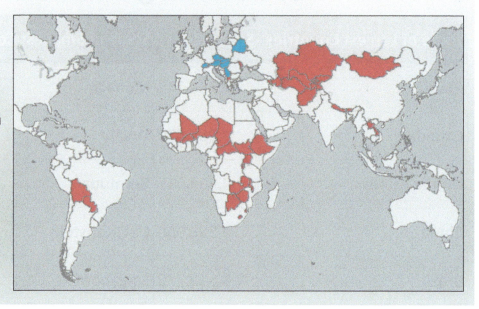

Figure 11.14. Landlocked countries. Explore this map at https://arcg.is/Xrb0y. Data source: World Bank.

but there is no incentive for the transit neighbor to improve those infrastructures merely to benefit the landlocked country. Then there are bribes and delays associated with crossing borders. Corrupt customs officers at ports and border crossings can demand bribes in exchange for goods passing through. And even when agents follow the rules, there is more red tape in the form of documents and clearances to move goods from port to a landlocked country. This both slows the flow of imports and exports and increases their costs. Ultimately, these barriers to trade limit investment, since companies are reluctant to put resources into places where the costs of moving goods will be higher. Furthermore, with fewer connections to the wider world and lighter flows of goods, people, and ideas, innovation tends to be limited in landlocked states.

Only a handful of landlocked states are highly developed, and all of them lie in Europe. Switzerland, Austria, Hungary, and others do just fine despite their lack of ocean access. This is due to the high quality of transportation infrastructure within Europe; a lack of corruption, and affluent, well-integrated markets within the region that landlocked countries can interact with.

Shape

Just as states come in all sizes, they come in all shapes as well. Some are compact, taking the shape of something resembling a circle. France, Uruguay, and Ethiopia have this form (figure 11.15). In theory, a *compact state* is easier to administer, since transportation linkages and administration from a central capital city can reach surrounding territory in all directions. *Elongated states*, in contrast, are long and thin. Chile is the most extreme example of an elongated state, stretching over 2,500 miles from north to south, with an average width of only about 100 miles. Panama and Norway are also described in this way. Unlike compact states, elongated state can be difficult to administer, since territories at each extreme can lie far from capital cities. Another

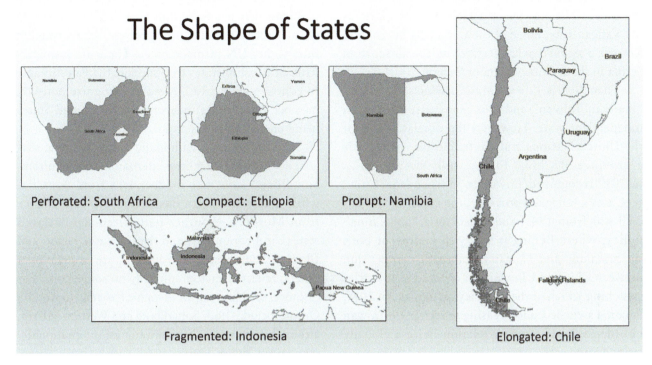

Figure 11.15. The shape of states. Data sources: Esri, HERE, Garmin, NGA, USGS.

type of state that can be difficult to administer is the fragmented state. This type of state is broken into many pieces, such as the multi-island states of Indonesia, with over 13,000 individual islands, and the Philippines, which consists of over 7,000 islands. *Perforated states* have "holes" in them that are occupied by other states. South Africa is a perforated state, encompassing the small, diamond-rich state of Lesotho. While this shape is not typically a problem for the perforated state, it can be for the smaller enclosed state, which faces the challenges of being landlocked.

The fifth common form for states is *prorupted* or *protruded*. These states have small pieces of territory that protrude out from the main mass of land like small arms. Thailand is a prorupted state, with a long southern protrusion along the Malay Peninsula. Often, the reason behind a prorupted state is to allow a small piece of land to access an important geographic feature. Namibia, for instance, has the Caprivi Strip, which reaches eastward to the Zambezi River. It was

incorrectly thought that the Zambezi River was navigable and would provide access to the eastern coast of Africa. However, it was later discovered that downriver were the Victoria Falls, making navigation impossible. In the case of the Democratic Republic of Congo, a small stretch of land reaches west, giving access to the Atlantic Ocean.

Go to ArcGIS Online to complete exercise 11.3: "Location: The economic consequences of being landlocked."

Number of countries in the world

After reading about the characteristics of states, you may ask, so how many countries are there? It seems like a straightforward question, but in reality, it comes down to who you ask. There are 193 member states of the United Nations, which many view as the official countries of the world. However, not all agree.

The United States recognizes 195 states, those of the United Nations, plus the Holy See, which holds the Vatican City, home of the Catholic Church, and Kosovo, a region that has declared independence from Serbia in the Balkans region of Europe.

Then there is Taiwan. Many products are labeled "made in Taiwan" and thus many assume it is an independent state. However, the official stance of the United Nations and United States is that it is a province of China. In fact, only twenty states in 2017 recognized Taiwan as independent. China sees it as a renegade province, the result of China's civil war from 1949, and insists on a "one China" policy, where Beijing is solely in charge of both the mainland and Taiwan. Diplomatic competition between China and Taiwan has resulted in countries switching which of the two they recognize. In 2017, Panama switched sides, ceasing to recognize Taiwan in order to build stronger relations with a globally rising China.

Recognition of Israel and Palestine is another area of intense debate among states. Israel, a member state of the United Nations, is not recognized by roughly thirty other UN member states. These are primarily Arab or Muslim states that support the Palestinians. In contrast, over 130 UN states recognize non-UN member Palestine as independent, the United States being one notable exception.

The list goes on, with many places claiming sovereignty but receiving little international recognition. Somaliland says it is independent from Somalia, while Western Sahara has declared independence from Morocco. Another half dozen self-declared states can be found around the Caucuses region and Ukraine. Some cases for sovereignty are stronger than others, reflected cartographically on some maps. For instance, the National Geographic basemap in ArcGIS Online includes both Somaliland and Western Sahara, although with notes that they are not internationally recognized (figure 11.16).

Self-Proclaimed States: Western Sahara and Somaliland

- **Neither is internationally recognized, yet their claims are significant enough to be shown on some maps.**

Figure 11.16. Self-proclaimed states: Western Sahara and Somaliland. Data sources: National Geographic, Esri, DeLorme, HERE, UNEP-WCMC, USGS, NASA, ESA, METI, NRCAN, GEBCO, NOAA, increment P Corp.

Exercise of power within states

Unitary and federal states

One of the key defining points of the modern nation-state is that the government regulates all activity within its formally recognized borders. This is often described as a monopoly on violence, where only the state has the right to enforce rules and exact punishment. Therefore, governments organize political control within their borders into two general categories: *unitary states* and *federal states*.

Most countries are unitary states, where power is delegated from a central government. In many cases, regulations and policies are made centrally and apply to all national territory. This can include everything from education standards, environmental regulations, labor rules, tax policy, and more. In some cases, power can be devolved to a subnational regional level, but the central government retains the right to revoke those rights. Regional political leaders can be appointed by the central government or elected to office by the local population. The United Kingdom's central government in London allows for substantial autonomy in Scotland, Wales, and North Ireland. Each of these countries has its own elected parliament or assembly and regulates a wide range of activities, including education, health and welfare policy, transportation, agriculture, and more. The British central government, in turn, retains control over issues such as immigration, foreign policy, and taxes.

A smaller number of countries are organized as federal states. In federal systems, there is a sharing of power between a national government and subnational governments. Unlike in unitary states where some regional autonomy is allowed, in federal systems, the division of power is codified as a permanent relationship that cannot be revoked by the national government. Typically, federal states are larger and/or multiethnic. For instance, seven of the eight largest countries in the world are organized as federal systems: Argentina, Australia, Brazil, Canada, India, Russia, and the United States. China is the exception in this group. By establishing subnational power structures, large and diverse areas are more easily managed. Diverse cultures, be they linguistic, religious, ethnic, or something else, are less likely to feel that a central authority is interfering with their rights. Switzerland, for instance, is a federal state divided into twenty-six cantons, each with its own constitution, legislature, and court system. Prior to Switzerland's formation as a single state in 1848, each of the cantons had existed as a sovereign entity, with one of four dominant languages: German, French, Italian, or Romansh. Thus, it was natural to maintain a high level of independence for each.

Nigeria in West Africa is another example of a federal state. Formed as part of Great Britain's colonial occupation in the region, it consists of over 250 distinct ethnic groups. It is religiously diverse as well, with about 50 percent Muslim, dominating the north, and 40 percent Christian, primarily in the south. Due to the difficulty of managing such diversity, the country was divided into three regions upon independence in 1960. Since then, there have been numerous revisions to the federal structure, with thirty-six states as of 2017. A federal structure has allowed for accommodation of cultural differences. For instance, the Christian south has a legal structure based on English common law, while Muslim northern states base their legal systems on Islamic Sharia law.

Authoritarianism and democracy

When studying states, one of the most important aspects is the relationship between rulers and citizens. Governments control activities within their borders, but there are different ways in which power can be exerted. One way of viewing the relationship between rulers and citizens is along the spectrum from *authoritarian* to *democratic*.

Authoritarian regimes concentrate power. This power can be in the hands of a single dictator or a single group of leaders or political party. It is not subject to any external controls, whether from a political opposition or laws. Dissent is not allowed, and control is typically enforced with state secret police. Sometimes, authoritarian regimes will paint a façade of freedom, through written constitutions, parliaments, or elections, but these hold no real power. For instance, the authoritarian state of North Korea incorporates the façade of liberty in its official name: The Democratic People's Republic of Korea. Likewise, presidential elections in Central Asia's Kazakhstan in 2015 gave nearly 98 percent of the vote to the incumbent president, hardly a sign of a free election.

On the other end of the spectrum is democracy. With democratic systems, power is shared among different competing groups. Political parties have the right to form, and elections are held without influence from the government in power. A free press is allowed and is not subject to government restrictions. The judiciary is independent and makes decisions based on the law, not political considerations. Civil liberties are upheld for the population, and while decisions are made by majority vote, minority rights are protected. The majority is held in check from a "winner takes all" view of power.

Of course, most states lie somewhere along the spectrum from pure authoritarianism to perfect democracy. The independent think-tank Freedom House makes an annual ranking of countries based on their degree of political rights and civil liberties, placing them along a spectrum from free to not free. This ranking is based on a number of variables, including a fair electoral process, allowance for political pluralism, transparent government decision making, freedom of the press and religion, the right to form political and labor groups, judicial independence, and protection of individual rights.

With the fall of the Soviet Union in 1991, the trend toward democracy made great strides. In 1986, 34 percent of countries were considered free, but by 2006, the number had risen to 47 percent. At the same time, not-free states fell from 32 to 23 percent. Gains were made in Eastern Europe and Central Asia, where Soviet authoritarianism ended, as well as in regions such as Latin America, where military governments ceded power to democratically elected ones (figure 11.17). But recent years, unfortunately, have seen a backslide. By 2016, the share of free countries had fallen to 45

percent from 47 percent, while non-free countries rose to 25 percent from 23 percent (figure 11.18). And, in many cases, even countries that did not slip to a lower category saw a decline in overall levels of political and civic freedom.

Some of the shift toward authoritarianism has been in countries where democratically elected leaders consolidated power and then stifled opposition. In regions across the globe, such as Turkey, Russia, Venezuela, Nicaragua, Tajikistan, and many others, laws were changed allowing leaders to remain in power well beyond previously established limits. In turn, they have managed to effectively remove opposition parties, vastly restrict press freedom, censure social media, and imprison rivals.

When the Soviet Union ended in 1991, there was much talk among political analysts about the triumph of liberal democracy. But it turns out that the celebrations were premature. Russia was considered partially free from 1991 to 2004 but then dropped to the not-free category after that. This drop corresponded with President Putin's consolidation of power. He was first elected in 2000, then reelected in 2004. In 2008, being ineligible to run for a third term, he took the role of Prime Minister, placing another "symbolic" candidate in the role of president. In 2012, he once again was elected president, and he won the 2018 presidential election with 76 percent of the vote, cementing a solid twenty years in power.

Putin has managed to retain power through a combination of media control, patronage, and intimidation. He has managed to destroy the free press, shutting down independent TV news outlets and exerting strong influence over those under state control. While the internet has proven harder to totally control, Russia under Putin has managed to successfully manipulate it to his benefits. There are official state-run "trolling factories," where false information is produced for blogs, video and newspaper comments sections, and social media posts. The idea is to create so much inaccurate and contradictory information that people decide no media can be trusted. Once people think that all of

Political Freedom and Civil Liberties
2016

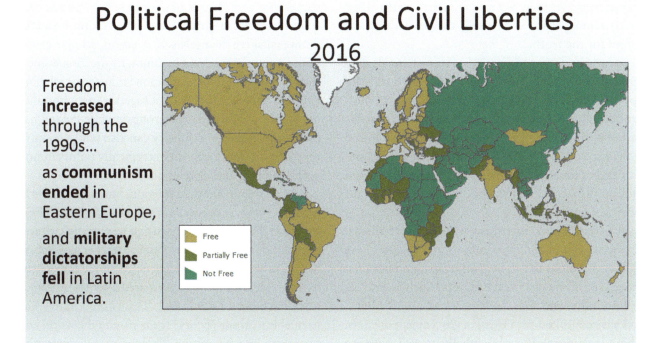

Freedom **increased** through the 1990s...

as **communism ended** in Eastern Europe,

and **military dictatorships fell** in Latin America.

- Free
- Partially Free
- Not Free

Figure 11.17. Political freedom and civil liberties, 2016. Data source: Freedom House.

Decline in Freedom
1973-2016

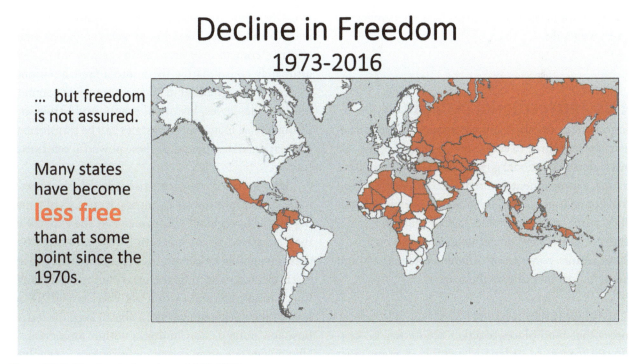

... but freedom is not assured.

Many states have become **less free** than at some point since the 1970s.

Figure 11.18. Decline in freedom, 1973–2016. Data source: Freedom House.

the information they receive is untrustworthy, then the government can say anything it wants without any regard for the truth.

Through patronage systems, Putin channels government contracts to allies and cuts off those who oppose him. This has created a loyal class of multimillionaire businesspeople who benefit from his authoritarian rule. Then there is intimidation and violence. Political opponents have been jailed, and journalists and other opposition figures have been shot on the street and even poisoned in places as far away as London.

Unfortunately for those who support liberal democracy, Putin and other authoritarian leaders have followers around the globe. Nationalist and populist groups in solidly democratic Europe, such as France, Britain, Italy, and Austria, have an affinity for Putin and have consulted with Russian government officials. China is gaining influence globally as well, offering investments in countries from Latin America to Africa and Asia in exchange for closer political ties. Some governments are happy to participate, since China does not push for human rights and transparency as conditions for investment, as many Western democratic governments do.

State and territorial stability

Nation-states function only as long as those within their borders agree to cooperate with each other and the established national government. A state can have a name, a government, international recognition, and fixed boundaries, yet still be subject to territorial instability. Some states are stable, lasting centuries, while others fracture into competing groups and civil war. These results come from different forces that bind countries together or tear them apart.

Centripetal forces

Forces that bind places together are known as *centripetal forces*. In order to be stable, a state must have a *raison d'être*, or reason for being. It must have

something that draws people to see members of a country as their in-group. In other words, the people of a state must see themselves as a nation, a single group with a common past and common future. Sometimes, the centripetal force is a common language or religion. Pakistan has Islam that binds it together (notwithstanding some militant terrorist groups), while Poland has the Polish language. Ethnicity can also bind a country together, although ethnicity is typically tightly bound with language and/or religion.

More important than religion, language, and ethnicity, however, are common values and ideals. These centripetal forces are larger than any single cultural group and can serve to tie people together as a nation despite other cultural differences. As stated previously, national identities are often created after state boundaries have been established. The United States has been able to absorb people from around the world and turn them into Americans. This successful integration is based on a common belief in ideas such as democracy and individual rights, which immigrants have absorbed and embraced for centuries. Great Britain was successful in building a British national identity from disparate groups in England, Wales, Scotland, and North Ireland. Likewise, French kings established effective control over areas with distinct languages, ultimately consolidating them into a French national identity. Through belief in and acceptance of common political ideologies, historical events, and heroes and through trust in institutions, states can be transformed into nation-states, bound by these powerful centripetal forces.

Devolution of power can also serve as a centripetal force in culturally diverse countries. It consists of granting power to regional governments at the expense of the central one. Be it a unitary or a federal state, when decision making is decentralized, geographically based resentments can be mitigated. Devolution has held the four nations of Great Britain together (as of 2018, at least). Likewise, Spain has been able to keep the restive Basque Country and Catalonia within its control by allowing its seventeen autonomous regions to run their own regional governments, with police, education and

Devolution: Great Britain and Spain

Decentralized power gives more control to local governments.

Great Britain has remained united in part by devolving power to Scotland, Wales, and North Ireland.

In Spain, this has placated the restive regions of Catalonia and the Basque Country.

Figure 11.19. Devolution: Great Britain and Spain. Data sources: Esri, HERE, Garmin, NGA, USGS.

health systems, environmental regulations, abilities to tax, and protection of regional languages (figure 11.19). Canada is divided into ten provinces, with Quebec recognized as an official nation with French as its official language. Similarly, in the United States, dense urban states on the East Coast, sparsely populated prairie states, Latino-heavy southern-border states, and many more, have been held together through the federal devolution of power to state governments.

Centrifugal forces

Centrifugal forces are those that tear a country apart. These can result in separatist movements, whereby a group of people in a region push to break away and form their own nation-state. Another term commonly used in this situation is *balkanization*, coined in reference to the Balkan Peninsula and the breakup of the Ottoman Empire into multiple nation-states in the early twentieth century.

Size and shape have the potential to act as centrifugal forces, with large countries being more difficult to integrate with transportation and communications networks and elongated or fragmented countries having similar problems. However, evidence that these factors play a substantial role in tearing countries apart is limited. Russia, Canada, the United States, China, and other large countries have proven to be very stable, as has Chile, the most elongated of states in the world. Both the Philippines and Indonesia, highly fragmented states, do have separatist movements, but neither has succumbed to them and broken apart.

The centrifugal pull caused by population diversity is the most significant cause of balkanization and separatist movements. When states fail at unifying people into a common national identity, regionally based minority groups may call for independence. This can be along a wide range of socioeconomic and cultural lines. Language, ethnicity, and religion are common lines of division, but education and standard of living can be another. Differing views of economic policy, political philosophy, and race have led to separatism as well. Often, multiple variables work together as centrifugal forces, so it can be difficult to say, for instance, that religion alone or ethnicity alone is the proximate cause of a separatist movement.

At this point, it is useful to return to the discussion of colonial-drawn state boundaries. As discussed previously, European colonial powers often drew boundaries without regard for cultural patterns as they existed on the ground. Thus, some cohesive groups were split into different states, while others, who may have been historical rivals, were lumped together into a single state. For instance, the country of Sudan was formed under British colonialism in the late nineteenth century. It contained a majority Arab Muslim population, but much of the south consisted of non-Arabs who practiced animist and Christian religions. Upon independence in 1956, the Arab majority attempted to impose its religion and culture, including Islamic law, throughout the entire country, leading to resistance among the people of the south. Civil wars between north and south raged off and on for decades and resulted in millions of deaths until South Sudan finally gained independence in 2011 (figure 11.20).

In Europe, the Balkans region erupted in conflict during the 1990s as the former Yugoslavia disintegrated along religious, ethnic, and linguistic lines. This conflict resulted in the deaths of over 100,000 people, the displacement of millions, and convictions of military and political leaders in the International Criminal Court for genocide. The single state of Yugoslavia, created after World War II, broke up into the nation-states of Bosnia and Herzegovina, Croatia, Slovenia, Macedonia, Serbia, and Montenegro. Kosovo currently seeks independence from Serbia.

Active European secessionist movements are also found in Great Britain and Spain (figure 11.21). Both have devolved power to regional authorities,

Centrifugal Forces and Balkanization

A lack of national unity can lead to the breakup of states.

Yugoslavia (left) and Sudan (right) both went through bloody civil wars before splitting into new nation-states based on linguistic, religious, and ethnic divisions.

Figure 11.20. Centrifugal forces and Balkanization Data sources: Esri, HERE, Garmin, NGA, USGS.

Figure 11.21. March for independence. Protesters in Barcelona supporting the secessionist movements of Scotland and Catalonia. Photo by ONiONA. Stock photo ID: 217160461. Shutterstock.com.

successfully maintaining their borders intact, yet pressures still exist. In Great Britain, Scotland, an independent nation until it was absorbed by Great Britain in the early 1700s, held an independence referendum in 2014, but a majority voted not to secede. In Spain, Catalonia, which contains the well-known city of Barcelona, has been seeking independence as well, partially on the basis of its Catalan language and partially on the basis of its higher standard of living; the people of Catalonia complain that the region sends much more tax revenue to Madrid than it receives from the central government. Spain has also faced many years of separatist activity from the Basque Country, a region with a distinct language and history. While quiet in recent years, Basque separatists have committed terrorist acts, including the murder of police officers and soldiers.

In much of the Middle East, the cradle of agricultural and urban civilization, territorial conflicts have shifted boundaries for millennia. In relatively recent history, the Ottoman Empire rose, drew new boundaries, then collapsed and was replaced with European colonial control and another set of boundaries. To this day, myriad groups are struggling for new nation-states tied to religious, linguistic, and ethnic identities. Most salient has been the conflict

in Iraq (figure 11.22). When the strongman Saddam Hussein was removed from power by US and coalition forces in 2003, it was assumed that the country would become a stable democracy in the region. However, a long history of conflict and mistrust between different culture groups made this ideal difficult to realize. The Shiite Muslim majority, who had been severely repressed under Hussein's rule, was happy to see him gone but then wanted to exercise its majority control. The Sunni Muslim minority population, who had enjoyed disproportionate power under Hussein, resisted Shiite attempts to take power. Then there were the Kurds, who had suffered greatly under the rule of Hussein and wanted to expand control of territory in the north of Iraq, possibly forming an independent Kurdistan with other Kurdish populations in Turkey, Syria, and Iran. Rather than achieving a peaceful democracy, these groups have fought bitter political and military battles. What was supposed to be a quick military intervention to remove a brutal dictator has transformed into civil conflict and political instability that has lasted well over a decade and resulted in over 250,000 deaths of civilians and combatants.

Border conflicts

Centrifugal forces can tear a country apart as a region attempts to break away and form a new nation-state. But there are other situations that can cause territorial instability and conflicts over borders. These can be broken down into four broad categories: *identity, demarcation, resources,* and *security*.

The case of identity relates most closely to the discussion on centrifugal forces. When a group sees itself as a nation distinct from the state it currently lives in, it can seek to break free. Sometimes its national identity straddles two or more countries, driving a movement that impacts multiple states. This is the case of the Kurds and their secessionist dreams of an independent Kurdistan, as mentioned above. In other situations, a region of one state will be incorporated, voluntarily or by force, and join another state. This process is referred to as *irredentism*. Typically, irredentism takes place

Centrifugal Forces in Iraq

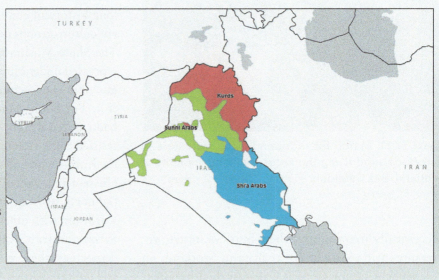

Shia Arabs, **Sunni Arabs**, and **Kurds** were ruled under Saddam Hussein's iron fist.

When he was removed by US and coalition forces in 2003, these groups began fighting for control of parts of the country, threatening the integrity of its boundaries.

There have been over **250,000 deaths** of civilians and combatants since then.

Figure 11.22. Centrifugal forces in Iraq. Data sources: Vogt, Manuel et al 2015 and Wucherpfennig et al, 2011.

when the incorporated region is viewed as an integral part of the larger state. It can be based on historical claims to territory, as when Argentina attempted unsuccessfully to take the British-controlled Falkland Islands (Islas Malvinas) in 1982 or based on cultural affinity, as with Russia and the Russian-speaking population of Ukraine (figure 11.23).

In the case of Ukraine, deep division lies between its western Ukrainian speaking half, which wants stronger relationships with Western Europe, and its eastern half, where Russian is the dominant language and residents prefer ties with Russia. Disagreement between the two halves rose to a point of crisis, and in 2014, Russia invaded and annexed the Crimean Peninsula. At the same time, Russia sent troops and military equipment to support pro-Russian rebels in eastern Ukraine (figure 11.24). In one especially tragic event, a commercial airliner flying over eastern Ukraine was shot down by poorly trained rebels using sophisticated Russian antiaircraft systems. Many of these

pro-Russian rebels desire unification of their region with Russia.

Border disputes can also arise from disagreements and uncertainty over demarcation. Many times, natural features are used to divide states, but detailed land surveys are not always available. For instance, a border may follow the peak of a mountain range, and countries may disagree exactly where the line lies. Another problem can come from the use of rivers as borders. While rivers can serve as a natural division between two states, they are not static features. With time, rivers can change course, potentially shifting borders. This was the case between Mexico and the United States along a segment of the Rio Grande near El Paso, Texas, and Ciudad Juarez, Mexico. In 1864, heavy rains caused the river's channel to shift south, adding about 700 acres to the United States. Over the years, roughly 5,000 Americans moved into this area, known as the Chamizal, and built homes and businesses. Mexico protested this loss of land, and the

Irredentism and the Russian Invasion of Ukraine

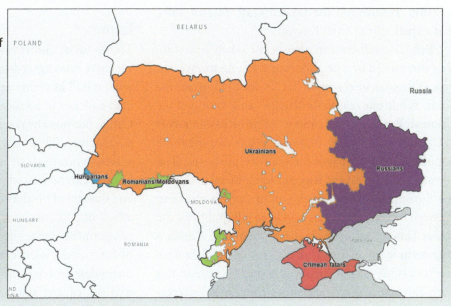

Many Russians believe parts of Ukraine rightfully belong to Russia.

In 2014, Russia invaded and annexed Crimea with its large Russian population and strategic Russian naval base.

Russia also provides military support to ethnic Russian separatist rebels in eastern Ukraine.

Figure 11.23. Irredentism and the Russian invasion of Ukraine. Data sources: Vogt, Manuel et al. 2015 and Wucherpfennig et al, 2011.

Figure 11.24. Destroyed airport in Donetsk, eastern Ukraine. In 2014–15, pro-Russian rebels and the Ukrainian military fought for control of the airport, ultimately leaving it destroyed. Photo by Denis Kornilov. Stock photo ID: 443034715. Shutterstock.com.

dispute continued until 1964, when much of the land was returned to Mexico and US residents were forced to relocate. The river is now channelized in cement and can no longer change location.

Conflict over natural resources is another reason behind border disputes. The Iraqi invasion of Kuwait in 1990 was partially related to oil resources and partly related to irredentism. From an irredentist standpoint, the Iraqi government had viewed Kuwait as an integral part of Iraq going back to the Ottoman Empire. But ultimately, it was a dispute over oil that led to the invasion. Iraq had major debts from its war with Iran and desperately needed oil revenue. It accused Kuwait of exceeding its OPEC (Organization of Petroleum Exporting Countries) quota, driving down world oil prices, and of slant drilling into Iraqi oil fields along its border. Iraq therefore invaded Kuwait, taking control of the country in a matter of hours. The international community rejected Iraq's invasion, resulting if the US-led Gulf War of 1991 that expelled Iraq from Kuwait.

Border conflicts can also occur when one state feels threatened by a neighboring state. Again, we can look at Russia's intervention in Ukraine to illustrate the point. While one reason for Russia's annexation of the Crimean Peninsula and support of pro-Russian rebels in Ukraine's east was to support the Russian-speaking population, another was for security purposes. The Ukrainian government was pursuing membership in the European Union, which would draw it closer to Western interests. Ukraine had been a Republic of the Soviet Union and upon its dissolution gave Russia a long-term lease on a naval base located in Ukrainian territory. One fear of Russia was that Ukrainian ties with Europe could threaten its base, a key strategic point of naval access from the Black Sea to the Mediterranean Sea and on to the Atlantic Ocean.

An ongoing border conflict related to control of natural resources and national security is taking place in the South China Sea. Here, China is making claims to nearly the entire sea, claiming economic rights to its resources and projecting its power deep into Southeast Asia. The United Nations Convention on the Law of the Sea states that countries have sovereign control twelve nautical miles beyond their coast and exclusive economic zones for 200 nautical miles. Within exclusive economic zones, a country has the right to fisheries, energy generation, mineral and oil extraction, or any other economic activity. China, in 1947, made claims to most of the South China Sea, delineating its claims with what is now known as the nine-dash line. Yet this line infringes on the 200-mile claims of countries such as Vietnam, Philippines, and Malaysia. To bolster its claims, China has claimed ownership of a series of islands and coral outcroppings in the sea, thus claiming all territory around them. On some of these islands, it is constructing thousands of acres of new land and building military facilities to show that it indeed has full control (figure 11.25). Conflict over

these disputed maritime borders threaten stability in the region, as thousands of commercial and military ships from numerous countries ply its waters.

Terrorism

Terrorism is, unfortunately, a topic of everyday news, yet there is no agreed-upon international definition of what it is. The term implies that a group or individual uses violence in an illegitimate way for political ends. Often, but not always, it is seen as committed by non-state actors. This view contrasts with the actions of states, whose use of inappropriate violence within borders can be viewed in terms of human rights abuses and outside of borders can be considered war. Despite a lack of consensus on a definition, the United States government defines terrorism as "the unlawful use of force and violence against persons or property to intimidate or coerce a government, the civilian population, or any segment thereof, in furtherance of political or social objectives." The key component of this definition is that an act has objectives aimed at some sort of political or social change.

Given that terrorism has political or social objectives, the reason behind it is some perceived injustice against a marginalized or oppressed group. Through terrorism, it is hoped that power or territory can be gained and the perceived injustice remedied. There are several broad categories that terrorism often falls into. With diffusion of the nation-state concept, some groups began using terrorism in struggles for national self-determination. It was used to oust European colonial powers, such as by Jewish paramilitaries against the British in Palestine. Across the Middle East and Africa, such as in Algeria, Kenya, and beyond, other nationalist groups also used guerrilla and terror tactics to strike against European colonial control. Terror has also been used by nationalist separatist movements. The Irish Republican Army (IRA) used bombings and assassinations in Great Britain from 1917 to the late 1990s, while Euskadi Ta Askatasuna (ETA, Basque Homeland and Liberty) did the same in a struggle for Basque independence in Spain. Back in the region of Palestine, the successful creation of the Jewish state

Chinese Island Building: Spratly Islands, South China Sea

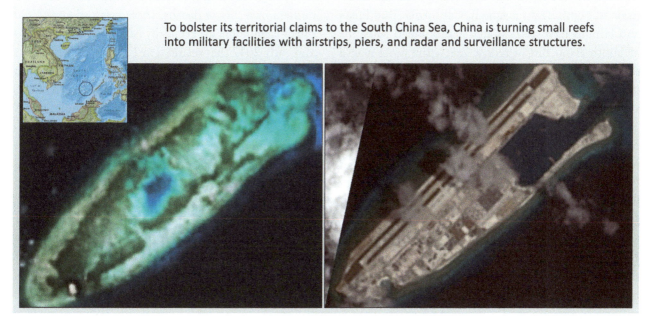

To bolster its territorial claims to the South China Sea, China is turning small reefs into military facilities with airstrips, piers, and radar and surveillance structures.

Figure 11.25. Chinese island building: Spratly Islands, South China Sea. Data sources: DigitalGlobe, National Geographic, Esri, DeLorme, HERE, UNEP-WCMC, USGS, NASA, ESA, METI, NRCAN, GEBCO, NOAA, increment P Corp.

of Israel created a new terrorist movement, that of the Palestinian Liberation Organization (PLO). The PLO used bombings, hijackings, and assassinations in its struggle for the independence of Palestinians under Israeli occupation.

In addition to nationalist terrorist movements, other groups are based on political ideology. In these cases, the goals are broader than the creation of a nation-state. Rather, they seek to change systems that they see as unfair or exploitative. Many have been tied to international Marxism and anti-imperialism, with the goal of destroying capitalist systems and their imperial control of developing states. In Latin America, Marxist groups such as the Revolutionary Armed Forces of Colombia (FARC) and Shining Path used bombings, murder, kidnapping, extortion, and hijacking in attempts to bring down the governments of Colombia and Peru. The 1970s and 1980s saw active Marxist terrorist groups in Europe and the

United States as well, with the Italian Red Brigades and the American Weather Underground. More recently, militant environmental and animal-rights groups have been active in the United States and Europe. These groups focus primarily on destroying property rather than harming people, yet the political and social intent of their actions places them in the terrorist category. Right-wing terrorist groups, such as white-supremacists and antigovernment militias, have been active in the United States and other countries as well. One of the worst acts of terrorism in the US was committed by an antigovernment "lone wolf" who blew up the Oklahoma City Federal Building in 1995, killing 168 people, including children in an on-site day-care center.

In recent years, another motivating factor has grown in influence: religion. All major religions have fundamentalist splinter groups that commit terrorism in the name of their faith. In the United States, this has involved bombings of abortion and reproductive health clinics by right-wing fundamentalist Christian militants. Also in the United States, members of the Jewish Defense League were arrested by the FBI in 2001 prior to bombing a mosque in California. Buddhists and Hindus have assassinated and bombed Muslims in South Asia. But in recent years, the most active religiously linked terrorist groups have been Islamic (figure 11.26). The deadliest groups globally in 2016 were Islamic State, followed by Boko Haram, the Taliban, and al Qaida. Based on these groups' regions of activity, the majority of deaths occurred in the Muslim world: Iraq, Afghanistan, Syria, Nigeria, and Pakistan. For instance, in 2016, seventeen of the twenty most deadly terrorist attacks took place in Afghanistan, Iraq, Nigeria, and Syria.

Looking specifically at the United States, both Islamists and far-right extremist groups have been the deadliest in recent decades. The largest death toll by Islamists was the September 11, 2001, attack that killed 2,996 people, while the largest far-right attack was the 1995 Oklahoma City Federal Building attack that killed 168 (figure 11.27). In addition to these outliers in terms of deaths, there were an additional thirty-eight homicide events by Islamists between 1990 and early 2017 that resulted in 136 deaths. Much less salient in the minds of many Americans is the death toll from far-right extremists. During this same time, these groups killed an additional 272 people in 178 homicide events. While Islamist extremists tended to target random victims in crowded places, more than half of the targets of far-right extremists were selected because of their religion, race or ethnicity, sexual orientation, or gender identity.

Regardless of the motivating factor driving terrorism, be it national self-determination, political ideology, or religion, there are clear geographic conditions that drive it. Most terrorist attacks occur in places with political terror and/or internal conflict.

Terrorism Deaths 2007-2016

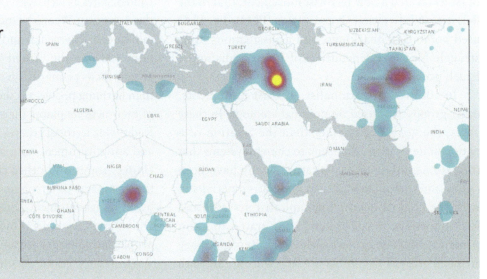

During this 10-year period, **over 50%** of all terrorism-related deaths occurred in **Iraq**, **Afghanistan**, **Syria**, **Nigeria**, and **Pakistan**.

Figure 11.26. Terrorism deaths, 2007–2016. Data source: National Consortium for the Study of Terrorism and Responses to Terrorism (START), 2017.

Figure 11.27. September 11, 2001, terrorist attack. The South Tower of the World Trade Center collapsing. On this day, 2,996 people were killed, the largest terrorist attack in US history. Photo by Dan Howell. Stock photo ID: 83242102. Shutterstock.com.

Political terror involves state-sanctioned killings, torture, disappearances, and political imprisonment by repressive governments. This was the case in Egypt and Turkey in the 2010s, when uprisings against the government resulted in political turmoil and hardhanded crackdowns on dissent. The human rights abuses that often accompany crackdowns such as these frequently result in further radicalization and easier recruitment for terrorist organizations. Likewise, high levels of terrorism are found in conflict zones, such as Afghanistan, Syria, and Iraq, where myriad factions use violence against civilians and the government in their struggle for power.

In countries with stable political situations, such as most of Europe and North America, as well as in relatively stable places such as Tunisia in North Africa, terrorism is more likely to arise from places with a lack of opportunity, low social cohesion, and alienation. In low-income neighborhoods, lower levels of education and fewer job opportunities leave many youths in this situation. This situation can be compounded by alienation from the majority culture when communities have large immigrant populations. For instance, in the European Union, first-generation immigrant youth have much higher levels of unemployment than nonimmigrant youth and thus struggle to feel that they are full members of society.

Terrorist organizations recruit from populations that feel under attack, either from a perceived threat against their religion, an animosity against the ruling government, or alienation and lack of opportunity. People in marginalized neighborhoods and in prisons and those with criminal histories are frequently targeted for recruitment. With people in these situations, terrorism can offer a feeling of companionship and a collective group identity that makes individuals feel more secure and powerful.

From a geographic standpoint, terrorism can be seen in terms of *nodes* and *conduits*. Between each node, groups must move money, people, weapons, and information. Recruitment, training, and ultimately execution of the terrorist act must be coordinated between linked nodes. At the center of networks are the *core nodes*. These are often located in places with weak government control. For instance, al Qaeda has been centered in Afghanistan and northern Pakistan, while Islamic State is based in Syria. The highest levels of strategy and planning are done in these core nodes. *Peripheral nodes* are those that undertake the attack and are thus located close to the target area. Between the core and peripheral nodes are *junction nodes* that link leadership's plans with those who will carry them out. They must recruit people for peripheral nodes, and then provide funding, training, and materiel directly to them. Both the junction and peripheral nodes are the most exposed to antiterrorism authorities, since they must work in the areas they plan to strike. From a counterterrorism perspective, the most effective strategy is to disrupt junction nodes, since they connect the core with those who will carry out attacks. If recruitment, training, and support from junction nodes is disrupted, peripheral nodes cannot succeed in carrying out attacks.

The problem with stopping so called lone wolf terrorists is that they are not tied to the typical node and conduit system. Rather, they are inspired by terrorist social media but act independently and without direct assistance from organized groups. But while it may be harder to prevent lone wolf attacks, the individuals who carry them out typically are less effective in causing damage and casualties because they work independently, without experienced trainers and sophisticated weaponry.

Go to ArcGIS Online to complete exercise 11.4: "Law of the sea: Global flash point in the South China Sea."

Supranational organizations: Cooperation among states

To mitigate threats to stability, countries frequently cooperate in terms of trade, national security, international dispute resolution, and other areas. For these reasons, countries voluntarily form supranational organizations, agreements between three or more states that grants some decision-making powers, and thus state sovereignty, to a larger body.

The United Nations
The United Nations is a supranational organization established in 1945 for the purpose of maintaining international peace and security and solving economic and social problems facing the world community. It consists of 193 states that meet each year in the General Assembly to debate issues of global concern. A smaller Security Council works to resolve conflicts between states peacefully and can impose economic sanctions or authorize use of force against countries if negotiations fail. In addition, the Economic and Social Council coordinates work on economic, social, and environmental issues, while the International Court of Justice adjudicates legal disputes over international law. The United Nations has additional specialty programs (table 11.2) that focus on many of the issues discussed in this book.

The European Union
Another supranational organization that makes the news on a regular basis is the European Union. Soon after World War II, European states began a process of integration, culminating in the formation of the European Union in 1992. In 2017, there were twenty-eight member countries, but the United Kingdom was in the process of leaving the union. This union performs many functions, but the most important are that it creates a single market for goods and services, allows for free movement of EU residents, and manages a single currency (figure 11.28).

With a single market, a unified set of rules and regulations means that products made in Spain can be sold in Germany, products made in France can be sold in Poland, and so on. The same is true for services and investment. No additional regulations can limit their sale within the union, and there are no customs or border restrictions on their movement. It also means that international trade agreements are made at an EU level, not at the individual country level. With 500 million residents, it is the world's largest single market and has trade agreements with dozens of countries outside the union. International trade increases economic growth, and the size of the EU market gives it the weight to promote a global trading system with rules that reflect EU values.

Free movement is another key function of the European Union. Residents of EU states can live and work in any other EU state. Thus, both Germany and the United Kingdom had over three million residents

United Nations: Select Programs

Program	Purpose
United Nations Development Programme	Reduce poverty and inequality.
United Nations High Commissioner for Refugees	Assist refugees.
World Food Programme	Eradicate hunger and malnutrition.
United Nations Population Fund	Promote reproductive health and family planning.
United Nations Human Settlements Programme	Promote sustainable human settlements and cities.
United Nations Environment Programme	Advocate for protection of the global environment.
World Bank	Finance poverty reduction projects.
World Health Organization	Advance physical and mental health and social well-being.
Joint United Nations Programme on HIV/AIDS	Stop and reverse the spread of HIV/AIDS.

Table 11.2. United Nations: Select programs. Data source: United Nations.

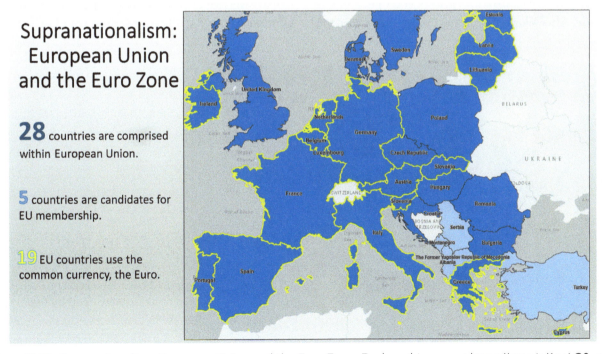

Supranationalism: European Union and the Euro Zone

28 countries are comprised within European Union.

5 countries are candidates for EU membership.

19 EU countries use the common currency, the Euro.

Figure 11.28. Supranationalism: European Union and the Euro Zone. Explore this map at https://arcg.is/1eeLG8. Data sources: Esri, HERE, Garmin, NGA, USGS.

from other EU countries in 2016, while Spain, France, Italy, and Switzerland had well over one million. These residents work in the full spectrum of jobs, from low-skill services in hotels and restaurants to high-paying positions in banking and technology. Free movement is also enhanced by the Schengen border-free area. This agreement allows for the free movement of people between member states without being subject to border controls. Free movement applies to all people, including non-EU tourists and businesspeople. The Schengen area consists of most, but not all, EU countries as well as a few non-EU countries.

In 1999, the EU Economic and Monetary Union issued the Euro currency, now used in twenty-six EU states (figure 11.29). This common currency replaced national ones, so there is no longer a French franc, a Spanish peseta, or an Italian lira. A single currency facilitates trade within the EU, making business transactions, travel, and shopping easier, as there is no need to convert currencies in each country.

The North Atlantic Treaty Organization

Supranational organizations also exist for purposes of defense. After World War II, the Cold War between the Soviet Union and Western democracies was heating up. In response, a group of European countries, plus the United States and Canada, formed the North Atlantic Treaty Organization (NATO) in 1949. Its purpose was to provide collective defense for member countries, many of which would be too small to defend themselves alone in a situation of conflict with the Soviets. The most well-known aspect of NATO is its Article 5, which states that an attack against one member is an attack against all. Thus, an attack on a relatively small country such as Denmark or Norway would be met with a counterattack by the militaries of much of the rest of Europe and North America. This was intended to dissuade the Soviet Union from expanding its influence further westward into Europe.

Article 5 has been invoked only once, after the September 11 terrorist attack by al Qaeda on the United States. Based on this article, NATO countries provided troops and support to counter al Qaeda and their Taliban protectors in Afghanistan. It also began counterterrorism sea patrols in the Mediterranean. NATO forces participated in peacekeeping roles in Kosovo in response to conflict in the Balkans and conducted air patrols to monitor Russian activity

Figure 11.29. Euro symbol. The Euro is the common currency used by twenty-six countries. Photo by Gordon Bell. Stock photo ID: 705327238. Shutterstock.com.

adjacent to Balkan states. Given the global nature of security risks, NATO has participated in counterpiracy activities off the Horn of Africa and assists African Union peacekeeping troops in Somalia. In each case, individual states relinquish some degree of sovereignty, potentially placing troops in harm's way for the benefit of all NATO members (figure 11.30).

Go to ArcGIS Online to complete exercise 11.5: "The challenges of European integration."

Figure 11.30. NATO exercise in Romania. British, US, Romanian, and Moldovan troops took part in this 2015 training exercise. NATO forces consist of troops from various national militaries that work in coordination for common defense. Photo by Studio 37. Stock photo ID: 645205543. Shutterstock.com.

References

Amnesty International. 2016. "Haiti/Dominican Republic: Reckless Deportations Leaving Thousands in Limbo." Amnesty International, June 15, 2016. https://www.amnesty.org/en/latest/news/2016/06/haiti-dominican-republic-reckless-deportations-leaving-thousands-in-limbo.

BBC News. 2016. "A Guide to Devolution in the UK." BBC News, September 18, 2016. http://www.bbc.com/news/uk-politics-35559447.

———. 2016. "Balkans War: A Brief Guide." BBC News, March 18, 2016. http://www.bbc.com/news/world-europe-17632399.

Bender, J. 2014. "Here Are the Self-Declared Nations You Won't See at the UN." Business Insider, September 24, 2014. http://www.businessinsider.com/the-self-declared-nations-you-wont-see-at-the-un-2014-9?op=1%2 F#public-of-somaliland-2.

Brantingham, P. J., G. E. Tita, M. B. Short, and S. E. Reid. 2012. "The Ecology of Gang Territorial Boundaries." Criminology 50, no. 3: 851–85. doi: https://doi.org/10.1111/j.1745-9125.2012.00281.x.

Council on Foreign Relations. n.d. "China's Maritime Disputes." Council on Foreign Relations InfoGuide. https://www.cfr.org/interactives/chinas-maritime-disputes?cid=otr-marketing_use-china_sea_InfoGuide#!/chinas-maritime-disputes?cid=otr-marketing_use-china_sea_InfoGuide.

Diener, Alexander C., and Joshua Hagen. 2012. Borders: A Very Short Introduction. New York: Oxford University Press.

The Economist. 2015. "Interiors: Why It's Better to Have a Coastline." The Economist, May 9, 2005. https://www.economist.com/news/americas/21650574-why-its-better-have-coastline-interiors.

Elder, M. 2014. "'Nothing Is True and Everything Is Possible,' by Peter Pomerantsev." New York Times book review, November 30, 2014. https://www.nytimes.com/2014/11/30/books/review/nothing-is-true-and-everything-is-possible-by-peter-pomerantsev.html.

European Commission. 2015. How the European Union Works: Your Guide to the EU Institutions. Luxembourg: EU Law and Publications.

European Commission Directorate-General for Trade. "EU Position in World Trade." European Commission. Updated October 2, 2014. http://ec.europa.eu/trade/policy/eu-position-in-world-trade.

Flint, C. 2017. *Introduction to Geopolitics*. London: Routledge, Taylor & Francis Group.

Freedom House. 2017. "Populists and Autocrats: The Dual Threat to Global Democracy." *Freedom in the World 2017*. https://freedomhouse.org/report/freedom-world/freedom-world-2017.

Hartshorne, Richard. 1950. "The Functional Approach in Political Geography." *Annals of the Association of American Geographers* 40, no.2: 95–130. doi: https://doi.org/10.1080/00045605009352027.

Ingraham, C. 2015. "This Is the Best Explanation of Gerrymandering You Will Ever See." *Washington Post*, March 1, 2005. https://www.washingtonpost.com/news/wonk/wp/2015/03/01/this-is-the-best-explanation-of-gerrymandering-you-will-ever-see/?utm_term=.a16ed8e7eca9.

Iraq Body Count (IBC). "IBC: The Public Record of Violent Deaths Following the 2003 Invasion of Iraq." https://www.iraqbodycount.org.

Jacobs, J. 2011. *The Death and Life of Great American Cities*. New York: Modern Library.

Klaas, B. 2017. "Gerrymandering Is the Biggest Obstacle to Genuine Democracy in the United States. So Why Is No One Protesting?" *Washington Post,* February 10, 2017. https://www.washingtonpost.com/news/democracy-post/wp/2017/02/10/gerrymandering-is-the-biggest-obstacle-to-genuine-democracy-in-the-united-states-so-why-is-no-one-protesting/?utm_term=.0bc6db3ea10c.

Legorano, G., and M. Force. 2017. "Middle-Class Catalans Drive Push for Independence." *Wall Street Journal*, October 1, 2017. https://www.wsj.com/articles/middle-class-catalans-not-workers-drive-push-for-independence-1507714203.

Lieberman, S. R. 1983. "The Evacuation of Industry in the Soviet Union During World War II." *Soviet Studies* 35, no. 1: 90–102. doi: https://doi.org/10.1080/09668138308411460.

Mellor, Roy E H. 2015. *Nation, State and Territory: A Political Geography*. New York: Routledge.

National Consortium for the Study of Terrorism and Responses to Terrorism (START). 2017. "Global Terrorism Database [data file]." http://www.start.umd.edu/gtd.

NATO. n.d. "What Is NATO?" https://www.nato.int/nato-welcome/index.html.

NPR. 2014. "50 Years Ago, A Fluid Border Made the US 1 Square Mile Smaller." NPR (September 25). https://www.npr.org/2014/09/25/350885341/50-years-ago-a-fluid-border-made-the-u-s-1-square-mile-smaller.

O'Leary, E., L. Cusack, R.-J. Bartunek, and A. Croft; 2017. "Factbox: How Catalan Autonomy Stacks Up Against Other Regions." *Reuters,* October 2, 2017. https://www.reuters.com/article/us-spain-politics-catalonia-devolution-f/factbox-how-catalan-autonomy-stacks-up-against-other-regions-idUSKCN1C727I.

Parkin, W., B. Klein, J. Gruenewald, J. Freilich, and S. Chermak. 2017. "Threats of Violent Islamist and Far-Right Extremism: What Does the Research Say?" National Consortium for the Study of Terrorism and Responses to Terrorism, February 22, 2017. http://www.start.umd.edu/news/threats-violent-islamist-and-far-right-extremism-what-does-research-say.

Popescu, Gabriel. 2011. *Bordering and Ordering the Twenty-first Century: Understanding Borders*. Lanham, MD: Rowman & Littlefield Publishers.

Sharma, Pradeep. 2007. *Economic Political Geography*. New Delhi: Discovery Publishing House.

Spruyt, H. 2011. "War, Trade, and State Formation." In *The Oxford Handbook of Political Science*. Oxford, NY: Oxford University Press.

United Nations High Commissioner for Refugees. 2017. "Statelessness and the Rohingya Crisis." *Refworld,* November 10, 2017. http://www.refworld.org/docid/5a05b4664.html.

US Department of State. n.d. "The Gulf War, 1991." US Department of State, Office of the Historian. https://history.state.gov/milestones/1989-1992/gulf-war.

———. 2016. "Country Reports on Terrorism 2016." https://www.state.gov/j/ct/rls/crt/2016/index.htm.

Vogt, Manuel, Nils-Christian Bormann, Seraina Rüegger, Lars-Erik Cederman, Philipp Hunziker, and Luc Girardin. 2015. "Integrating Data on Ethnicity, Geography, and Conflict: The Ethnic Power Relations Dataset Family." *Journal of Conflict Resolution* 59, no. 7: 1327–42. doi: https://doi.org/10.1177/0022002714567948.

The World Bank. 2014. *Improving Trade and Transport for Landlocked Developing Countries*. Washington, DC: The World Bank.

Wucherpfennig, Julian, Nils Weidmann, Luc Girardin, Lars-Erik Cederman, and Andreas Wimmer. 2011. "Politically Relevant Ethnic Groups across Space and Time: Introducing the GeoEPR Dataset." *Conflict Management and Peace Science* 28, no.5:423–37. doi: https://doi.org/10.1177/0738894210393217.

Chapter 12

Humans and the environment–pollution and climate change

In China, students at expensive private schools play on fields enclosed by massive tents with air filters to keep out smog. In Japan, the 2011 Fukushima nuclear power plant disaster forced nearly 500,000 people to evacuate the surrounding area due to radiation leaks. Contaminated water kills hundreds of thousands of people each year. New York, Houston, and other cities around the world are flooded by massive hurricanes, and record breaking wildfires scorch the southwestern United States, Portugal, and other regions as the climate changes and brings drought, irregular precipitation, and extreme storms. Clearly, we face a wide range of environmental impacts: pollution of our air, water, and land and global uncertainty brought about by a changing climate (figure 12.1).

Figure 12.1. Air pollution in Beijing, China. Photo by testing. Stock photo ID: 354568241. Shutterstock.com.

As humans impact and modify our planet, many argue that we have entered a new geologic epoch, one that is less dependent on natural changes in the earth and atmosphere and more dependent on humans. This human-induced epoch is known as the *Anthropocene*, a stage in which human actions have drastically altered the earth. People have transformed landscapes by altering plant and animal compositions, cutting and filling mountains and valleys, altering the flow of rivers, and changing the global climate. But while many people see these changes as a condition of modern human society, in fact, humans have been moving the planet into the Anthropocene ever since *Homo sapiens* migrated out of Africa.

Over twenty thousand years ago, as humans settled Eurasia, Australia, and the Americas, they immediately began altering the natural environment. With fire they cleared forests, through hunting they contributed to extinctions of megafauna, and with their migrations they carried invasive plants and animals to new places. The process of environmental transformation accelerated with the advent of agriculture roughly ten thousand years ago. Forests and grasslands were cleared on a massive scale for wheat and rice cultivation. Additional clearing resulted as pastoralism diffused across the continents via domesticated animals such as cows, goats, and sheep. Some researchers even argue that deforestation and increasing methane emissions from these actions were the first stage in human-induced climate change. As humans further expanded via the seas, they brought environmental change with them to ecologically isolated islands, such as Cyprus in the Mediterranean, Tonga in the Pacific, and many other places. Islands were dramatically altered as humans introduced new plant and animal species and cleared forests for human uses. In many cases, these new species quickly decimated native ones. Bird species went extinct, as did mammals, lizards, flowers, and other native life. Often, overall biological diversity was drastically reduced after human arrival.

With urbanization and the expansion of global trade, human impacts on the environment accelerated. Bronze Age people around the Mediterranean cleared native forests for orchards, such as olives, grapes, and figs. In the Americas, the Mayan people and others cleared tropical forests as their populations grew as well. As early as the thirteenth and fourteenth centuries, there is evidence that Northern and Western Europeans were overexploiting fish reserves.

Overall, it can be said that there are no pristine natural landscapes on the earth, and there likely have not been any for thousands of years. Humans are part of the ecosystem and modify it just as other species do, although arguably on a greater scale. Because of the substantial impacts we make on the environment, it is essential to understand how they happen and how to mitigate them to ensure healthy ecosystems for ourselves and other life on earth.

Pollution

As human populations have grown and settlement patterns have become denser, pollution of the air, water, and land has become of greater concern. Pollution occurs when a substance is added to the environment at a quicker pace than it can be dispersed, decomposed, or diluted. Smoke from a fire pit that quickly disperses into the air may not be considered pollution, but when smoke from myriad fires—whether from chimneys, garbage incineration, factories, vehicles, wood-burning stoves, or other sources—is emitted in an urban area and accumulates over the city, it becomes pollution. The same holds true with water. A small amount of fertilizer runoff that is quickly diluted in a large river poses no problem, but when larger amounts flow into water bodies, they can change the water's chemistry and kill native flora and fauna. When most products were made of wood, animal skins, and other organic materials that quickly decompose, land pollution was not an issue. However, with plastics and other synthetic materials, which take hundreds if not thousands of years to break down, impacts on the land can be substantial. For instance, animals are injured and killed when they consume these materials or get tangled in them and are unable to move. Likewise, synthetic chemicals that decompose slowly can accumulate in soils, making them toxic to humans who live or farm there. Because of the harm that pollution causes to humans and other life, many companies, institutions, and government agencies are working to reduce it and make our world a cleaner place to live.

Air pollution

Contaminated air can be one of the most visible types of pollution. It is often easy to see a brownish haze lying over cities, and peoples' bodies can quickly react to high levels of air pollution, which causes itchy eyes, runny noses, and impaired breathing. Air pollution impacts a large segment of the world's population. Each year, over 80 percent of people in urban areas are exposed to air pollution levels that exceed World Health Organization standards. This exposure causes a wide range of health problems, including an increased risk of stroke, heart disease, lung cancer, asthma, and more. More than three million people annually die prematurely from these causes. But the impact from air pollution is not spatially even. Low- and middle-income countries are disproportionately impacted: fully 98 percent of urban residents in those places are exposed to unhealthy air, while just 56 percent of urban residents in high-income countries are.

Sources of air pollution

Various contaminants contribute to air pollution. In the United States, six pollutants that can harm human health and the environment are regulated: ground-level ozone, particulate matter, carbon monoxide, lead, sulfur dioxide, and nitrogen dioxide. Many of these pollutants come from the burning of fossil fuels. Carbon monoxide comes largely from cars, trucks, and other vehicles and machinery, while nitrogen dioxide comes from these sources as well as from power plants. Sulfur dioxide comes primarily from power plants and other industrial facilities. Lead, on the other hand, comes from ore and metals processing, aircraft that use leaded fuel, waste incinerators, utilities, and lead-acid battery producers. Particulate matter consists of tiny particles that, when inhaled, can cause serious health problems. These particles can come from construction sites, fields, fires, and unpaved roads, but most are formed by chemical reactions of sulfur dioxide and nitrogen oxides. Ozone is not directly emitted from any source. Rather, it forms when nitrogen oxides react with volatile organic compounds. This process occurs when emissions from power plants, vehicles, and other industrial sources react in the presence of sunlight.

Temporal trends

As private motor vehicle use has diffused to a wider proportion of the world's population and as manufacturing continues to use more energy and produce more goods, air pollution globally has increased. World Bank measurements of PM 2.5 ambient particulate matter (particles measuring less than 2.5 microns that can penetrate deep into the respiratory system and cause severe health damage) show that at the world scale, exposure levels increased by over 11 percent between 1990 and 2015. However, this increase was not spatially uniform (figure 12.2). By income, both low-income and high-income countries saw declines in air pollution exposure, while increases were found in the middle-income countries. Increases in middle-income countries are attributable to expanding manufacturing economies and growing wealth. As these countries develop, they shift from agricultural economies to typically dirtier industrial ones. At the same time, growing affluence means that more people can afford exhaust-spewing

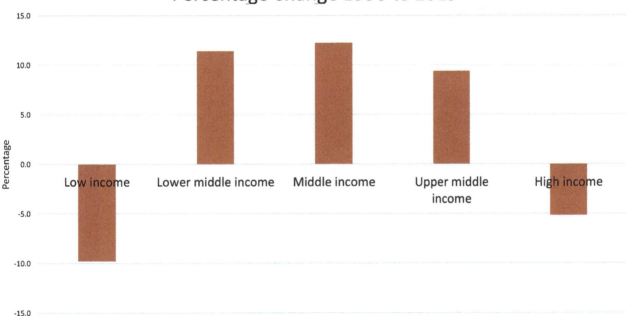

Figure 12.2. Particulate matter exposure, percentage of change, 1990 to 2015, by country income level. Data source: World Bank.

private vehicles (figure 12.3). With more factories, cars, and motorbikes in middle-income countries, air pollution levels rise. This can be seen in lower-middle-income Bangladesh, which saw a 40 percent increase in particulate matter air pollution between 1990 and 2015, the result of rapid expansion of the garment industry in recent years. And, of course, China, an upper-middle-income country, has gone through one of the most significant industrial transformations in the history of the world. As a result, its levels of air pollution increased by over 20 percent since 1990.

The picture is brighter for the high-income United States. Nationally, particulate matter (PM 2.5) levels fell by an impressive 42 percent between 2000 and 2016. Declines were found in all regions of the country, generally a positive sign of improving environmental quality. However, the greatest decline was found in the central part of the US, including states such as Ohio, Illinois, Indiana, and West Virginia. In this region, a decline in air pollution is likely tied to deindustrialization, meaning that cleaner air came at the expense of economic vitality and employment.

Nevertheless, the overall improvement in air quality is something to be celebrated. It may be hard to imagine, but notoriously polluted cities across the country once had even dirtier air. For example, Houston's particulate matter declined 26 percent between 2000 and 2015. During the same time, Pittsburgh's fell by 42 percent, while that of Cleveland and Los Angeles went down by 48 percent.

Spatial patterns

As of 2015, air pollution impacted those in middle-income countries to the greatest degree (figure 12.4). Levels of particulate matter in lower-middle-income countries was 3.4 times greater than in high-income ones, while upper-middle-income countries were 2.5 times larger. This can be seen in figure 12.5, where upper-income countries in North America, Western Europe, and parts of the Pacific have relatively low levels of particulate pollution. Higher levels of pollution are found in poorer regions, such as in much of Latin America and especially in parts of Africa, the Middle East, and Asia. As mentioned previously, many middle-income countries have seen substantial growth in industry and private vehicle use, which contributes to poor air quality. Another factor that needs to be highlighted in places that have not seen substantial manufacturing growth is dust from agricultural and pastoral uses. In places such as the Sahel region of sub-Saharan Africa, farming and grazing is replacing natural forests and grasslands. This leaves more soil exposed, which is picked up by wind and contributes to high levels of particulate pollution. In fact, dust

Figure 12.3. Traffic jam in Jakarta, Indonesia. Increasing private vehicle use in middle-income countries has contributed to worsening air pollution. Photo by AsiaTravel. Stock photo ID: 494677234. Shutterstock.com.

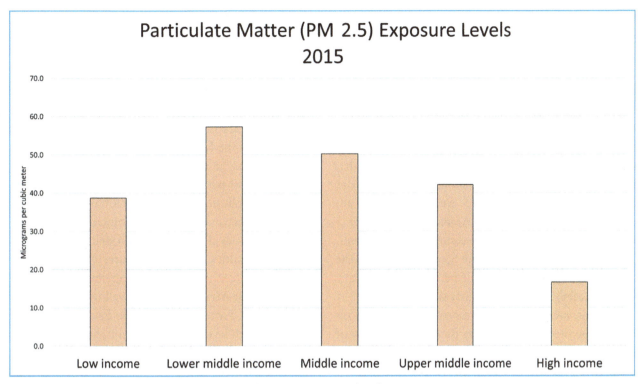

Figure 12.4. Particulate matter exposure, 2015, by county income level. Data source: World Bank.

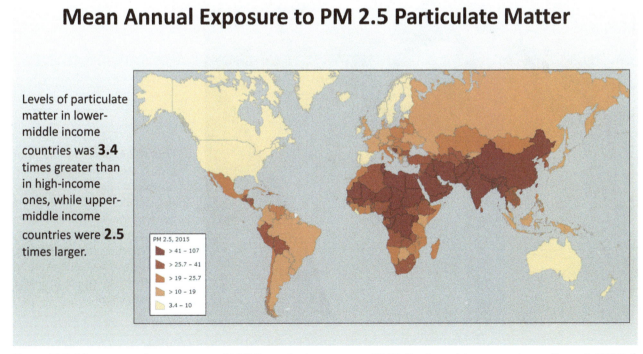

Figure 12.5. Mean annual exposure to PM 2.5 particulate matter, 2015. Explore this map at https://arcg.is/19LKvf. Data source: World Bank.

kicked up by wind often creates massive sandstorms, some of which are so large they cross the Atlantic and impact air quality in the United States (figure 12.6).

The World Health Organization recommends that annual mean levels of PM 2.5 particulate pollution fall at 10 µg/m³ (micrograms per cubic meter) or less. However, at a national scale, few countries meet this standard (figure 12.5). The United States and Canada, as well as a few European countries, Australia, New Zealand, and a handful of other small countries fell within these limits in 2015. Among the worst are oil-producing Qatar and Saudi Arabia, both with particulate levels over ten times the recommended maximum. The industrializing countries of South Asia exceed the level by over seven times, while manufacturing-heavy China exceeds the level by nearly six times.

Levels such as these have profound impacts on human health. Air pollution increases the risk of stroke and heart disease, contributes to pulmonary obstruction diseases and lung cancer, and causes respiratory infections in children. These conditions too often result in early death. Deaths from air pollution are less common in high-income countries and most prevalent in parts of Asia and Africa (figure 12.7). The lowest rates are found in Sweden, Australia, and New Zealand, where less than 1 death per 100,000 people is attributed to air pollution. In the United States, the rate is a relatively modest 7 deaths per 100,000. At the high end, however, rates are much greater. The highest rate is found in North Korea where well over 200 people per 100,000 die prematurely due to air pollution. North Korea has a substantial industrial sector but is technologically backward, resulting in very dirty factories as well as inadequate safety and health-care infrastructure. Other countries, such as India, China, and Afghanistan in Asia and much of West Africa have rates well over 100 deaths per 100,000. Notably, the Middle East, a place with relatively high levels of air pollution, has lower death rates than would be expected. It appears that the combination of exceedingly poor air quality along with inadequate health care leads to higher death rates.

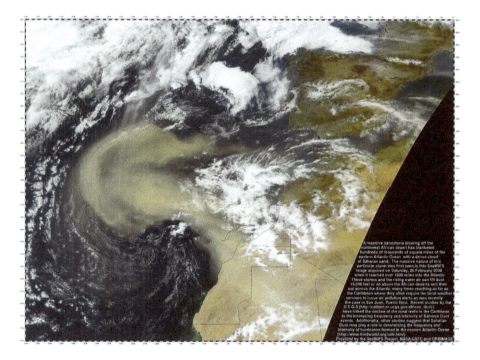

Figure 12.6. Satellite image of a dust storm blowing off the northwest coast of Africa. Land degradation from farming and grazing leaves soils exposed and contributes to wind-blown particulate air pollution. Sometimes the quantities are so large that they impact air quality in North America. Image by NASA.

It should be noted that the death rates discussed here and shown in figure 12.7 are age-standardized. As you'll recall from the chapter on population, death rates tend to be higher in places with more elderly and lower in places with more youth. For this reason, air pollution death rates between countries have been standardized to account for each country's age structure. Using age standardization, rates for Sweden, with a larger elderly population, will not be misleadingly too high, and Afghanistan, with a larger youth population, will not misleadingly be too low.

While national-scale data helps illustrate global impacts of air pollution, there is great spatial variation within countries. Levels of pollution and health impacts depend on physical characteristics of the landscape, such as topography and weather, as well as proximity to sources of pollution. Places in valleys or largely enclosed by hills often trap contaminants, resulting in higher levels of air pollution than found in places with flat and open topography. Wind can contribute to cleaner air by dispersing it, while still weather conditions can allow contamination to accumulate, worsening air quality. Furthermore, places that lie downwind of pollution sources will have higher levels of pollution than those that are upwind. In the Los Angeles region, for instance, sea breezes typically push smog inland from the coasts, which then accumulates in cities at the base of the San Gabriel Mountains. Likewise, cities such as Mexico City and Santiago, Chile, both surrounded by mountains, have consistently poor air quality (figure 12.8).

Proximity to pollution sources also plays a powerful role in levels of air pollution and impacts on human health. Transportation routes and facilities are

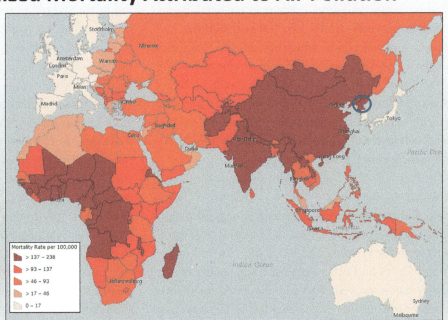

Figure 12.7. Age-standardized mortality rate attributed to household and ambient air pollution (per 100,000 population). Explore this map at https://arcg.is/19LKvf. Data source: World Health Organization.

Figure 12.8. Smog against the Andes Mountains in Santiago, Chile. Air pollution is often more severe in places where mountains inhibit its dispersal. Photo by Alexmillos. Stock photo ID: 289832411. Shutterstock.com.

among the most severe producers of air pollution in many locales, increasing the risk of cancer and other cardiorespiratory illnesses. Health risks are substantial within 300 feet of a freeway but can still impact health within 1,000 feet and beyond, depending on wind patterns. The same holds true for truck-heavy distribution centers, rail yards, and port facilities. Oil refineries, power plants, and other industrial facilities are also important emitters of contaminants, leading to increased rates of health impacts for those living close to or downwind from these locations. Especially in terms of particulate matter, agricultural regions can have high levels of air pollution, even when industry and vehicle emissions are limited.

Solutions

Air pollution has increased globally alongside manufacturing, motor vehicle use, and removal of natural vegetation for farming and livestock, but within the United States and other upper-income countries, air quality has improved. This provides hope for people in poorer regions that face serious health impacts from contaminated air. Solutions exist, but they cost money and are unlikely to be implemented voluntarily, so government regulation is typically required.

Air pollution regulations gained traction in the 1940s and 1950s in North America and Europe, as manufacturing and automobile use was expanding rapidly. Myriad stories show the severity of the air pollution crisis, from a "gas attack" one summer day in 1943 when visibility in Los Angles was reduced to three blocks, leaving residents with stinging eyes and throat-scraping sensations, to London's killer smog that took 4,000 lives in 1952. Regulations initially began at the local and state levels but ultimately were embraced on a national level, such as with the US Clean Air Act of 1970.

A wide range of rules have been implemented in upper-income countries to tackle air pollution, many of which have been resisted by those who see them as an inconvenience or as harming economic opportunity. At the household level, regular municipal trash service replaced backyard incineration of garbage. Other regulations on consumer products, such as paints and even barbecue starter liquids, have had positive impacts on air quality. Strict limits on burning coal and wood for heating homes have also been established in many cities. On farms, the use of "smudge pots," where used motor oil and old tires are burned to prevent frost in orchards, has been eliminated.

With automobiles and other vehicles, a wide range of solutions have been implemented. For instance, in many US cities, sleeves on gas fuel pump nozzles prevent vapors from leaking into the air. Lead has been eliminated from gasoline, and reformulated fuels burn more cleanly. Catalytic converters on vehicles reduced tailpipe emissions substantially. With time, more electric vehicles will reduce emissions along roadways, but overall air quality will improve only if power plants produce electricity in a clean manner.

Factories and power plants have been required to meet emissions standards via better operations practices and clean-air technologies. Some fuels are cleaner than others; for example, natural gas is cleaner than coal. Smokestack emissions have been reduced when regulators require using what lawyers refer to as "best practicable technology" or "best available technology," varying standards that account for economic and/or technological feasibility. Of course, technological solutions improve with time, meaning that legal definitions such as these can result in ongoing improvement of air quality.

One interesting solution to air pollution that gives companies flexibility in meeting standards is the use of pollution credits, or emissions trading. This works by setting emissions targets for industries, say, in terms of total output of sulfur dioxide and nitrogen oxide. Industries can install emissions-reducing technology to reach these goals, and if they exceed the goals, they can sell pollution credits to another facility that has not met the goals. By selling credits, a company can offset some of the cost of clean-air technology. At the same time, companies that do not meet standards can buy pollution credits. This may be more cost effective than installing new clean-air technologies, giving an important level of flexibility to some companies. The key to this type of program is that the total level of allowed emissions within a region is reduced each year. Air quality is thus improved but without rigidly regulating how much each individual company can emit per year.

Go to ArcGIS Online to complete exercise 12.1: "US air pollution: Where is it bad and where is it improving?" and exercise 12.2: "Air pollution exposure at a local scale."

Water pollution

In 1969, the Cuyahoga River in Ohio caught fire. It was at least the ninth time in a century that the river was in flames, the result of industrial and domestic waste from steel mills, chemical plants, and sewage discharges. While damage from the fire was less than in previous ones—a 1912 fire killed five dockworkers, and a 1952 fire caused millions of dollars in damages—this one was caught on film and broadcast on national television. Shock from these images helped lead to the formation of the US Environmental Protection Agency in 1970 and the federal Clean Water Act in 1972. As with air pollution, by the 1960s, Americans were calling for laws to regulate and control contaminants that were poisoning water for both humans and ecosystems.

Great gains have been made in much of the developed world since then, but water pollution continues to negatively impact lakes, rivers, and oceans. And as with other types of pollution, contaminated water in developing countries poses an even greater threat to human health and ecosystems.

Sources of water pollution

Water pollution comes from five major sources: agriculture, industry and energy production, mining, water-system infrastructure, and human waste. Agricultural production employs large amounts of inputs, such as nitrogen and phosphate fertilizers and pesticides. These chemicals seep into groundwater and run off into water bodies, causing a wide range of impacts to ecosystems. For instance, nitrogen and phosphorus feed algae, which can overtake native plants and deplete dissolved oxygen used by fish. Agricultural runoff can also increase the salinity of water as high-salt soils erode and wash into rivers, streams, and lakes. Runoff of this type also can increase sediment loads in water, harming fish, plants, and other aquatic life.

Factories and power plants are other significant sources of water pollution (figure 12.9). Discharge from these sources often includes toxic metals, such as lead and mercury; toxic chemicals, including solvents, pesticides, and asbestos; phosphorus and nitrogen

Figure 12.9. Industrial water discharge pipe in Antwerp, Belgium. Photo by IndustryAndTravel. Stock photo ID: 367273355. Shutterstock.com.

nutrients; and suspended matters, including particulates and sediments. All of these pollutants, when discharged in sufficient quantities, can harm wildlife and contaminate water used by humans for drinking, bathing, and other purposes. In addition, water is sometimes used by industrial facilities and power plants for cooling purposes. During this process, heat is transferred from the facility to the water, which is then discharged with a temperature well above the ambient temperature of the water body, harming aquatic life.

Mine runoff is another source of water pollution. As rocks and minerals are extracted from mines, toxic compounds such as lead, copper, arsenic, and zinc, as well as sulfur, can leach into nearby water bodies and subsurface groundwater. The mining process also uses mercury and other toxic chemicals to extract minerals, which can further contaminate water when it escapes from the mine site. As with agriculture, erosion and sedimentation from mines also cause ecological damage.

A less commonly known source of water pollution comes from water-system infrastructures, such as dam construction and irrigation systems. In the case of dams, natural water flows are altered. This can lead to changes in sedimentation flows, which prevent the natural movement of nutrients downstream. Altered water flow patterns also impact fish and other native life that thrive in fluctuating river flows. Likewise, irrigation systems draw water from natural water bodies for use on farms. This can substantially reduce the quantity of water that flows downstream, reducing plant and animal life.

Lastly, an important source of water pollution comes from human waste. Globally, over 80 percent of wastewater is discharged without treatment. However, there is wide variation between upper- and lower-income countries. In high-income countries, roughly 70 percent of municipal and industrial wastewater is treated, but this figure falls to just 8 percent in low-income countries. Meanwhile, over 2.4 billion people do not have access to improved sanitation, while one billion practice open defecation. This means that in

some places, human feces can flow directly into surface and groundwater.

Temporal trends

Global datasets for water quality over time are limited, making general conclusions about water pollution trends more difficult than for air pollution. Nevertheless, temporal trends appear to reflect those seen with air quality: some improved water quality in the developed world and declining quality in much of the developing world.

Nutrient loads of nitrogen and phosphates have increased globally, promoting the growth of oxygen-depleting algae that damage fisheries and other ocean and fresh water ecosystems. For example, in the case of coastal areas, the number of "dead zones," places with oxygen levels that no longer support most marine life, roughly doubled each decade between 1910 and 2010. These are mostly found at river mouths, where agricultural runoff from upstream accumulates and suffocates marine life.

In addition to nitrates, which come largely from agriculture, industrial and mining activity discharges myriad toxic compounds. In recent decades, as manufacturing shifted from developed countries, which often have stricter environmental controls, to developing countries with weaker controls, water pollution has become of greater concern. In China, over 80 percent of water wells in parts of the country are unfit for drinking or bathing due to industrial and agricultural contamination. Because of this, many cities must draw water from groundwater reservoirs hundreds or thousands of feet below the surface, where toxic contaminants have not yet percolated. China's massive industrialization during the past few decades makes it an extreme case, but similar patterns can be found in other less-developed manufacturing countries as well.

The greatest threat to water quality and human health in developing countries comes from untreated human waste and wastewater. As developing countries continue to urbanize, human waste is increasingly concentrated in densely populated areas. Given that over 80 percent of wastewater goes untreated, the volume of sewage discharged into rivers, lakes, and oceans has increased in many places, threatening water supplies and making many places unfit for swimming or bathing.

In contrast, evidence shows that in the developed world, some types of water quality have improved. In the affluent European countries of the Netherlands, Belgium, and Denmark, for example, nitrogen balances per hectare of agricultural land fell substantially between 1990 and 2004. Likewise, environmental regulations in the United States, including the Clean Water Act, have greatly improved water quality in some areas. Toxic components such as mercury, lead, DDT, and PCBs all saw declines since the 1970s, although the rate of decline has slowed in recent years. Certain pesticides found in water bodies have fallen as well due to government restrictions and/or better agricultural management techniques. Household wastewater, one of the greatest threats to human health, is nearly universally treated in the most affluent countries today. In Canada, for instance, about 60 percent of the population was connected to wastewater treatment facilities in 1990, but by 2010, it had risen to around 85 percent. The United States has improved as well, but to a lesser degree, reaching 75 percent coverage by 2012. Affluent countries in Europe have even higher rates of treated sewer connections: over 95 percent in Germany and Spain and over 80 percent in France, to name a few.

With nutrient loads of nitrogen and phosphates, progress is more limited. For instance, 30 percent of agricultural streams in the United States still contain nitrogen levels higher than those recommended by the EPA for human consumption. Likewise, nitrate transport downriver to the Gulf of Mexico was 10 percent higher in 2008 than in 1980. Part of the reason for less progress in this area is that early pollution controls focused more on *point-source* pollution rather than on *nonpoint-source* pollution. Point-source pollution controls regulate discharges from fixed locations, such as drainage pipes at factories and power plants. Nonpoint-source pollution comes from dispersed locations, such as agricultural and urban runoff. More recent regulations are tackling nonpoint-source pollution, but major gains are yet to be seen.

Spatial trends

Flaming rivers and toxic plumes of chemical effluent tend to grab the headlines and can be of concern in localized areas, but the greatest waterborne threat to human health comes from untreated human waste. Fecal contamination of water sources comes from both open defecation (no latrine) and untreated sewage wastewater. Rainwater runoff washes fecal material from open defecation into rivers and lakes, while much untreated sewage is discharged directly into water bodies. This pollutes water with pathogens that cause serious diseases such as diarrhea, cholera, trachoma, and schistosomiasis. It is estimated that over two million people die annually from diarrhea alone, three-quarters of whom are children under the age of five.

Globally, the largest number of people using open defecation is in India, where over 500 million people lack access to toilet facilities. Other countries in Asia with large numbers include Indonesia, Pakistan, Nepal, China, and Cambodia. In sub-Saharan Africa, the other region with substantial numbers, Ethiopia, Nigeria, Sudan, Niger, Burkina Faso, and Mozambique stand out. Figure 12.10 shows the percentage of people with access to improved sanitation facilities in 2012. Regionally, sub-Saharan Africa has the lowest proportion, followed by South Asia. For instance, only 10 percent or less of the populations of South Sudan and Niger have access, while in India, about 40 percent do. Rates are relatively high in Latin America and the Caribbean, but there is substantial variation. In Argentina and Chile, nearly all of the population has improved sanitation, but just over one-quarter of the population has access in Haiti, while less than half does in Bolivia. In upper-income regions of North America and the European Union zone, on the other hand, around 99 percent of the population has access to improved sanitation.

Access to improved sanitation has a clear spatial relationship with mortality from exposure to unsafe water. In figure 12.11, it is clear that sub-Saharan Africa and South Asia are the epicenter of water-borne disease deaths, along with Haiti in the Western Hemisphere. In India, a person is over forty-five times

Proportion of Population with Improved Sanitation

Over 500 million: The number of people in India with no improved sanitation facility.

South Asia and sub-Saharan Africa have the lowest proportion of population with access.

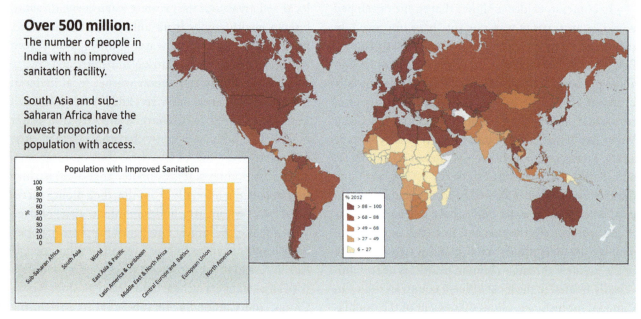

Figure 12.10. Population with improved sanitation. Explore this map at https://arcg.is/1GbLT9. Data source: World Health Organization.

Mortality Rate from Unsafe Water

In **India**, a person is over **450 times** more likely to die from waterborne disease than in the United States or Canada. This number rises to a **1,700 times** greater risk in countries such as **Central African Republic**, the **Democratic Republic of the Congo**, and **Angola**.

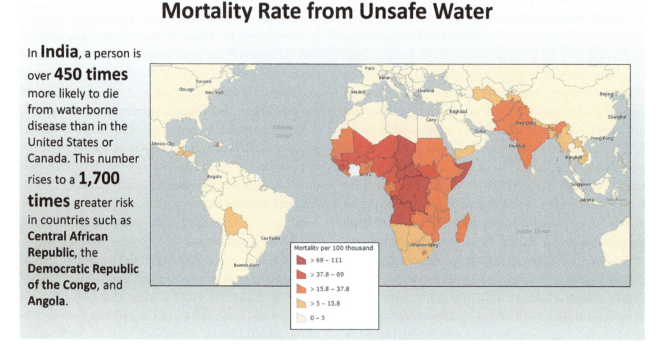

Figure 12.11. Mortality rates from unsafe water. Explore this map at https://arcg.is/1GbLT9. Data source: United Nations.

more likely to die from waterborne disease than in the United States or Canada. This number rises to a shocking 170 times greater risk in countries such as Central African Republic, the Democratic Republic of the Congo, and Angola.

Solutions

Solutions to water pollution vary by type and level of available technology. Point-source pollution in upper-income countries has been reduced substantially through government-established standards and implementation of filters on effluent. For instance, in the United States, the Clean Water Act established the National Pollutant Discharge Elimination System (NPDES), which requires permits to discharge sewer, chemical and biological wastes, radioactive materials, solid waste, and other pollutants into water bodies. In developing countries, this type of control is less common, as the costs of technology, as well as political will and effective enforcement of regulations, is more limited.

Nonpoint-source pollution is less well controlled in both developed and developing countries. Reducing runoff from agriculture and livestock land can be accomplished in different ways. With geospatial technologies such as remote sensing, farmers can now map fertilizer requirements with great detail, allowing them to use just the right amount of fertilizer on individual patches of land. Less fertilizer use means less runoff into nearby water bodies. Other solutions include planting certain cover crops that recycle excess nitrogen from the soil; planting buffers of trees, shrubs, or grasses around fields to absorb runoff; managing drainage; and reducing field tillage. In the case of urban runoff, rains wash oil, animal waste, and garbage into storm drains. Often, these drains are not connected to municipal wastewater treatment plants and instead run directly into lakes, rivers, and oceans. Connecting storm drains to treatment facilities can help reduce this type of nonpoint-source pollution.

Managing human waste, the deadliest of all water pollutants, has been largely accomplished in upper-income countries, but costs and even culture

have made progress much slower in developing ones. Construction of toilet facilities, sewerage systems, and treatment plants is prohibitively expensive for many poor countries, resulting in only a small proportion of waste being properly treated. But culture, surprisingly, can also inhibit the use of proper toilet facilities. In India, many rural residents view open-defecation as healthy and convenient, so much so that in rural areas, 40 percent of households with a working toilet had at least one resident who did not use it. The government has set out to build over 100 million new toilets in rural India, but problems persist: they are often used as storage rooms once they are built and government workers leave. Without a concerted education campaign on the health benefits of toilets, construction alone is unlikely to be effective.

Despite financial and cultural challenges, human waste treatment is being approached in innovative ways. Toilets connected to expensive municipal sewerage systems and treatment plants are not the only solution. Lower-cost pit latrines and compost systems can also be effective. In Bolivia, feces is collected from dry toilets and then composted with worms and converted into fertilizer for use on potato fields. In Tanzania, private collectors pump waste from latrine pits and haul it to municipal treatment facilities, eliminating the need for costly sewerage systems. As long as human waste is properly contained, transported, and treated, it does not matter if it is done through a modern sewerage and treatment facility or a simpler pit and compost system.

Go to ArcGIS Online to complete exercise 12.3: "Water pollution: Sanitation facilities and human health."

Land/solid waste pollution

In the late 1970s, President Jimmy Carter declared environmental emergencies at Love Canal in Niagara Falls, New York. At this location, a school and hundreds of homes had been built directly on top of a landfill where Hooker Chemicals and Plastics Corporation (now

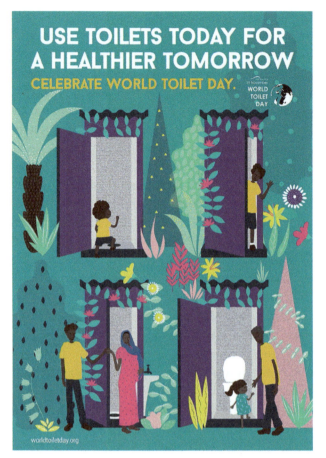

Figure 12.12. Informational poster for World Toilet Day. The United Nations runs this campaign to promote the healthy disposal of human waste. Image from United Nations.

Occidental Chemical Corporation) had disposed of over 21,000 tons of hazardous chemicals from 1942 to 1953. As a result, by the 1970s, residents were reporting chemical residues, foul odors, and increased rates of cancer and other health problems. Ultimately, nearly 1,000 families were evacuated from the site, and partially as a result, in 1980, the US government passed what is commonly known as the *Superfund law* (Comprehensive Environmental Response, Compensation, and Liability Act [CERCLA]).

But hazards do not come only from buried waste materials. In 2017 alone, massive garbage refuse dumps have caught fire and collapsed onto surrounding homes, killing hundreds in Sri Lanka, Ethiopia, and the Philippines. These tragedies resulted from developing countries' inability to safely dispose of growing urban waste combined with housing shortages that result in squatter settlements on undesirable land.

As with air and water pollution, land pollution can cause serious health impacts and negatively impact ecosystems from improper disposal of solid waste materials. Various types of filters in factories, water sanitation facilities, and power plants help keep contaminants from being discharged into the air and water, but ultimately, the filtered material must be disposed of somewhere. The same is true for solid waste material from households, construction sites, and manufacturing plants. In poor countries, much waste is dumped randomly in gullies, ditches, and water bodies, where it contributes to human health problems and harms wildlife. As countries develop, more waste is collected by municipal governments. This material typically ends up in open dumps, sanitary landfills, and hazardous waste disposal facilities, with varying levels of environmental protection.

Sources of solid waste

Solid waste comes from a variety of sources. We are all aware of our own solid waste production through residential disposal of paper, cardboard, plastics, paints, electronics, and other items. Then there is industrial solid waste from manufacturing and power plants, such as packaging and food waste, hazardous chemical wastes, and ashes. Commercial and institutional waste from businesses, schools, hospitals, and government buildings includes additional plastics, paper, glass, metals, biohazard waste, and e-waste. Construction and demolition waste from development sites includes wood, steel, concrete, dirt, bricks, and other such material. Municipal services from street cleaning, landscaping, and wastewater treatment facilities results in waste such as plant and tree trimmings and solid sewage sludge.

Temporal trends

Solid waste production has increased with population, urbanization, and economic development. More people obviously means more consumption of goods and greater use of sanitary facilities. But even more important than the number of people is where they live and how affluent they are. Urban residents tend to be wealthier than their rural counterparts. They purchase more goods and dispose of more paper, plastics, glass, and other materials. In fact, urban residents produce about twice as much waste as rural ones. Likewise, as countries develop economically, people consume more. Increased affluence leads to more electronic goods, more prepackaged foods, more clothing and household goods, and so on. As the world becomes even more urban and affluent, the production of solid waste is likely to increase. Currently, about 1.3 billion tons of municipal solid waste are created each year, but this is likely to increase to 2.2 billion tons by 2025.

The silver lining to this trend is that upper-income countries tend to recycle more and use more efficient production techniques that can minimize solid waste leftovers. This can partially offset increases in waste production. However, many countries have yet to pass through middle-income levels, and recycling is not at the point where it replaces most solid waste.

Spatial trends

By region, the greatest amount of waste is produced by the mostly rich countries of the OECD (Organization for Economic Co-operation and Development), which includes most of Western Europe, the United States and Canada, Australia and New Zealand, Japan and South Korea, Chile and Mexico, among others (table 12.1). These countries produce twice as much waste per capita as middle-income regions in Eastern and Central Asia, Latin America, and the Middle East. At the low end are poorer regions, such as sub-Saharan Africa and South Asia. These countries produce roughly one-quarter the amount of waste per capita as the OECD countries.

In terms of pollution, collection rates are possibly more important than the amount of waste produced. High-income countries collect roughly 98 percent of municipal solid waste, using methods that are mechanized, efficient, and frequent. Waste is typically left in containers that can be quickly and easily picked up by refuse collection trucks at least weekly. In contrast, low-income countries collect about 41 percent of waste. Even this smaller amount can consume a

Solid Waste Production by Region

	kg/per capita/day
OECD	2.2
Eastern and Central Asia	1.1
Latin America	1.1
Middle East and North Africa	1.1
East Asia and Pacific	0.95
Sub-Saharan Africa	0.65
South Asia	0.45

Table 12.1. Solid waste production by region. Kilograms per capita per day. Data source: World Bank.

large portion of a municipal budget, as waste is placed curbside in small bags that are easily ripped open and scattered by dogs and informal recycling-waste pickers (figure 12.13). As would be expected, countries of the OECD collect the most waste, while those of South Asia and sub-Saharan Africa collect the least. Uncollected waste presents risks to health, spreading pathogens and increasing disease-carrying rodent and insect populations. It can also clog drainage systems, increasing contact between people and polluted rain runoff.

After collection, methods of disposal determine the degree of contamination and health impacts caused by solid waste. At the lowest level are semi-controlled or controlled open dumps. These consist of land set aside for the placement of solid waste, which have limited to no controls on leachate contamination, toxic dust, or vermin. While this type of disposal is uncommon in high-income regions, up to one-third of waste can end up in these in low- and middle-income regions. Nearby residents are exposed to contaminated air and water as well as diseases carried by insects and animals that feed in the dump. Health impacts can be especially dire for informal

workers, some of whom live at the dump, who pick through the waste in search of materials to reuse or recycle (figure 12.14).

The next level of waste disposal facilities includes controlled landfills. Landfills differ from open dumps in that they include liners that contain leachate and remove at least some for treatment. They also get compacted by heavy machinery and are covered daily with soil to control toxic dust. Moving up to an even higher level of disposal are sanitary landfills. These are the most well designed, containing and treating nearly all leachate, thus posing minimal risk to air, water, and nearby residents. Some sanitary landfills are designed specifically for hazardous waste, and they include double liners and more stringent monitoring and controls. In high-income regions, the single largest destination for solid waste is to controlled or sanitary landfills.

Solutions

Contamination from solid waste disposal in the developed world no longer poses the risk it once did, but improper disposal is the norm in many lower-income countries. The obvious solution for lower-income

Figure 12.13. Uncollected garbage in Harare, Zimbabwe. Over half of solid waste goes uncollected in low-income regions, posing threats to human health and wildlife. Photo by Cecil Bo Dzwowa. Stock photo ID: 752490544. Shutterstock.com.

Figure 12.14. *Catadores* (rubbish pickers) at an open dump in Rio de Janeiro, Brazil. In many developing countries, poor residents make a living by picking through waste from their more affluent compatriots. Photo by A. Paes. Stock photo ID: 755069146. Shutterstock.com.

Figure 12.15. Sanitary landfill in Turin, Italy. Sanitary landfills include liners to prevent leachate contamination of water and are capped with soil daily to prevent toxic dust dispersion. Photo by Mike Dotta. Stock photo ID: 409903858. Shutterstock.com.

countries is to improve collection and to build more controlled landfills (figure 12.15). This can be a challenge, however, since municipal funds are limited and collection alone can consume 80 to 90 percent of a city's solid waste budget.

Another solution, for both rich and poor countries, is to reduce, reuse, and recycle, so that less material needs to be disposed of in the first place. Many high-income countries recycle 20 percent of their solid waste, with Singapore recycling fully 60 percent of its waste. The European Union, in 2010, recycled 35 percent of household waste, with a mandated target of 50 percent by 2020. In low- and middle-income countries, data is limited, since most recycling is done by informal waste pickers on streets and in dumps, making estimates difficult to calculate.

The reuse of goods is likely more common in lower-income countries, where limited incomes prevent people from buying and discarding goods at the same rates as in upper-income countries. Aside from lowering incomes, improving reuse rates requires a change in cultural attitudes to make it more acceptable to buy used clothing and to purchase durable items rather than disposable ones. Metal silverware instead of plastic at work, reusable water bottles instead of disposable

plastic, refurbished electronics over new ones—these and similar measures can help to limit the production of solid waste.

Recycling and reuse help to reduce the amount of goods purchased and disposed of in the first place. Reduction can also be achieved in manufacturing and shipping of goods through more efficient processes that use fewer raw materials and create less waste material, reduce packaging, and so on.

Where contamination has already occurred, land can be cleaned and rehabilitated. The Superfund program in the United States, spurred by the tragedy in Love Canal, forces the parties responsible for contamination to either clean up sites or reimburse the government for cleanup work done by the Environmental Protection Agency. Under this program, over 390 sites have been cleaned and removed from the Superfund list; however, over 1,300 sites remain (figure 12.16). Obviously, much more work needs to be done, but cleanup can take years and even decades. In fact, dozens of sites on the Superfund National Priorities list were identified in the early 1980s.

Go to ArcGIS Online to complete exercise 12.4: "Land pollution: Superfund sites."

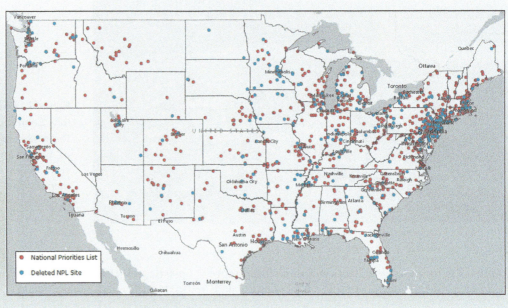

Over 1,300:
The number of
National Priority
List Superfund
sites. These are
toxic sites that
remain to be
cleaned up.

Nearly 400:
The number of
sites that have
been cleaned and
deleted from the
Superfund list.

Figure 12.16. Superfund sites. Explore this map at https://arcg.is/1Gymay. Data source: US EPA.

Climate change

A winter blizzard buries the northeastern United States with record snowfall, hurricanes strike the Gulf Coast and cause millions of dollars in flood damage, and the western United States sees the driest year on record. People often use cases such as these to support or deny the existence of climate change, specifically, the phenomena of a warming planet. But that is to confuse weather with climate. Weather describes short-term behavior of the atmosphere, while climate reflects long-term patterns. To use a probably imprecise analogy, weather is like your mood and climate is like your personality. And just as one's personality tends to change over time, all evidence is that the earth's climate is changing as well. However, it is impossible to say any single weather event, such as a large snowstorm, hurricane, or drought, is due to a changing climate. Throughout history, there have been extreme weather events, even before evidence of warming. What climate

change does is increase the probability of these events. Another analogy relates to baseball. If a player using steroids hits a home run, did the steroids cause the home run? Most likely, the player hit home runs even before taking steroids. But what steroids do is increase the probability of hitting a home run. The same goes for climate change. A single category 5 hurricane may or may not be due to climate change. But what is certain is that climate change increases the probability of category 5 hurricanes.

Evidence of global warming

The scientific consensus is that the earth's climate is changing: it is warming and the warming is primarily due to human activity. The cause is a rapidly increasing level of carbon dioxide and other gases in the earth's atmosphere, the result of the burning of fossil fuels, as well as the clearing of land for agriculture (figure 12.17). These gases occur naturally and, by

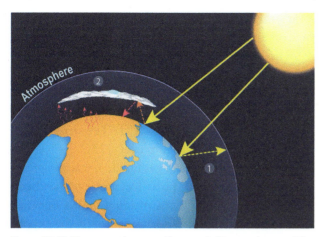

Figure 12.17. The greenhouse effect and climate change. The earth's natural greenhouse gases trap some of the sun's energy and keep the planet warm, allowing life to thrive (1). Human burning of fossil fuels is increasing greenhouse gases, causing more of the sun's energy to be trapped, warming the earth beyond historic norms (2). Image by Designua. Stock vector ID: 527285524. Shutterstock.com.

trapping some of the sun's energy, are what allow life to thrive on our planet. However, human burning of fossil fuels is increasing greenhouse gases significantly, causing more of the sun's energy to be trapped and warming the earth beyond historic norms.

Ice core data combined with more recent atmospheric measurements illustrate how carbon dioxide (CO_2) levels have risen dramatically in recent time. For over 400,000 years, CO_2 levels rose and fell but never below 300 parts per million (figure 12.18). Since the advent of the Industrial Revolution, however, carbon dioxide levels have accumulated rapidly, rising by one-third and passing 400 parts per million by 2017. These levels are unprecedented and are contributing to dramatic changes in the earth's climate.

NASA presents a series of datasets that illustrate how CO_2 is changing our planet. As atmospheric CO_2

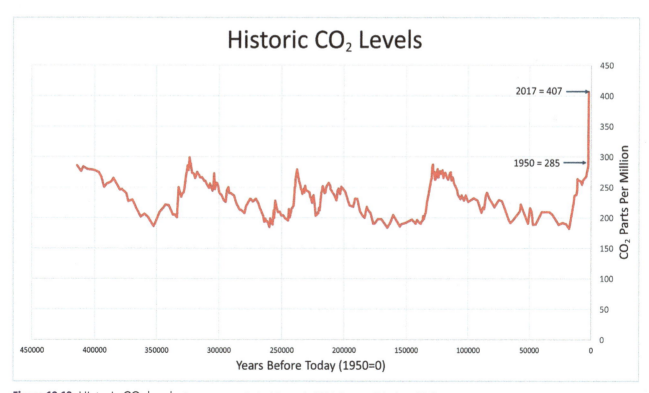

Figure 12.18. Historic CO_2 levels. Data sources: Petit, J.R., et al., 2001; Tans and Keeling, 2018.

has risen, so too have global temperatures. Figure 12.19 illustrates how temperature has changed compared to the 1951–80 mean. The long-term trend is upward, but the greatest increase has occurred in the last thirty-five to forty years. In fact, sixteen of the seventeen warmest years from the past century have occurred since 2001.

Some argue that this warming is due more to natural solar irradiance than to human-induced increases in greenhouse gases. This theory is plausible, but the evidence shows that solar irradiance is not a significant contributor. Since the mid-1700s, energy from the sun has been constant or has increased only slightly. If it was indeed a contributing factor, temperatures would rise in all layers of the atmosphere. However, only surface layers of the atmosphere are warming, where greenhouse gases trap heat, while upper layers have actually cooled. Furthermore, climate models cannot account for the level of heating we are seeing with solar irradiance alone. Only by including greenhouse gases can the changes be accounted for.

This warming is causing significant declines in polar ice cover. Arctic sea ice cover in September, which is the month that it reaches minimum extent, has been steadily declining since measurements in the late 1970s (figures 12.20 and 12.21). The same pattern is seen with Antarctic ice (figure 12.22).

Because of greenhouse gas buildup, warming temperatures, and melting polar ice, sea level has been rising. It occurs as millions of square miles of polar ice melt and return to the sea and through the expansion of sea water as it warms. Satellite measures since 1993 show that the global mean sea level has risen by roughly eighty millimeters, or a little over three inches (figure 12.23). Coastal tide gauge measurements from the 1880s indicated that sea level has risen eight inches since then.

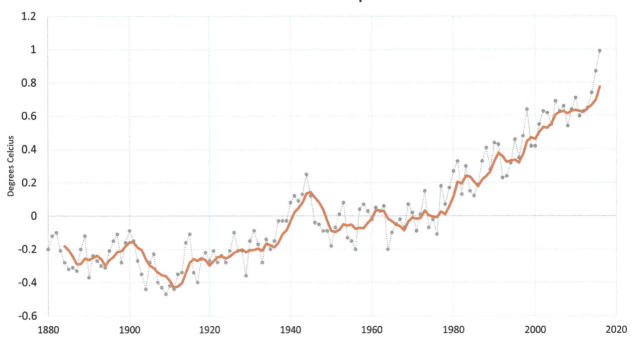

Figure 12.19. Temperature difference from mean. Solid line represents five-year moving average. Data source: NASA/GISS.

September Arctic Ice Extent

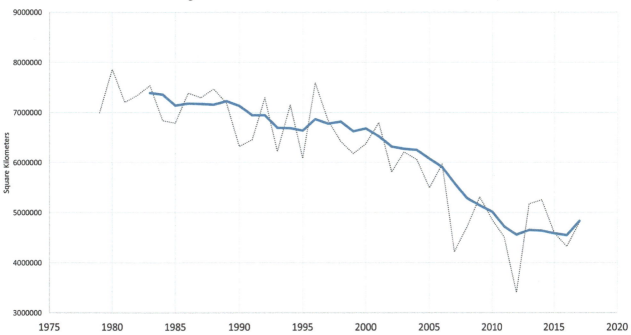

Figure 12.20. September Arctic ice extent. Data sources: NASA and National Snow and Ice Data Center.

Arctic September Sea Ice Coverage: 1984 and 2012

The 1984 image (above) represents the average minimum extent for 1979-2000.

The downward trend in the minimum extent is evident in the 2012 image (above), which was approximately half that of 1984.

Figure 12.21. Arctic September sea ice coverage: 1984 and 2012. Image source: NASA Scientific Visualization Studio.

Figure 12.22. Antarctic ice mass: Change from 2002. Data source: NASA.

Figure 12.23. Global mean sea level increase: 1993–2017. Data source: NASA.

Effects

Temperature, precipitation, and extreme weather

Climate change impacts many aspects of life on earth. Temperatures are expected to continue rising as human-produced greenhouse gases continue to be emitted. With rising temperatures, heat waves will become more common. Already, multi-month extreme heat has been on the rise since reliable records began in 1895. Furthermore, extreme heat days that occurred once every twenty years are projected to occur once every two to three years in much of the United States.

Precipitation will be altered as well, with many wet places getting wetter and many dry places getting drier. Generally, much of the North and Northeast regions of the US will see more precipitation, while the Southwest will see less. As another example, in Africa, countries along the already dry Mediterranean will likely see less precipitation, while tropical regions along the equator will see more. In drier regions, increasing heat and lack of precipitation will contribute to a greater frequency and intensity of wildfires. Extreme storms with heavy downpours will become more frequent as well, as warm air holds more water vapor than cooler air. These downpours will cause more flood events. Heavy precipitation and flooding will also become more common in areas at risk of hurricanes. Already since the 1980s, there has been a substantial increase in intensity, frequency, and duration of hurricanes in the Atlantic, the result of warming ocean waters.

Impacts on people, cities, and economies

Changes in temperature and precipitation will impact many aspects of human society, including numerous themes covered in this book. Air pollution is likely to increase along with climate change. Increasing numbers of hot, sunny days will contribute to greater ground-level ozone pollution, while wildfires and dust will add to particulate pollution. Health impacts from dirty air will be compounded by extreme heat days that contribute to deaths and hospitalizations from heat stroke and related conditions. Water pollution increases with flooding, whereby pathogens are picked up by floodwaters and carried into contact with people on streets and sidewalks and in homes, schools, and businesses.

As you'll recall, humans are now an urban species, with over 50 percent of all people living in cities and towns. Yet most urban settlements, along with their manufacturing and service sector economies, have been built under climatic conditions of a cooler world. Increasing temperatures will put pressure on electrical grids as they struggle to power a greater number of air conditioners for a greater number of hours. Extreme downpours from heavy storms can bring down power lines and cut off electricity. Likewise, flooding threatens myriad infrastructure systems. Rising water from heavy downpours, hurricanes, and storm surges enhanced by rising sea levels can swamp highways and subways, overflow sewerage systems, and short-circuit electrical systems running everything from apartment and office building elevators to factory and oil refinery installations.

Many argue that Hurricane Sandy, which struck the New York metropolitan region in 2012, was just a sample of what is to come for many coastal urban areas. During that event, 100 people died in the metropolitan region as fourteen-foot storm surges swamped homes and businesses. Flooding and wind cut electricity to 8.5 million customers, leaving many Manhattanites in high-rise buildings without heat or light and unable to get out by elevator. Subway tunnels flooded, and tens of thousands of people were displaced from their homes.

As seen with New York's subways, urban transportation infrastructure is also at risk from climate change (figure 12.24). Fuel supplies are threatened from flooding and storm events that damage refineries and distribution networks; bridges and roads can be destroyed by heavy flooding; and coastal ports, highways, and airports can be swamped by rising sea levels and storm surges. Incredibly, heat waves are also preventing commercial airliners from taking off. In Phoenix, Arizona, temperatures close to 120 degrees heat the air to a point where its density is too low for airplanes to get lift. Late-afternoon flights have been canceled and rescheduled for times when the

Figure 12.24. Flooded underpass from Hurricane Sandy in New York. Photo by Kobby Dagan. Stock photo ID: 117337669. Shutterstock.com.

temperature is lower. As global temperatures rise, this problem will affect airports in regions with extreme heat ever more frequently.

Outside of crops grown in greenhouses, the spatial distribution of agriculture is highly dependent on climate. Changes in climate will substantially impact the type and quantity of crops and livestock produced around the world. With greater variation in temperature and precipitation, through heat waves, drought, floods, and other extreme events, agricultural output is likely to become much more unpredictable. Year-to-year variations will increase, with good output some in seasons and poor output in others. This will make food prices more volatile, putting low-income consumers at risk when agricultural production falls and prices rise. Warmer temperatures will also increase the prevalence of weeds, insects, and plant diseases, requiring greater use of pesticides and herbicides to keep them under control.

Agricultural impacts will vary by region. In North America, yields may increase in northern regions, but overall agricultural output is expected to decline by midcentury. For instance, in the West and Southwest, an increase in the number of dry days will reduce crop and livestock production. A lack of chilling will also negatively impact fruit and nut trees, an important source of California's agricultural economy, as well as plum and cherry trees in the Northeast.

In Africa, a similar pattern is likely. Overall production of key crops such as cereals will fall, although specific regions may see some increases in crop yields (figure 12.25). Heat will also increase food spoilage, given that proper transportation and refrigeration technology is more limited than in richer regions. In Asia, rice production, the key staple for the region, is likely to decline, but wheat production could increase in Pakistan. Projections for Latin American agriculture are mixed. Southeastern regions of South America, such as southern Brazil, Uruguay, and Argentina, could see warmer and wetter conditions that improve production, while wider variation in precipitation in Central America will likely harm key crops such as coffee and maize.

Political instability, conflict, and migration

In terms of political instability and conflict, some argue that climate change increases violence and threatens the stability of states. As the climate changes, existing societal organization becomes stressed. This becomes evident when cities and other human settlements, as well as agricultural, manufacturing, and service economies, face warmer temperatures, shifting patterns of precipitation, greater volatility in storm activity, and

Figure 12.25. Outside Koulomboutej village, Niger. Agriculture will become more difficult in arid landscapes such as the Sahel region, which crosses Africa south of the Saharan desert. Photo by Giulio Napolitano. Stock photo ID: 158008097. Shutterstock.com.

rising seas. What once worked well may no longer function, while mismanagement of resources can become quickly unmasked.

For instance, in the dry Sahel region of Africa, there may be more ethnic conflict as rainfall variability forces a switch from mixed-crop farming to livestock ranching and a concomitant increase in struggles for grazing land. In North Africa, Syria and Egypt have already seen climate-induced conflict. Syria once prided itself for its production of wheat and cotton, but declining winter precipitation led to its worst-ever drought from 2007 to 2010. Combined with inefficient agricultural practices, this led to agricultural and pastoral devastation and the displacement of almost two million people. This occurred even before Syria's bloody civil war. Egypt, on the other hand, has always been extremely dependent on wheat imports to feed its population. When drought and heatwaves struck Russia and China in 2010–11, limited supplies of imported wheat drove prices of bread in Egypt up by 300 percent. Food riots resulted, contributing to anger against the leadership of now-deposed president Hosni Mubarak. In the South China Sea, warming waters are shifting fish populations to new areas farther north. This is disrupting important fishing grounds for countries such as Vietnam and the Philippines and contributing to tension over China's claims to much of the sea. Throughout the world, changes such as these are contributing to instability and conflict and are likely to accelerate as climate further changes.

As has happened throughout history, when stresses become too great in a place, people seek lives in new locations. Climate change is likely to increase push forces to migrate, as economic livelihoods and settlements become unsustainable. This will be felt more strongly in developing countries, where resources to mitigate impacts, be they sea walls or irrigation systems, are financially out of reach. Migration flows will lead to urban areas, as people flee unsustainable rural communities where agriculture has declined, and from vulnerable cities prone to flooding. Receiving cities will struggle in both the developed and developing world to absorb large migrant flows from what some are referring to as *climate refugees*.

Broadly, there are three types of areas most at risk. First are dryland regions with variable precipitation. These are places where water for agriculture and human consumption is already scarce. As these places become even drier, with longer periods of drought and intermittent downpours and flooding, hundreds of millions of people will face pressure to migrate. Africa's Mediterranean north and Sahel regions, as well as dry regions of northern India and southeast Pakistan, western China, coastal Peru, and others, will likely see climate-driven pressures to migrate.

The second type of region most at risk includes heavily populated low-lying coasts in areas of tropical cyclones. Over 200,000 people fled New Orleans after category 5 Hurricane Katrina in 2005, while the number of Honduran migrants caught at the US border rose significantly after Hurricane Mitch struck in 1998. In addition to the southeastern United States and tropical Latin America, other places that face regular flooding from events such as these include Bangladesh in the Indian Ocean and the Philippines in the Pacific, among others.

Third, low-lying atolls in the Pacific and Indian oceans are threatened. Islands such as Kiribati, the Maldives, Seychelles, Solomon Islands, and Micronesia often consist mostly of land just a few meters above sea level. With rising oceans and more extreme storms, much of the land area of these islands will be under water or flooded so often as to be uninhabitable (figure 12.26). The risk is becoming such a concern that New Zealand has considered creating a special refugee visa for displaced Pacific Islanders.

A changing climate will reshuffle where humans can best survive and thrive. Economic activity and settlement patterns will have to adjust to these changes, as crops once suited to some places can no longer thrive, water infrastructure for cities requires more extensive distribution networks and flood control, and manufacturing and service businesses reorient where products are made and where consumers live. Managing these changes properly will cost lots of money, while neglect

Figure 12.26. Malé, Maldives. Islands such as this are at risk from rising sea levels and extreme storms. The average elevation of the Maldives is four feet, with a high point of just eight feet. Photo by Chumash Maxim. Stock photo ID: 159659003. Shutterstock.com.

will likely lead to conflict, political instability, and disruptive migration flows.

Solutions

In recent years, there has been strong political disagreement over the extent and cause of climate change. Regardless, the science clearly shows that it is significant and that it is caused by humans, with at least 97 percent of actively publishing climate scientists in agreement. Models still vary in how much temperatures will rise, to what degree precipitation will change, how high sea levels will rise, and which regions will be most negatively impacted, but the impacts will be real and substantial in all scenarios. With that said, however, it is clearly fair to debate the best response to climate change. Creative and potentially useful solutions are being proposed from people of diverse political orientations. This is where open debate should be encouraged.

Solutions to climate change can be broken down into two categories. First is *mitigation*. Mitigation refers to reducing levels of carbon dioxide and other greenhouse gases in the atmosphere. Second is *adaptation*. This refers to reducing the vulnerability of populations, such as from flooding, heatwaves, and reduced agricultural output.

Mitigation

The long-term and permanent solution to climate change is through mitigation: the reduction of carbon dioxide and other greenhouse gases to levels that can be absorbed naturally by plants, trees, and oceans. The bad news, as shown earlier in this chapter, is that carbon dioxide emissions have increased steadily over the years. But there is good news as well. Between 1963 and 2008, studies show that carbon dioxide emissions per unit of energy produced in the United States has fallen. This tells us that total emissions are increasing along with population and economic growth, but we are becoming more efficient in our energy production over time. If we can continue to become even more efficient, emissions can ultimately be reduced to sustainable levels.

To push forward remedies to climate change, 174 countries have signed the *Paris Agreement*, a nonbinding platform whereby signatories promise to reduce carbon emissions so that global temperatures in this century rise less than 2 degrees Celsius (3.6 degrees Fahrenheit) above preindustrial levels and to aim for temperature increases of only 1.5 degrees Celsius (2.7 degrees Fahrenheit). How this is to be achieved is not stipulated, giving countries wide leeway to develop mitigation policies.

Absorption of carbon dioxide is heavily influenced by land-use change. As people clear forests and other vegetation for cities, rangeland, and farms, the amount of vegetation available to absorb gases decreases. Reforestation, on the other hand, increases absorption. The problem is that overall, land-use change is resulting in less carbon-absorbing vegetation, not more. Reforestation programs and better land-use planning can help, but more mitigation strategies are required.

This means that mitigation must rely heavily on reductions of emissions. Different approaches are being used, and there are vibrant debates as to which is the most cost effective. In places such as California and parts of the northeastern United States, carbon pricing via cap and trade programs are being used. As with air and water pollution, cap and trade programs set a cap for the total amount of carbon that can be emitted within an area. Companies are then issued emissions permits, which they can use to emit carbon, or they can reduce their emissions and sell the remaining credits to other companies. Over time, permits are removed from the program, thus reducing the amount of permissible carbon emissions.

Other government regulations and standards are being used to reduce carbon emissions as well. Fuel standards for cars and trucks force manufacturers to produce and sell more fuel-efficient vehicles. In the United States, passenger cars in 1980 traveled 24.3 miles per gallon, but by 2014, this number had increased to 36.4 miles per gallon. In government building construction, LEED (Leadership in Energy and Environmental Design) certification has made buildings more energy efficient, requiring less air conditioning and heating. Government-certified Energy Star appliances, such as washers, dryers, and refrigerators, are much more efficient than in the past, while various environmental standards have increased energy efficiency in manufacturing processes.

Government subsidies also impact carbon emissions. Many governments have helped subsidize energy production or consumption via tax breaks, rebates, subsidized loans, and other tools. In all cases, a government subsidy means that the government pays part of the cost of the energy. China, along with big oil states such as Iran, Saudi Arabia, Russia, and Venezuela, provides heavy subsidies to greenhouse gas–producing fossil fuels. Many countries subsidize cleaner renewable energy, such as solar, wind, and geothermal, to jump-start its use.

Opponents of climate change policies argue that the costs of government intervention, be it a tax on carbon via cap and trade, regulations that mandate automobile fuel efficiency, or subsidies for renewable energy, are too high. From a purely economic standpoint, they argue that government spending on these types of mitigation strategies will divert money from other uses, whether private investment or government spending on schools, roads, and the military.

Some point out that technological change and free markets are leading to the use of climate-friendly energy sources anyway. Between 2009 and 2016, the unsubsidized total lifetime cost of building and running wind energy facilities has fallen by 66 percent. During the same time, the cost of utility-scale soar projects has fallen by 85 percent (figure 12.27). This has made them competitive with conventional generating technologies such as coal and natural gas. From 2010 to 2016, the greatest addition to net energy capacity came from renewable energy sources, primarily solar and wind, resulting in substantial growth of renewable electricity output (figure 12.28). Worryingly, coal also increased substantially, but that will change moving toward 2040. By that time, renewables will be even more prevalent, while the role of coal will fall.

Figure 12.27. Solar farm in Qinghai province, China, a country that has been taking the lead in solar power production, accounting for a substantial amount of new installed capacity. Photo by lightrain. Stock photo ID: 636564641. Shutterstock.com.

The question remains whether market and technological solutions will be fast enough to prevent catastrophic warming or if further governmental regulation will be required to speed up reductions in greenhouse gas emissions. By 2022, 30 percent of energy will be produced globally by renewables, up from 24 percent in 2016. However, coal use will still dominate. And while coal growth is expected to fall by 2040, what is needed even more are declines in its use.

Renewable Electricity Output

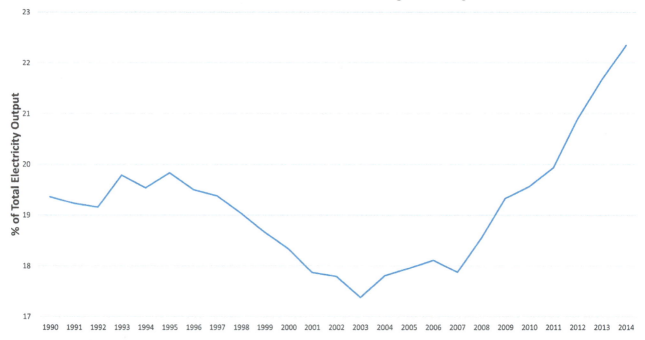

Figure 12.28. Renewable electricity output as a percentage of total electricity output. Renewable energy sources have been providing a steadily increasing share of electricity at a global scale since the turn of the century. Data source: World Bank.

Adaptation

Earth is already warming, with corresponding changes in temperature and precipitation and more numerous extreme weather events. Consequently, adaptation to climate change is necessary, even if greenhouse gas emissions are ultimately brought under control.

Adaptation will take many forms. Land-use planning for homes and businesses must consider rising sea levels and larger areas at risk for flood. Infrastructure and transportation networks must be made more resilient to storms and flooding. Water systems must be expanded to conserve water from irregular precipitation patterns, while agricultural regions will have to change the types of crops grown and install more widespread irrigation systems. Seawalls and storm surge barriers will be needed in vulnerable coastal communities (figure 12.29). In some cases, relocation

Figure 12.29. The Maeslantkering storm-surge barrier in the Netherlands. These massive arms swing shut to block heavy storm surges from flooding Rotterdam. Expensive adaptations such as these may be necessary to protect New York and other vulnerable coastal cities from rising sea levels and more severe storms. Photo by GLF Media. Stock photo ID: 590781014. Shutterstock.com.

of populations will be necessary. Heatwaves will put pressure on electrical grids. Public health will be at risk, so adaptation must include emergency planning for everything from flood rescue to air-conditioned community cooling centers.

Most governmental adaptation in the United States is occurring at the local level. In Chicago, the city is promoting green rooftops to reduce urban heat. New York and other cities are updating flood insurance maps requiring properties formerly considered safe to buy policies. In the Miami region, regulations are discouraging new developments in areas prone to rising seas and flooding. Phoenix, Arizona, and other cities are focusing on public health by establishing community cooling centers that open during heatwaves.

While some people argue over the significance of climate change, many hard-nosed corporate managers are already adapting their businesses to reduce risks from a warming planet. Coca-Cola has invested in water projects around the world to ensure reliable water for its beverages in a time of uncertain precipitation. ConAgra Foods is diversifying its supply chain so that tomatoes and other produce come from multiple suppliers to reduce the risk of shortages from drought or storm damage. Pacific Gas and Electric Company is upgrading its grid to accommodate higher demand from air conditioning and is working to improve reservoirs and canals to ensure a steady flow of water for hydroelectric power as the Sierra Nevada snowpack declines. Many other companies are taking similar actions to evaluate risks to their physical infrastructure, transportation and supply-chains, and customer needs.

All mitigation and adaptation strategies cost money. Carbon pricing for cap and trade raises the price of fossil fuels, which are the most commonly used sources of energy today. Rising prices on everything from gasoline for our cars to electricity for our offices and factories can harm economic growth. Government rules and regulations do the same by imposing costs on producers who use fossil fuels. Subsidies and research and development for cleaner renewable energy divert

money from investment by private companies or from other government programs, be it education, health care, or military spending. Adapting infrastructure to withstand flooding and extreme weather will also be very expensive, as will building sea walls, irrigation and water-supply conduits, and flood-control projects. But if we choose not to pay today, the cost will be even greater in the future as impacts from a warming climate become even more severe. The time of climate change denial is over; hard decisions must now be made on the best way to tackle this global threat to human society.

Go to ArcGIS Online to complete exercise 12.5: "Global carbon emissions."

References

BBC News. 2017. "Phoenix Flights Cancelled Because It's Too Hot for Planes." *BBC News,* June 20, 2017. http://www.bbc.com/news/world-us-canada-40339730.

Bhada-Tata, Perinaz, and Daniel A. Hoornweg. 2012. "What a Waste?: A Global Review of Solid Waste Management." *Urban Development Series Knowledge Papers; No. 15.* Washington, DC: World Bank Group. http://documents.worldbank.org/curated/en/302341468126264791/What-a-waste-a-global-review-of-solid-waste-management.

Boivin, N. L., M. A. Zeder, D. Q. Fuller, A. Crowther, G. Larson, J. M. Erlandson, T. Denham, and M. D. Petraglia. 2016. "Ecological Consequences of Human Niche Construction: Examining Long-Term Anthropogenic Shaping of Global Species Distributions." *Proceedings of the National Academy of Sciences.* http://www.pnas.org/content/113/23/6388.full.

Buckley, C., and V. Piao. 2016. "Rural Water, Not City Smog, May Be China's Pollution

Nightmare." *The New York Times,* April 12, 2016. https://www.nytimes.com/2016/04/12/world/asia/china-underground-water-pollution.html?_r=0.

California Air Resources Board. 2017. *Air Quality and Land Use Handbook.* California Environmental Protection Agency Air Resources Board. https://www.arb.ca.gov/ch/landuse.htm.

Center for Climate and Security. 2017. *Epicenters of Climate and Security: The New Geostrategic Landscape of the Anthropocene.* Washington, DC: Center for Climate and Security. https://climateandsecurity.org/epicenters.

Duggan, B. 2017. "Death Toll Rises in Ethiopian Trash Dump Landslide." *CNN,* March 15, 2017. https://www.cnn.com/2017/03/15/africa/ethiopia-trash-landslide-death-toll/index.html.

Dunbar, B. 2005. "What's the Difference Between Weather and Climate?" NASA, February 1, 2005. https://www.nasa.gov/mission_pages/noaa-n/climate/climate_weather.html.

European Environment Agency. 2013. "Highest Recycling Rates in Austria and Germany—But UK and Ireland Show Fastest Increase." Press release, March 3, 2013. https://www.eea.europa.eu/media/newsreleases/highest-recycling-rates-in-austria.

Intergovernmental Panel on Climate Change (IPCC). 2014. *Fifth Assessment Report—Impacts, Adaptation and Vulnerability.* Geneva: IPCC. https://www.ipcc.ch/report/ar5/wg2.

International Energy Agency. 2017. "Energy Subsidies by Country, 2016." https://www.iea.org/weo/energysubsidies.

———. "Renewables 2017: A New Era for Solar Power." https://www.iea.org/publications/renewables2017.

Jaquith, C. 2016. *Levelized Cost of Energy Analysis 10.0.* New York: Lazard. https://www.lazard.com/perspective/levelized-cost-of-energy-analysis-100.

Larsen, M. C., P. A. Hamilton, and W. H. Werkheiser. 2013. "Water Quality Status and Trends in the United States." In *Monitoring Water Quality.* Washington, DC: USGS Publications Warehouse RSS.

NASA. 2014. "Global Climate Change." https://climate.nasa.gov.

NBC News and Reuters. 2017. "Sri Lanka Garbage Dump Disaster: 100 Feared Dead after Trash Pile Collapse." *NBC News.com,* April 17, 2017. https://www.nbcnews.com/news/world/sri-lanka-trash-dump-disaster-100-feared-dead-after-trash-n747231.

Palaniappan, M., P. Gleick, L. Allen, M. Cohen, J. Christian-Smith, and C. Smith. 2010. *Clearing the Waters: A Focus on Water Quality Solutions.* Oakland, CA: Pacific Institute.

Patel, A. 2014. "Why Many Indians Can't Stand to Use the Toilet." *Wall Street Journal,* October 8, 2014. https://blogs.wsj.com/indiarealtime/2014/10/08/why-many-indians-cant-stand-to-use-the-toilet/?mg=prod%2Faccounts-wsj.

Petit, J. R., and Jean Jouzel. 2001. "Vostok Ice Core Data for 420,000 Years." *IGBP Pages/World Data Center for Paleoclimatology Data Contribution Series,* No.2001-076. NOAA/NGDC Paleoclimatology Program, Boulder, CO.

Roque, P. 2000 "31 Filipinos Killed in Garbage Dump Collapse." *ABC News,* July 10, 2000. http://abcnews.go.com/International/story?id=83209&page=1.

South Coast Air Quality Management District (AQMD). n.d. "REgional CLean Air Incentives Market (RECLAIM)." Accessed February 12, 2017. http://www.aqmd.gov/home/programs/business/about-reclaim.

———. n.d. "The Southland's War on Smog: Fifty Years of Progress Toward Clean Air (through May 1997)." Accessed February 16, 2017. http://www.aqmd.gov/home/research/

publications/50-years-of-progress#The%20 Arrival%20of%20Air%20Pollution.

Tans, P., and R. Keeling. 2018. "Trends in Atmospheric Carbon Dioxide." National Oceanic and Atmospheric Administration/Earth System Laboratory. https://www.esrl.noaa.gov/gmd/ccgg/trends.

Taylor, L. 2017. "New Zealand Considers Visa for Climate 'Refugees' from Pacific Islands." *Reuters*, November 17, 2017. https://www.reuters.com/article/us-newzealand-climatechange-visa/new-zealand-considers-visa-for-climate-refugees-from-pacific-islands-idUSKBN1DH1JB.

United Nations Educational, Scientific and Cultural Organization (UNESCO). 2012. *UN World Water Development Report, Managing Water Under Uncertainty and Risk.* Paris: UNESCO. http://www.unesco.org/new/fileadmin/MULTIMEDIA/HQ/SC/pdf/WWDR4%20Volume%201-Managing%20Water%20under%20Uncertainty%20and%20Risk.pdf.

———. 2017. *UN World Water Development Report, Wastewater: The Untapped Resource.* Paris: UNESCO. http://www.unesco.org/new/en/natural-sciences/environment/water/wwap/wwdr/2017-wastewater-the-untapped-resource.

United Nations Framework Convention on Climate Change. 2017. "The Paris Agreement." http://unfccc.int/paris_agreement/items/9485.php.

US Environmental Protection Agency. 2017. "Love Canal Site Profile." https://cumulis.epa.gov/supercpad/SiteProfiles/index.cfm?fuseaction=second.Cleanup&id=0201290#bkground.

———. "Criteria Air Pollutants." https://www.epa.gov/criteria-air-pollutants.

———. 2018. "Superfund." https://www.epa.gov/superfund.

US Global Change Research Program. 2014. *National Climate Assessment.* Washington, DC: US Global Change Research Program. https://nca2014.globalchange.gov/report.

The World Bank. 2017. *Atlas of Sustainable Development Goals 2017.* Washington, DC: World Bank Group.

World Health Organization. 2016. *Ambient Air Pollution: A Global Assessment of Exposure and Burden of Disease.* Geneva, Switzerland: World Health Organization.

Index

About the author

J. Chris Carter is a professor of geography at Long Beach City College, where he has over fifteen years' experience teaching human geography and geographic information systems. He earned his PhD in the joint San Diego State University/University of California–Santa Barbara geography doctoral program, where his research focused on economic and urban change in Latin America. In addition to Long Beach City College, he has taught at the University of La Serena, Chile, and California State University, Long Beach. His areas of interest lie in the spatial patterns of urban, demographic, social, and economic change in the United States, Latin America, and beyond.